Lecture Notes in Mathematics 1639

Editors:
A. Dold, Heidelberg
F. Takens, Groningen

Springer
Berlin
Heidelberg
New York
Barcelona
Budapest
Hong Kong
London
Milan
Paris
Santa Clara
Singapore
Tokyo

Karel Dekimpe

Almost-Bieberbach Groups:
Affine and
Polynomial Structures

Springer

Author

Karel Dekimpe*
Katholieke Universiteit Leuven
Campus Kortrijk
Universitaire Campus
B-8500 Kortrijk, Belgium
e-mail: Karel.Dekimpe@kulak.ac.be

* Postdoctoral Fellow of the Belgian National Fund for Scientific Research (N.F.W.O.)

Cataloging-in-Publication Data applied for

Die Deutsche Bibliothek - CIP-Einheitsaufnahme

Dekimpe, Karel:
Almost Bieberbach groups: affine and polynomial structures /
Karel Dekimpe. - Berlin ; Heidelberg ; New York ; Barcelona ;
Budapest ; Hong Kong ; London ; Milan ; Paris ; Santa Clara ;
Singapore ; Tokyo : Springer, 1996
  (Lecture notes in mathematics ; 1639)
  ISBN 3-540-61899-6
NE: GT

Mathematics Subject Classification (1991): Primary: 20H15, 57S30
                                          Secondary: 20F18, 22E25

ISSN 0075-8434
ISBN 3-540-61899-6 Springer-Verlag Berlin Heidelberg New York

The use of general descriptive names, registered names, trademarks, etc. in this
publication does not imply, even in the absence of a specific statement, that such
names are exempt from the relevant protective laws and regulations and therefore
free for general use.

Typesetting: Camera-ready TeX output by the author
SPIN: 10479900   46/3142-543210 - Printed on acid-free paper

Preface

The reader taking a first glance at this monograph might have the (wrong) impression that a lot of topology/geometry is involved. Indeed, the objects we study in this book are a special kind of manifold, called the infra–nilmanifolds. This is a class of manifolds that can, and should, be viewed as a generalization of the flat Riemannian manifolds. However, the reader familiar with the theory of the flat Riemannian manifolds knows that such a manifold is completely determined by its fundamental group. Moreover, the groups that occur as such a fundamental group can be characterized in a purely algebraic way. More precisely, a group E is the fundamental group of a flat Riemannian manifold if and only if E is a finitely generated torsion free group containing a normal abelian subgroup of finite index. These groups are called Bieberbach groups. It follows that one can study the flat Riemannian manifolds in a purely algebraic way.

This group theoretical approach is also possible for the infra–nilmanifolds, which are obtained as a quotient space under the action of a group E on a simply connected nilpotent Lie group G, where E acts properly discontinuously and via isometries on G. (If G is abelian, then this quotient space is exactly a flat Riemannian manifold). The fundamental group of an infra–nilmanifold is referred to as an almost–Bieberbach group. It turns out that much of the theory of Bieberbach groups extends to the almost–Bieberbach groups. Thus for instance, a group E is the fundamental group of an infra–nilmanifold if and only if E is a finitely generated torsion free group containing a normal nilpotent subgroup of finite index.

The aim of this book is twofold:

1. I wish to explain and describe (in full detail) some of the most important group–theoretical properties of almost–Bieberbach groups.

I have the impression that the algebraic nature of almost–Bieberbach groups is far from well known, although many of their properties are just a straightforward generalization of the corresponding properties of the Bieberbach groups. On the other hand, I do not claim to be a specialist of Bieberbach (or more general crystallographic) groups and so a lot more of the theory of Bieberbach (crystallographic) groups still has to be generalized. I hope therefore that this book might stimulate the reader to help in this generalization.

2. I also felt there is a need for a detailled classification of all almost–Bieberbach groups in dimensions ≤ 4. We will see that an infra-nilmanifold is completely determined by its fundamental group. So my classification of almost–Bieberbach groups can also be viewed as a classification of all infra–nilmanifolds of dimensions ≤ 4. I myself use the tables of almost–Bieberbach groups not really as a classification but as an elaborated set of examples or "test cases" for new hypotheses. I hope that, one day, they can be of the same value to you too.

I tried to write this monograph both for topologists/geometers as for algebraists. Therefore, I made an effort to keep the prerequisites as low as possible. However, the reader should have at least an idea of what a Lie group is. Also, a little knowledge of the theory of covering spaces can be helpful now and then. From the algebraic point of view, I assume that the reader is fairly familiar with nilpotent groups and that he is acquainted with group extensions and its relation to cohomology of groups.

Although this work is divided into eight chapters, there are really three parts to distinguish.

1. In the first part (Chapter 1 to Chapter 3), we define almost–crystallographic and almost–Bieberbach groups. We spend a lot of time in providing alternative definitions for them. Also we show how the three famous theorems of L. Bieberbach on crystallographic groups can be generalized to the case of almost–crystallographic groups. These first chapters could already suffice to let the reader start his own investigation of almost–crystallographic groups.

2. Chapter 4 forms a part on its own. It deals mainly with my own field of interest, namely the canonical type representations. These are representations of a polycyclic–by–finite group (in our situation always virtually nilpotent), which respect in some sense a given

filtration of that group. We discuss both affine and polynomial representations and present some nice existence and uniqueness results. The reason for considering polycyclic–by–finite groups is natural in the light of Auslander's conjecture.

3. The last part of this monograph (Chapter 5 to Chapter 8) describes a way to classify almost–Bieberbach groups. We also give a complete list of all almost–Bieberbach groups in dimensions ≤ 4, which were obtained using the given method. Moreover, we show how it was possible to use these tables and find in a pure algebraic way some topological invariants (e.g. Betti numbers) of the corresponding infra–nilmanifolds.

Finally, I would like to say a few words of thanks. To Professor Paul Igodt who introduced me to the world of infra–nilmanifolds and who proposed me to investigate the possibility of classifying the almost–Bieberbach groups. I am also grateful to Professor Kyung Bai Lee, since I owe much of my knowledge on almost–Bieberbach groups to him. But most of all I must thank my wife Katleen, for her encouragement when I was doing mathematics in general and especially for her support and practical help when I was writing this book.

Karel Dekimpe,
Kortrijk, August 19, 1996

Contents

Chapter 1

Preliminaries and notational conventions

1.1 Nilpotent groups

In this first chapter we discuss the fundamental results needed to understand this book. Our primary objects of study are virtually nilpotent groups. Remember that a group G is said to be virtually $\mathcal{P}$, where $\mathcal{P}$ is a property of groups, if and only if G contains a normal subgroup of finite index which is $\mathcal{P}$.

Although we assume familiarity with the concept of a nilpotent group, we recall some special aspects of this theory in order to fix some notations.

Let N be any group, then the upper central series of N

$$Z_*(N): \; Z_0(N) = 1 \subseteq Z_1(N) \subseteq \cdots \subseteq Z_i(N) \cdots$$

is defined inductively by the condition that

$$Z_{i+1}(N)/Z_i(N) = Z(N/Z_i(N))$$

where $Z(G)$ denotes the center of a group G. The group N is said to be nilpotent if the upper central series reaches N after a finite number of steps, i.e. there exists a positive integer c such that $Z_c(N) = N$. If c is the smallest positive integer such that $Z_c(N) = N$, we say that N is c–step nilpotent or N is nilpotent of class c.

Another frequently used central series is the lower central series. This series uses the commutator subgroups of a group N. We use the convention that the commutator $[a, b] = a^{-1}b^{-1}ab$ for all $a, b \in N$. Conjugation in N with a is indicated by $\mu(a)$. Sometimes we use $b^a = a^{-1}ba = \mu(a^{-1})(b)$.

The lower central series of N is the central series

$$N = \gamma_1(N) \supseteq \gamma_2(N) \supseteq \cdots \supseteq \gamma_i(N) \supseteq \cdots$$

where the i–fold commutator subgroups $\gamma_i(N)$ are defined inductively by the formula

$$\gamma_{i+1}(N) = [N, \gamma_i(N)].$$

The groups we are interested in are the finitely generated torsion free nilpotent groups for which it makes sence to consider central series

$$N_* : \ 1 = N_0 \subseteq N_1 \subseteq N_2 \subseteq \cdots \subseteq N_c = N$$

with torsion free quotients N_i/N_{i-1} for $1 \le i \le c$. We will refer to such a central series N_* as a torsion free central series.

Given such a torsion free central series, there exists integers $k_i \in \mathbb{Z}$ such that $N_i/N_{i-1} \cong \mathbb{Z}^{k_i}$. We write $K_i = \sum_{j \ge i} k_j$. We also write K for K_1, which is the rank or Hirsch number of N.

A set of generators

$$\{a_{1,1}, a_{1,2}, \ldots, a_{1,k_1}, a_{2,1}, \ldots, a_{2,k_2}, a_{3,1}, \ldots, a_{c,k_c}\}$$

of N will be called compatible with N_* iff

$$\forall i \in \{1, 2, \ldots, c\}: \ a_{1,1}, a_{1,2}, \ldots, a_{i,k_i} \text{ generate } N_i.$$

It is clear at once that any torsion free central series of N admits a compatible set of generators. Such a compatible set of generators may be obtained in the following way: First we choose k_1 generators of N_1, say $a_{1,1}, a_{1,2}, \ldots, a_{1,k_1}$. Then we complete this set to a set of generators for N_2. So we have to choose elements $a_{2,1}, \ldots a_{2,k_2}$. We continue this way and finally we find the last k_c generators $a_{c,1}, \ldots, a_{c,k_c}$. Any element $n \in N$ can now be written uniquely in the form

$$n = a_{c,1}^{x_{c,1}} a_{c,2}^{x_{c,2}} \ldots a_{1,k_1}^{x_{1,k_1}} \in N, \text{ for some } x_{i,j} \in \mathbb{Z}.$$

This shows that we may identify n with its coordinate vector

$$(x_{1,1}, x_{1,2}, \ldots, x_{1,k_1}, x_{2,1}, \ldots, x_{i,j}, \ldots, x_{c,k_c}) \in \mathbb{Z}^K.$$

For all torsion free finitely generated nilpotent groups N, the upper central series determines a torsion free central series, while in general the lower central series fails to have torsion free factors. However, we can alter the lower central series slightly in order to get a torsion free central series. To explain this we need the concept of the isolator:

Definition 1.1.1 *(see also [56], [59])*
*Let G be a group. For H a subgroup of G, the **isolator** of H in G
(sometimes called the **root set**) is defined by*

$$\sqrt[G]{H} = \{g \in G \parallel g^k \in H \text{ for some } k \geq 1\}.$$

In general, the isolator of a subgroup H in G doesn't have to be a subgroup
itself. E.g. if $H = 1$ then $\sqrt[G]{H}$ is exactly the set of torsion elements of G,
which needn't be a group in general. We will only need the isolator of a
commutator subgroup.

Lemma 1.1.2 *Let G be any group. Then,*

1. *$\forall k \in \mathbb{N}_0$: $\sqrt[G]{\gamma_k(G)}$ is a characteristic subgroup of G.*

2. *$\forall k \in \mathbb{N}_0$: $G/\sqrt[G]{\gamma_k(G)}$ is torsion free.*

3. *$\forall k, l \in \mathbb{N}_0$: $[\sqrt[G]{\gamma_k(G)}, \sqrt[G]{\gamma_l(G)}] \subseteq \sqrt[G]{\gamma_{k+l}(G)}$.*

For the proof of this lemma we refer the reader to [56, page 473].

It follows that for any finitely generated, torsion free c-step nilpotent
group N the series

$$\sqrt[N]{\gamma_{c+1}(N)} = 1 \subseteq \sqrt[N]{\gamma_c(N)} \subseteq \cdots \subseteq \sqrt[N]{\gamma_2(N)} \subseteq \sqrt[N]{\gamma_1(N)} = \sqrt[N]{N} = N$$

is a torsion free central series. We will refer to this series as **the adapted
lower central series**.

For any group G, the groups $\sqrt[G]{\gamma_i(G)}$ can be determined by means of a
universal property. Write $\tau_i(G) = G/\sqrt[G]{\gamma_{i+1}(G)}$ and denote the canonical
projection of G onto $\tau_i(G)$ by p. The group $\tau_i(G)$ is the biggest possible
torsion free quotient of G, which is nilpotent of class $\leq i$. Formally

Lemma 1.1.3 Universal property of $\tau_i(G)$.
*Let G be any group and suppose that N is a torsion free nilpotent group
of class $\leq i$. Given a group homomorphism $\varphi : G \to N$, there exists a
unique morphism $\psi : \tau_i(G) \to N$ such that $\varphi = \psi \circ p$. I.e. the following
diagram commutes:*

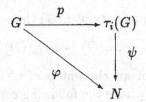

<u>Proof:</u> As N is nilpotent of class $\leq i$, all $i+1$–fold commutators are mapped trivially. So the morphism φ factors through $G/\gamma_{i+1}(G)$. Also, as N is torsion free, the characteristic subgroup $\tau(G/\gamma_{i+1}(G))$ consisting of all torsion elements of $G/\gamma_i(G)$ is mapped trivially. Therefore, there is a factorization

$$\varphi : G \to G/\gamma_{i+1}(G) \to (G/\gamma_{i+1}(G))/\tau(G/\gamma_{i+1}(G)) \to N. \qquad (1.1)$$

But $\sqrt[G]{\gamma_{i+1}(G)}$ consists exactly of those elements which are mapped into the set of torsion elements $\tau(G/\gamma_{i+1}(G))$ under the canonical projection of G onto $G/\gamma_{i+1}(G)$. So,

$$\tau(G/\gamma_{i+1}(G)) = \sqrt[G]{\gamma_{i+1}(G)}/\gamma_{i+1}(G)$$

from which it follows that the factorization (1.1) mentioned above can be written as:

$$\varphi : G \xrightarrow{\ p\ } \tau_i(G) \xrightarrow{\ \psi\ } N.$$

This establishes the existence of the map ψ. The uniqueness is obvious. ∎

The above proposition determines the subgroup $\sqrt[G]{\gamma_{i+1}(G)}$ of G completely. For suppose there exists another normal subgroup A of G (together with a canonical projection $q : G \to G/A$), such that any morphism $\varphi : G \to N$ as above can be written in the form $\varphi = \psi' \circ q$. In this case let N be equal to $G/\sqrt[G]{\gamma_{i+1}(G)}$ and let $\varphi = p$. It is obvious that the map ψ' in this case maps the coset gA onto $g\sqrt[G]{\gamma_{i+1}(G)}$, for all $g \in G$. By reversing the roles, we obtain a morphism $\psi : G/\sqrt[G]{\gamma_{i+1}(G)} \to A : g\sqrt[G]{\gamma_{i+1}(G)} \to gA$, which is the inverse of ψ'. Therefore, the groups A and $\sqrt[G]{\gamma_{i+1}(G)}$ coincide. As an application of the above universal property we find:

Lemma 1.1.4 *Let G be any group.*
For all $j \geq i$, there is a canonical isomorphism

$$\tau_i(\tau_j(G)) \cong \tau_i(G).$$

It follows that for $\overline{G} = \sqrt[G]{\gamma_{j+1}(G)}$

$$\sqrt[G/\overline{G}]{\gamma_{i+1}(G/\overline{G})} = \sqrt[G]{\gamma_{i+1}(G)}/\overline{G}.$$

<u>Proof:</u> By the universal property of $\tau_i(G)$ there is a canonical morphism

$$c_1 : \tau_i(G) \to \tau_i(\tau_j(G))$$

which maps the coset of an element g of G onto the coset of $g\overline{G}$ in $\tau_i(\tau_j(G))$. Conversely, we have the following commutative diagram

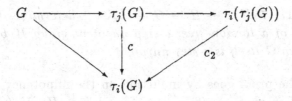

where

1. the non labeled arrows are canonical projections onto a quotient group.

2. c is induced by the fact that $\sqrt[G]{\gamma_{j+1}(G)}$ is contained in $\sqrt[G]{\gamma_{i+1}(G)}$ $(j \geq i)$.

3. c_2 is obtained by the universal property of $\tau_i(\tau_j(G))$.

It is clear now that c_1 and c_2 are each others inverse. The last claim of the lemma, concerning the equality of the two subgroups of $G/\overline{G}$, follows from the comments preceding this lemma.

∎

The following technical lemmas will be needed at special occasions during our treatment of almost–crystallographic groups.

Lemma 1.1.5 *Let H be a torsion free, normal subgroup of finite index in a group G. Assume $z \in Z(H)$ and $x \in G$ such that $[x, z] \neq 1$. Then any commutator of the form $[x, [\ldots [x, [x, z]]\ldots]$ is not trivial.*

<u>Proof :</u> Consider the sequence $(c_i)_{i \in \mathbb{N}}$ in $Z(H)$ defined by $c_0 = z$ and $c_{i+1} = [x, c_i]$. We proceed by induction. Assume $c_i \neq 1$ and $c_{i+1} = 1$ for $i \geq 1$. If $[G : H] = m$, $x^m \in H$ and hence it commutes with c_{i-1}. A trivial computation shows that

$$1 = [x^m, c_{i-1}] = \prod_{j=1}^{m} [x, c_{i-1}]^{x^{m-j}} = \prod_{j=1}^{m} c_i^{x^{m-j}} = c_i^m.$$

Since H is torsion free $c_i = 1$, which is a contradiction.

∎

Lemma 1.1.6 *If $0 \to H \to G \to K \to 1$ determines G as a central extension of an abelian group H by a group K which is nilpotent of class $\leq c$, then G is nilpotent of class $\leq (c + 1)$.*

The proof is straightforward and left to the reader.

Lemma 1.1.7 *If* $1 \to H \to G \to \mathbb{Z}_k \to 1$ *determines* G *as a nilpotent extension of a torsion–free, c-step nilpotent group* H *by a finite cyclic group, then* G *itself is c-step nilpotent.*

<u>Proof :</u> The proof goes by induction on the nilpotency class of H. So, assume H is abelian. Take $g \in G$ and $h \in H$. If $[g,h] \neq 1$, lemma 1.1.5 implies that G is not nilpotent. Consequently, it follows that the extension is a central one. Having chosen a section $s : \mathbb{Z}_k \to G$, elements $g \in G$ can be written as $hs(x^l)$ ($h \in H$, x is a generator of $\mathbb{Z}_k$ and $0 \leq l \leq k - 1$). Now, it is clear that if g_1 and g_2 are in G, $[g_1, g_2] = [s(x^{l_1}), s(x^{l_2})]$. Since $s(x)^l$ and $s(x^l)$ belong to the same coset of H, it follows that $[g_1, g_2] = [s(x)^{l_1}, s(x)^{l_2}] = 1$, so G is abelian.

Now assume H is of class c. Lemma 1.1.5 implies that $Z(H) \subset Z(G)$ when G is nilpotent. Consider the short exact sequence $1 \to H/Z(H) \to G/Z(H) \to \mathbb{Z}_k \to 1$. Here, $H/Z(H)$ is torsion–free $(c-1)$-step nilpotent. By induction $G/Z(H)$ itself is $(c-1)$-step nilpotent. Apply lemma 1.1.6 to the extension $1 \to Z(H) \to G \to G/Z(H) \to 1$ and deduce that G is nilpotent of class $\leq c$.

■

Lemma 1.1.8 *Let* G *be any group and suppose that* T *is a torsion free normal subgroup of* G, *while* F *is a finite normal subgroup of* G. *Then*

$$[T, F] = 1.$$

<u>Proof:</u> Let $t \in T$ and $f \in F$, then

$$[t, f] = \underbrace{t^{-1} f^{-1} t}_{\in F} f = t^{-1} \underbrace{f^{-1} t f}_{\in T} \in T \cap F = 1$$

■

Lemma 1.1.9 *Let* φ *be any automorphism of* $\mathbb{Z}^k$. *If there exists a subgroup* A *of finite index in* $\mathbb{Z}^k$ *on which* φ *is the identity, then* φ *is the identity automorphism.*

<u>Proof:</u> φ can be represented by an invertible matrix M with integral entries. Seen as an element of $\mathrm{Gl}(n, \mathbb{R})$, this matrix represents a linear mapping leaving fixed a generating set (i.e. A) of the real vector space $\mathbb{R}^n$. This implies that M is the $n \times n$–identity matrix.

■

1.2   Nilpotent Lie groups

Although we intend to keep the topological/geometrical background needed to understand this book as small as possible, we need at least some
knowledge concerning nilpotent Lie groups. In fact, most of the stuff we
will use can be found in the magnificent paper of A.I. Mal'cev [51]. We
refer to this paper for all the proofs of the claims we make here.

Throughout this section G will denote a connected and simply connected nilpotent Lie group. We use $\mathfrak{g}$ to indicate the Lie algebra of G.
This Lie algebra $\mathfrak{g}$ has the same dimension and nilpotency class as G.
Moreover, in the case of connected and simply connected nilpotent Lie
groups it is known that the exponential map $\exp : \mathfrak{g} \to G$ is bijective. We
denote its inverse by log. The exponential map earns its name because of
the fact that for matrix groups/algebras the exponential map is indeed
given by the exponentiation of matrices. I.e. $\exp(A) = \sum_{n=0}^{\infty} \frac{A^n}{n!}$.
If H is another connected and simply connected nilpotent Lie group, with
Lie algebra $\mathfrak{h}$, then we have the following properties:

- For any morphism $\varphi : G \to H$ of Lie groups, there exists a unique
  morphism $d\varphi : \mathfrak{g} \to \mathfrak{h}$ (differential of φ) of Lie algebras, making the
  following diagram commutative:

$$
\begin{array}{ccc}
G & \xrightarrow{\varphi} & H \\
\log \downarrow \uparrow \exp & & \log \downarrow \uparrow \exp \\
\mathfrak{g} & \xrightarrow{d\varphi} & \mathfrak{h}
\end{array}
\qquad (1.2)
$$

- Conversely, for any morphism $d\varphi : \mathfrak{g} \to \mathfrak{h}$ of Lie algebras, there
  exists a unique morphism $\varphi : G \to H$ of Lie groups, making the
  above diagram commutative.

In G, it makes sense to speak of a^x where $a \in G$ and $x \in \mathbb{R}$. (E.g.
consider the one-parameter subgroup of G passing through a). A formal
definition may look as follows:

Definition 1.2.1

$$
a^x = \exp(x \log a), \quad \forall a \in G, \ \forall x \in \mathbb{R}.
$$

The definition satisfies all the expected conditions:

1. $a.a^{-1} = a^{-1}.a = 1$; $a^0 = 1$,

2. $a^n = \underbrace{a.a\ldots a}_{n \text{ times}}$, if $n \in \mathbb{N}$,

3. $\left(a^{\frac{1}{n}}\right)^n = a$, if $n \in \mathbb{N}$,

4. $(a^x)^y = a^{xy}$,

5. $a^x.a^y = a^{x+y}$,

6. If $\varphi : G \to H$ is a morphism between connected and simply connected nilpotent Lie groups, then $\varphi(a^x) = (\varphi(a))^x$.

We give a proof of this last property, using the commutative diagram (1.2):

$$
\begin{aligned}
\varphi(a^x) &= \varphi(\exp x \log a) \\
&= \exp(d\varphi(x \log a)) \\
&= \exp(x \, d\varphi(\log a)) \\
&= \exp(x \log(\varphi(a))) \\
&= (\varphi(a))^x .
\end{aligned}
$$

∎

We also mention the famous Campbell–Baker–Hausdorff formula:

$$\forall A, B \in \mathfrak{g}: \ \exp(A).\exp(B) = \exp(A * B), \qquad (1.3)$$

where

$$A * B = A + B + \frac{1}{2}[A, B] + \sum_{m=3}^{\infty} C_m(A, B).$$

Here $C_m(A, B)$ stands for a rational linear combination of m–fold Lie brackets in A and B. Since our Lie algebras are nilpotent, the sum involved in $A * B$ is always finite. As an immediate consequence of this formula, one sees that

$$\forall a, b \in G: \ log(a.b) = \log a * \log b.$$

Of major importance to us, is the concept of a uniform lattice of G.

Definition 1.2.2 *Let G be a connected and simply connected nilpotent Lie group. A uniform lattice of G is a uniform discrete subgroup, i.e. a discrete subgroup with compact quotient, N of G.*

We remark that not all connected and simply connected nilpotent groups admit lattices.

One of the nicests results of Mal'cev is the "unique isomorphism extension property"

Theorem 1.2.3 *Let G and H be two connected and simply connected nilpotent Lie groups. Suppose moreover that N and M are uniform lattices of G and H respectively. Then any isomorphism $\varphi : N \to M$ extends uniquely to an isomorphism of Lie groups of G onto H.*

In case we use this property for $M = N$, we also say "the unique automorphism extension property ".

V.V. Gorbacevič ([33]) generalized this theorem as follows:

> Let N be a uniform lattice of a simply connected, connected nilpotent Lie group G and let H be an arbitrary simply connected, connected nilpotent Lie group. Then any morphism $\varphi : N \to H$ extends uniquely to a morphism $G \to H$.

Mal'cev also describes all possibilities of uniform lattices in a connected and simply connected nilpotent Lie group.

Theorem 1.2.4 *Any lattice N of a connected and simply connected nilpotent Lie group G is a finitely generated torsion free nilpotent group. Conversely, for any torsion free finitely generated nilpotent group N there exists (up to isomorphism) exactly one connected and simply connected nilpotent Lie group G containing N as a uniform lattice. We refer to this G as the* **Mal'cev completion** *of N.*

Let N be a torsion free and finitely generated nilpotent group with a torsion free central series N_* as in the previous section. Suppose moreover that

$$\{a_{1,1}, a_{1,2}, \ldots, a_{1,k_1}, a_{2,1}, \ldots, a_{c,k_c}\}$$

is a set of generators compatible with N_*. Then the elements

$$\{A_{1,1} = \log(a_{1,1}), A_{1,2} = \log(a_{1,2}), \ldots, A_{c,k_c} = \log(a_{c,k_c})\}$$

form a basis for the Lie algebra $\mathfrak{g}$ of the Mal'cev completion G of N. It follows that any element g of G can be written uniquely in the form

$$g = a_{c,1}^{x_{c,1}} a_{c,2}^{x_{c,2}} \ldots a_{1,k_1}^{x_{1,k_1}}. \text{ for some } x_{i,j} \in \mathbb{R}.$$

Any element can thus be identified with a coordinate vector

$$(x_{1,1}, x_{1,2}, \ldots, x_{1,k_1}, x_{2,1}, \ldots, x_{i,j}, \ldots, x_{c,k_c}) \in \mathbb{R}^K.$$

Moreover, such an element belongs to the lattice N if and only if its coordinate vector belongs to $\mathbb{Z}^K$. Using the Campbell–Baker–Hausdorff formula one proves that multiplication in G (and so in N) is given by a polynomial function of these coordinates.

Finally we want to mention the following properties:

Lemma 1.2.5 *Let N be a uniform lattice of a connected and simply connected nilpotent Lie group G with Lie algebra $\mathfrak{g}$. Then,*

1. $\gamma_i(N)$ *and* $\sqrt[N]{\gamma_i(N)}$ *are uniform lattices of* $\gamma_i(G)$,

2. $Z_i(N)$ *is a uniform lattice of* $Z_i(G)$,

3. $\exp(\gamma_i(\mathfrak{g})) = \gamma_i(G)$,

4. $\exp(Z_i(\mathfrak{g})) = Z_i(G)$.

Here the upper and lower central series of a Lie algebra $\mathfrak{g}$ are defined analogously to the upper and lower central series of a group. We also find an alternative description of the isolator subgroup.

Lemma 1.2.6
Let N be a finitely generated, torsion free nilpotent group with Mal'cev completion G, then $\forall i : \sqrt[N]{\gamma_i(N)} = N \cap \gamma_i(G)$.

<u>Proof:</u> Using the universal property of the groups $\tau_i(N)$ (see also the comments after lemma 1.1.3) it is enough to proof that for any morphism $\varphi : N \to M$ where M is a torsion free nilpotent group of class $\leq i$ there exists a unique morphism $\psi : N/(N \cap \gamma_i(G))$ making the following diagram commutative.

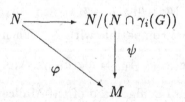

It is obvious that ψ, if it exists, is unique. To prove the existence of ψ, consider the finitely generated, torsion free nilpotent group M' of class $\leq i$ given by $M' = \varphi(M)$. Denote the Mal'cev completion of M' by H. By the result of V.V. Gorbacevič, mentioned above, we know that the morphism φ extends (uniquely) to a homomorphism $\tilde{\varphi} : G \to H$.

Moreover, as H is nilpotent of class $\leq i$, $\tilde{\varphi}$ factors through $G/\gamma_{i+1}(G)$. The map ψ is now obtained as a restriction of $\tilde{\varphi}$ to the subgroup

$$N \cdot \gamma_{i+1}(G)/\gamma_{i+1}(G) = N/(N \cap \gamma_{i+1}(G)).$$

This shows that $\sqrt[N]{\gamma_{i+1}(N)} = N \cap \gamma_{i+1}(G)$. ∎

Chapter 2

Infra–nilmanifolds and Almost–Bieberbach groups

2.1 Crystallographic and Bieberbach groups

Let G be any connected and simply connected Lie group. We denote by $\mathrm{Aut}\,(G)$ the group of automorphisms of G (as a Lie group). The semidirect product $G \rtimes \mathrm{Aut}\,(G)$ acts on G in a canonical way by

$$\forall g, x \in G, \ \forall \alpha \in \mathrm{Aut}\,(G): \ {}^{(g,\alpha)}x = g\alpha(x).$$

We denote the group $G \rtimes \mathrm{Aut}\,(G)$ by $\mathrm{Aff}(G)$ and refer to it as the group of affine diffeomorphisms of G. If $G = \mathbb{R}^n$ the abelian Lie group, then $\mathrm{Aff}(G) = \mathbb{R}^n \rtimes \mathrm{Gl}\,(n, \mathbb{R})$ the usual affine group (with the usual action on $\mathbb{R}^n$).

In this book we will be concerned with discrete subgroups Π of $\mathrm{Aff}(G)$ which act properly discontinuously on G. If, moreover, the action of Π on G is free (i.e. Π is torsion free), then the quotient space $\Pi \backslash G$ is a manifold.

Let us look at a concrete case and suppose G is a connected and simply connected abelian Lie group. So G equals $\mathbb{R}^n$ for some n. It is well known that the orthogonal group $O(n)$ is a maximal compact subgroup of $\mathrm{Gl}\,(n, \mathbb{R})$. A uniform discrete subgroup Π of $\mathbb{R}^n \rtimes O(n) \subseteq \mathrm{Aff}(\mathbb{R}^n)$ is called a **crystallographic group** of dimension n. If Π is torsion free, Π is said to be a **Bieberbach group** . These groups are well known by the work of L. Bieberbach. Standard references are [60] and [13]. We mention three famous theorems on Bieberbach groups (in fact on crystallographic groups) which were published around 1911 and 1912, by L. Bieberbach

and ,G. Fröbenius ([6], [7], [32]). The order in which the theorems are
listed is not the historical order in which they appeared. We will shortly
sketch the history of these theorems after there statements.

Theorem 2.1.1 First Bieberbach theorem

Let Π be an n-dimensional crystallographic group, then $\Gamma = \Pi \cap \mathbb{R}^n$ is a
lattice of $\mathbb{R}^n$ and Π/Γ is finite.

This means that for an n-dimensional crystallographic group Π, the trans-
lational part $\Gamma = \Pi \cap \mathbb{R}^n$ of Π is a free abelian group isomorphic with $\mathbb{Z}^n$,
such that the vector space spanned by Γ is the whole space $\mathbb{R}^n$. If $\Pi = \Gamma$,
then Π is torsion free and the manifold $_\Pi\backslash\mathbb{R}^n$ is an n–dimensional torus.
For more general torsion free Π we see that the manifold $M = {}_\Pi\backslash\mathbb{R}^n$
inherits the flat Riemannian structure of $\mathbb{R}^n$ (as Π is a group of distance
preserving transformations) and so M is a compact flat Riemannian manifold. Moreover, all compact flat Riemannian manifolds are obtained in
this way. Thus the Bieberbach groups are exactly the fundamental groups
of the compact flat Riemannian manifolds.

Theorem 2.1.2 Second Bieberbach theorem

Let Π and Π' be two n–dimensional crystallographic groups. If $\psi : \Pi \rightarrow \Pi'$
is an isomorphism, then there exists an element $\alpha \in \mathrm{Aff}(\mathbb{R}^n)$ such that
$\forall \gamma \in \Pi : \psi(\gamma) = \alpha\gamma\alpha^{-1}$.

From this, one deduces that two flat Riemannian manifolds with isomorphic fundamental groups are "affinely" diffeomorphic and so, a flat
Riemannian manifold is up to a well understood diffeomorphism completely determined by its fundamental group.

Theorem 2.1.3 Third Bieberbach theorem *or* Fröbenius' Theorem

Up to conjugation in $\mathrm{Aff}(\mathbb{R}^n)$, there are only finitely many n–dimensional
crystallographic groups.

This implies that, up to affine diffeomorphism, there are only finitely
many n–dimensional flat Riemannian manifolds.

These theorems constitute an answer to one of the famous Hilbert
Problems, number 18 to be precise. The first Bieberbach theorem is indeed historically the first theorem which L. Bieberbach proved. After
this theorem L. Bieberbach could show that there where only finitely
many crystallographic groups of a given dimension, up to **isomorphism**.

It was G. Fröbenius who reacted that one should not consider crystallographic groups up to isomorphism, but up to affine equivalence. Fröbenius then also proved theorem 2.1.3 as stated above. Inspired by the work of Fröbenius, L. Bieberbach then showed that an isomorphism between two crystallographic groups is always induced by an affine conjugation, i.e. he proved, as last, the Second Bieberbach Theorem. So please be aware that the ordering of the Bieberbach theorems is not a historical ordering.

The algebraic structure of a crystallographic group is well know, one part due to the first Bieberbach theorem and the converse part due to H. Zassenhaus ([62]). We summarize this in the following theorem:

Theorem 2.1.4 Algebraic characterization
Let Π be a crystallographic group of dimension n, then $\Gamma = \Pi \cap \mathbb{R}^n$ is the unique normal, maximal abelian subgroup of Π.
Conversely, let Q be an abstract group such that Q contains a normal, maximal abelian group Γ of finite index, with $\Gamma \cong \mathbb{Z}^n$, then there exists a monomorphism $\varphi : Q \to \mathrm{Aff}(\mathbb{R}^n)$ such that $\varphi(Q)$ is a crystallographic group.

We will refer to such a Q as an abstract crystallographic (Bieberbach) group and even as a crystallographic (Bieberbach) group.

2.2 Almost–crystallographic groups

As an immediate generalization of the crystallographic groups, we look what happens in case G is a connected, simply connected nilpotent Lie group. Again we consider a maximal compact subgroup C of $\mathrm{Aut}\,(G)$. A uniform discrete subgroup E of $G \rtimes C$ is called an almost–crystallographic group (abbreviated in the sequel as AC–group). A torsion free AC–group is called an almost–Bieberbach group (abbreviated as AB–group) and the quotient-space $_E\backslash G$ is now called an infra–nilmanifold. (In case $E \subset G$, $_E\backslash G$ is said to be a nilmanifold). So the nilmanifolds take over the role of the tori and the infra–nilmanifolds the role of the flat Riemannian manifolds. Of course, the dimension of an AC–group is the same as the dimension of the corresponding Lie group. The adjective "almost" is inpired by the geometrical fact that the infra–nilmanifolds are exactly the almost flat manifolds. We are not going into detail about this geometrical point of view, but refer to [35], [31] and [57] for more information.

All three Bieberbach theorems have been generalized to the nilpotent case.

The first theorem was generalized by L. Auslander and published in 1960 ([1]). The formulation is as follows:

Theorem 2.2.1 Generalizing the first Bieberbach theorem
Let $E \subseteq G \rtimes \mathrm{Aut}\,(G)$ be an n–dimensional almost–crystallographic group. Then $E \cap G$ is a uniform lattice of G and $E/(E \cap G)$ is finite.

Although the proof of this theorem is not so difficult (but it is rather long), the techniques used do not appear elsewhere in this book and therefore we present this theorem here without proof. The interested reader should consult [1].

In his paper, Auslander also proves that $N = E \cap G$ is the (unique) normal maximal nilpotent subgroup of E. (In fact, Auslander only shows that N is maximal amongst all **normal** nilpotent groups, but his arguments can be used to prove that N is maximal nilpotent amongst all groups. This was remarked by K.B. Lee and F. Raymond in [48]). An alternative proof will be presented later on. We still call N the translational part of E. The finite quotient $F = E/N$ is often referred to as the holonomy group of E. The theorem above shows that the Lie group G is precisely the Mal'cev completion of the translational group $E \cap G$.

L. Auslander also formulated a generalization for the second Bieberbach theorem. However, this formulation was incorrect (even for Bieberbach groups) and a counter example was presented in ([48]). A correct generalization of this second theorem is also given in [48] by K.B. Lee and F. Raymond:

Theorem 2.2.2 Generalizing the second Bieberbach theorem
Let $\psi : E \to E'$ be an isomorphism between two almost–crystallographic groups (of a fixed Lie group G), then ψ can be realized as a conjugation by an element of $\mathrm{Aff}(G)$.

This theorem implies that two infra–nilmanifolds with isomorphic fundamental groups are affinely diffeomorphic. This theorem is of significant importance since it states that a classification of infra–nilmanifolds can be done on the group level.

We wrote "of a fixed Lie group G" between parentheses, since an isomorphism between two AC–groups is only possible if the corresponding simply connected, connected nilpotent Lie groups are isomorphic. Indeed, as the translational subgroups of E and E' are the unique normal and maximal nilpotent subgroups N and N' of E and E' respectively, it follows that any isomorphism $\psi : E \to E'$ restricts to an isomorphism of N

onto N'. By the unique isomorphism extension property mentioned in the previous chapter, we know that ψ induces indeed an isomorphism between the two Mal'cev completions.

To prove theorem 2.2.2, we first define a notion of 1-cohomology with non-abelian coefficients. Let A and B be any groups. Suppose that a morphism $\varphi : A \to \text{Aut}(B)$ is given. A map $f : A \to B$ is called a 1-cocycle (from A to B, with respect to φ) if and only if

$$\forall x, y \in A : f(xy) = (\varphi(x)f(y))f(x).$$

Two 1-cocycles f and g are equivalent if $f(x) = (\varphi(x)(b^{-1}))g(x)b$ for all $x \in A$. It is readily seen that this relation is an equivalence relation. The set of equivalence classes is denoted by $H^1_\varphi(A, B)$ and is non empty since it contains at least one class, namely the class of the map $f : A \to B : x \mapsto 1$. If B is an abelian group, this notion of 1-cohomology coincides with the usual notion.

The following lemma generalizes the fact that $H^1(F, \mathbb{R}^n) = \{1\}$, for all finite groups F.

Lemma 2.2.3 *Let F be any finite group and suppose that G is a connected, simply connected nilpotent Lie group. Then, for any morphism $\varphi : F \to \text{Aut}(G)$, we have that*

$$H^1_\varphi(F, G) = \{1\}.$$

<u>Proof:</u>
We proceed by induction on the nilpotency class c of G. If $c = 1$, $G = \mathbb{R}^n$ and the lemma is well known in this case. Any automorphism $\varphi(x)$ ($x \in F$) restricts to an automorphism of $Z(G)$ (which we denote by the same symbol) and induces an automorphism $\bar\varphi$ of $G/Z(G)$. We denote the canonical projection of $g \in G$ in $G/Z(G)$ by $\bar g$.

Suppose a 1-cocycle $f : F \to G$ is given. Then there is an induced 1-cocycle $\bar f : F \to G/Z(G)$ (with respect to $\bar\varphi$) given by $\bar f(x) = \overline{f(x)}$. By the induction hypothesis, we know that f is cohomologuous to 1 and so there exists an element $a \in G$ such that

$$\bar f(x) = \bar\varphi(x)(\bar a^{-1})\bar a = \overline{\varphi(x)(a^{-1})a}, \; \forall x \in F.$$

Therefore, we can define a map

$$g : F \to Z(G) : x \mapsto \varphi(x)(a^{-1})a(f(x))^{-1}.$$

A trivial computation shows that this map g is a 1–cocycle with respect to φ (restricted to $Z(G)$). However, as $Z(G) \cong \mathbb{R}^k$ for some k, this 1–cocycle is cohomologuous to 1. So, there exists a $z \in Z(G)$ such that $g(x) = \varphi(x)(z^{-1})z$, $\forall x \in F$. This implies that $\varphi(x)(z^{-1})z = \varphi(x)(a^{-1})a(f(x))^{-1}$, from which it follows that

$$\forall x \in F: \ f(x) = \varphi(x)((az^{-1})^{-1})az^{-1}.$$

In other words, f is equivalent to 1.

■

Proof of theorem 2.2.2:
As explained above, the isomorphism $\psi : E \to E'$ induces an isomorphism $\nu : N \to N'$ between the translation subgroups, which extends (uniquely) to an automorphism of G, which we also denote by ν. Remark that

$$\forall n \in N : \psi(n,1) = (\nu(n),1) = (1,\nu)(n,1)(1,\nu)^{-1}.$$

This implies that, after conjugating with $(1,\nu)$, we may (and do) suppose that ψ induces the identity map on the translational subgroups.

Let $(g,\alpha) \in E$ and denote $\psi(g,\alpha) = (h,\beta)$. For all $(n,1) \in N$

$$(g,\alpha)(n,1)(g,\alpha)^{-1} = (g\alpha(n)g^{-1},1).$$

After applying ψ to both sides of the above equality, we find that

$$(h\beta(n)h^{-1},1) = (g\alpha(n)g^{-1},1) \Rightarrow \beta(n) = \mu(h^{-1}g)\alpha(n)$$

from which it follows that β is completely determined by α, g and h (on N and thus on G by the unique automorphism extension property). As h is determined (via ψ) by g and α, we can define a map

$$f : E \to G : (g,\alpha) \mapsto f(g,\alpha) = g^{-1}h.$$

Moreover, by comparing $f(g,\alpha)$ with $f(g',\alpha)$, we see that f is independent of g and so, f is in fact a map of the finite group $F = E/(E \cap G)$ into G. This f is a 1–cocycle with respect to $\varphi : F \to \operatorname{Aut}(G)$ defined by mapping the coset of (g,α) to α.

The previous lemma implies that f is cohomologuous to 1. Let a be an element of G, for which $f(g,\alpha) = \alpha(a^{-1})a$, then

$$\psi(g,\alpha) = (gf(g,\alpha), \mu((f(g,\alpha))^{-1})\alpha) = (a, \mu(a)^{-1})(g,\alpha)(a,\mu(a)^{-1})^{-1}.$$

This finishes the proof of theorem 2.2.2.

■

2.3   How to generalize the third Bieberbach theorem

The third theorem is not as straightforward to generalize as the other two. This is caused by the fact that in the abelian case all n–dimensional tori are diffeomorphic, while in the nilpotent case there are infinitely many, non–homeomorphic nilmanifolds in dimensions ≥ 3. In fact, there can be infinitely many simply connected, connected nilpotent Lie groups in a given dimension. The following considerations are due to K.B. Lee ([45]). For example, one can consider the three dimensional Heisenberg group H:

$$H = \left\{ \begin{pmatrix} 1 & x & y \\ 0 & 1 & z \\ 0 & 0 & 1 \end{pmatrix} \mid\mid x, y, z \in \mathbb{R} \right\}. \tag{2.1}$$

For any integer $k > 0$, we obtain a uniform lattice Γ_k of H, which is the subgroup generated by:

$$a = \begin{pmatrix} 1 & 0 & 0 \\ 0 & 1 & 1 \\ 0 & 0 & 1 \end{pmatrix} \quad b = \begin{pmatrix} 1 & 1 & 0 \\ 0 & 1 & 0 \\ 0 & 0 & 1 \end{pmatrix} \quad c = \begin{pmatrix} 1 & 0 & \frac{1}{k} \\ 0 & 1 & 0 \\ 0 & 0 & 1 \end{pmatrix}.$$

It is easy to see that

$$\Gamma_k \; : \; < a, b, c \mid\mid [b,a] = c^k, \; [c,a] = [c,b] = 1 > .$$

Since $H_1(\Gamma_k, \mathbb{Z}) = \Gamma_k / [\Gamma_k, \Gamma_k] = \mathbb{Z}^2 \oplus \mathbb{Z}_k$, we may conclude that the manifolds $\Gamma_k \backslash H$ are pairwise non–homeomorphic. So, it is certainly not exact to claim that there are only finitely many infra–nilmanifolds in each dimension. Another possible way to look at the third Bieberbach theorem is the following: each torus covers only finitely many flat Riemannian manifolds. So we try to generalize the third theorem by fixing a nilmanifold and by looking at the infra–nilmanifolds covered by it. But $\Gamma_1 \backslash H$ is a k–fold covering of $\Gamma_k \backslash H$, and so, one nilmanifold covers infinitely many other nilmanifolds. Therefore, K.B. Lee introduces the notion of an essential covering. Let $M = {}_N \backslash G$ be a nilmanifold and $M' = {}_E \backslash G$ an infra–nilmanifold. A covering $p : M \to M'$ is said to be essential iff the induced map on the level of the fundamental groups is so that $p_*(N) = E \cap G$. This implies that E and N have "the same" translational part, namely N.

A correct formulation, due to K.B. Lee, for the generalization of the third theorem is:

Theorem 2.3.1 *There are, up to an affine diffeomorphism, only finitely many infra–nilmanifolds which are essentially covered by a fixed nilmanifold.*

We postpone the proof of this theorem for a while. In fact we are going to translate the above formulation into a group theoretic language. The fact that this algebraic formulation is really equivalent to the topological one will become clear in the sequel of this book.

Definition 2.3.2
Let N be a finitely generated, torsion free nilpotent group. A group extension $1 \to N \to E \to F \to 1$, with F finite and in which N is a maximal nilpotent subgroup of E will be called "essential".

At this point we whish to recall the notion of the Fitting subgroup of a polycyclic–by–finite group (e.g. see [59]).

Definition 2.3.3 *Let Γ be a polycyclic–by–finite group. Then the Fitting subgroup of Γ, denoted by $\mathrm{Fitt}\,(\Gamma)$ is the unique maximal normal nilpotent subgroup of Γ.*

Using this terminology, we see that the extension $1 \to N \to E \to F \to 1$, with N torsion free finitely generated nilpotent and F finite, is essential if and only if $N = \mathrm{Fitt}\,(E)$ and there is no nilpotent subgroup of E, strictly containing N.

It is easy to see that a covering of an infra–nilmanifold by a nilmanifold is essential if and only if the induced extension of the fundamental groups is essential. Therefore it will be enough to prove the generalization of the third theorem on the group level, which can be stated as follows:

Theorem 2.3.4
Let N be a finitely generated, torsion free nilpotent group. Then there are, up to isomorphism, only finitely many (essential) extensions of the form $1 \to N \to E \to F \to 1$ in which N is the maximal nilpotent subgroup and F is finite.

In his proof, K.B. Lee used a counting principle based on a group cohomology argument. As the example in the next section will show, this principle is unsuitable for this context and hence should be replaced. In the section thereafter we develop a different proof of the theorem, now based on the finiteness of the number of conjugacy classes of subgroups in an arithmetic group, a result which goes back to the work of A. Borel and Harish-Chandra ([8]; see also [59]).

2.4 The first proof of the generalized third Bieberbach theorem revisited

From a conceptual point of view the idea behind the proof in [45] is that of reducing the classification of AC-groups having a fixed N as maximal nilpotent subgroup, to the known classification of crystallographic groups. It is this approach which opens the way we will follow to classify explicitly the AC-(and AB-) groups in dimensions 3 and 4. However, as we will point out in an example, it is unsuitable for counting the isomorphism types of the AC-groups obtained and consequently, it should be revised as an argument to the proof of theorem 2.3.4.

Let us again focus on the situation of a group E containing a finitely generated, torsion free, c-step nilpotent group N of finite index; i.e. we consider an extension (not necessarily essential)

$$1 \to N \to E \overset{j}{\to} F \to 1 \tag{2.2}$$

with $j : E \to F : x \mapsto j(x) = \bar{x}$.

Since the isolator of $\gamma_2(N)$ in N is normal in E, we obtain a short exact sequence

$$1 \to \dfrac{N}{\sqrt[N]{\gamma_2(N)}} \to \dfrac{E}{\sqrt[N]{\gamma_2(N)}} \to F \to 1. \tag{2.3}$$

$N/{\sqrt[N]{\gamma_2(N)}}$ is a finitely generated free abelian group, and hence it becomes an F-module via a morphism $\varphi_{N,E} : F \to \mathrm{Aut}(N/{\sqrt[N]{\gamma_2(N)}})$.

Let us write Γ for the isolator of $\gamma_c(N)$ in N. Given the canonical projection $E \to E/\Gamma$, we will write x_Γ for the image of an element $x \in E$. Remark that Γ is contained in $Z(N)$. In the extension

$$1 \to N/\Gamma \to E/\Gamma \to F \to 1. \tag{2.4}$$

N/Γ is torsion free, nilpotent of class $c - 1$ and of finite index in E/Γ. Similarly to (2.3), we obtain

$$1 \to \dfrac{N/\Gamma}{{}^{N/\Gamma}\!\sqrt{\gamma_2(N/\Gamma)}} \to \dfrac{E/\Gamma}{{}^{N/\Gamma}\!\sqrt{\gamma_2(N/\Gamma)}} \to F \to 1. \tag{2.5}$$

By lemma 1.1.4 there is a canonical isomorphism

$$\tau : \dfrac{E}{\sqrt[N]{\gamma_2(N)}} \to \dfrac{E/\Gamma}{{}^{N/\Gamma}\!\sqrt{\gamma_2(N/\Gamma)}}$$

such that the following diagram commutes :

$$
\begin{array}{ccccccccc}
1 & \to & \dfrac{N}{\sqrt[N]{\gamma_2(N)}} & \to & \dfrac{E}{\sqrt[N]{\gamma_2(N)}} & \to & F & \to & 1 \\
 & & \downarrow \tau' & & \downarrow \tau & & \| & & \\
1 & \to & \dfrac{N/\Gamma}{\sqrt[N/\Gamma]{\gamma_2(N/\Gamma)}} & \to & \dfrac{E/\Gamma}{\sqrt[N/\Gamma]{\gamma_2(N/\Gamma)}} & \to & F & \to & 1
\end{array}
$$

Now it is clear that, with respect to the induced action $\varphi_{N/\Gamma,E/\Gamma}$ of F on

$$
\dfrac{N/\Gamma}{\sqrt[N/\Gamma]{\gamma_2(N/\Gamma)}} \, , \tau' \text{ is an isomorphism of } F\text{-modules.}
$$

Keeping all this in mind, we establish a relation between the action $\varphi_{N,E}$ and the maximal nilpotency of N in E:

Proposition 2.4.1 $grp(N \cup \{x\})$ *is nilpotent* $\Leftrightarrow \bar{x} \in \ker(\varphi_{N,E})$.

<u>Proof:</u> Let $x \in E$ such that $\varphi_{N,E}(\bar{x}) = 1$. Then $\forall n \in N$, $x^{-1}nxn^{-1} = r_n \in \sqrt[N]{\gamma_2(N)}$. We use induction (on the nilpotency class of N) to prove that $grp(N \cup \{x\})$ is nilpotent.

The abelian case is immediate. Assume N is of class $c \geq 2$. Since τ' is an isomorphism of F-modules, $\varphi_{N/\Gamma,E/\Gamma}(\bar{x}) = 1$. Hence, by induction, $grp(N/\Gamma \cup \{x_\Gamma\})$ is nilpotent and of class $c - 1$ (see lemma 1.1.7). We are going to show that

$$
1 \to \Gamma \to grp(N \cup \{x\}) \to grp(N/\Gamma \cup \{x_\Gamma\}) \to 1
$$

is a central extension. Remember that $\Gamma \subseteq Z(N)$ and thus, the only thing left to check is whether $[x, \Gamma] = 1$. By lemma 1.1.9 it is sufficient to prove that $x \in C_E(\gamma_c(N))$.

So, assume $n \in N$ and take $m \in \gamma_{c-1}(N)$. Since $grp(N/\Gamma \cup \{x_\Gamma\})$ is nilpotent and $m_\Gamma \in Z(N/\Gamma)$, it follows by lemma 1.1.5 that $[x_\Gamma, m_\Gamma] = 1$, or equivalently, $x^{-1}mx = mz$ with $z \in \Gamma$. Then $x^{-1}[n, m]x = [r_nn, mz]$ $(r_n \in \sqrt[N]{\gamma_2(N)})$. An easy computation now shows that

$$
x^{-1}[n, m]x = [r_n, z]^n[n, z][r_n, m]^{nz}[n, m]^z = [n, m]
$$

implying that $x \in C_E(\gamma_c(N))$. This shows that $grp(N \cup \{x\})$ is nilpotent.

Conversely, suppose $x \in E$ such that $grp(N \cup \{x\})$ is nilpotent. Again we use induction on the class of N. The abelian case is trivial. Assume N is of class c $(c \geq 2)$. So, $grp(\{x_\Gamma\} \cup N/\Gamma)$ is nilpotent. By induction, we deduce that $\varphi_{N/\Gamma, E/\Gamma}(\bar{x})$ is the identity. The desired result follows at once by the isomorphism τ' of F-modules above. ∎

From the above proposition, we deduce a key observation, first made by K.B. Lee in [45] allowing to start the reduction process mentioned above. For another proof (using Lie group theory) we refer the reader to the original paper of K.B. Lee.

Lemma 2.4.2 *Let N be a finitely generated torsion free $c-$step nilpotent group. Define $Z = \sqrt[N]{\gamma_c(N)}$. Then, for any finite group F, we have that*

$$1 \to N \to E \to F \to 1 \text{ is essential}$$

$$\Updownarrow$$

$$1 \to N/Z \to E/Z \to F \to 1 \text{ is essential.}$$

Proof: First of all note that N/Z is a torsion free nilpotent group. It follows directly from proposition 2.4.1, that N is maximal nilpotent in E if and only if N/Z is maximal nilpotent in E/Z. ∎

Related to this lemma, notice the following commutative diagram:

$$
\begin{array}{ccccccccc}
& & 1 & & 1 & & & & \\
& & \downarrow & & \downarrow & & & & \\
& & Z & = & Z & & & & \\
& & \downarrow & & \downarrow & & & & \\
1 & \longrightarrow & N & \longrightarrow & E & \longrightarrow & F & \longrightarrow & 1 \\
& & \downarrow & & \downarrow & & \| & & \\
1 & \longrightarrow & N/Z & \longrightarrow & E/Z & \longrightarrow & F & \longrightarrow & 1 \\
& & \downarrow & & \downarrow & & & & \\
& & 1 & & 1 & & & &
\end{array}
\qquad (2.6)
$$

With this observation in mind, it is natural that induction is involved to reach the crystallographic case. Indeed, assume N is of nilpotency class $c > 1$. Then a group E having the normal subgroup N as Fitting subgroup of finite index, induces a group E/Z in which N/Z is maximal nilpotent and of finite index. Assume there are only finitely many isomorphism types for E/Z and fix one of these, say E/Z. We look for all extensions of E/Z by Z which give rise to a group containing N as the maximal nilpotent normal subgroup. Since Z is central in N, the action of E/Z on Z factors through F. Consequently there are only finitely many E/Z-module structures $\varphi : E/Z \to \operatorname{Aut} Z$ to consider. For each of them, there is a restriction morphism res: $H^2_\varphi(E/_Z, Z) \to H^2(N/_Z, Z)$. An extension $< E >$ in $H^2_\varphi(E/_Z, Z)$ will contain N as maximal nilpotent normal subgroup if and only if its restriction res $(< E >)$ determines a group which is isomorphic to N. So far, there is no problem. However, to decide on isomorphism types, group cohomology is not the best instrument. Consequently, the argument that res only maps a finite number of elements onto the class of N, although correct, is unsecure with respect to an isomorphism type classification. Indeed, there might be more than one class in $H^2(N/_Z, Z)$ representing a group isomorphic to N and it is not clear why a group $< E >$ in the inverse image under res of one class should be isomorphic to some group in the inverse image of an other class. In fact, as our example will show, there will often be infinitely many different cohomology classes, all representing N, and infinitely many of them will lie in the image of res.

Among the examples we know of to illustrate this problem, the following one is of minimal dimension (in fact in dimension 3 this problem does not occur). It also shows that the final counting argument in the proof of theorem 2.3.4 as given in [45] does not count all essential extensions.

Example. Take

$$N \; : < a, b, c, d \,|\, [b, a] = d, \text{ all other commutators trivial} > .$$

One verifies easily that in this case $Z = \text{grp}\{d\}$. Consequently, $N/_Z \cong \mathbb{Z}^3$. Now, consider the 3-dimensional crystallographic group $E/_Z = \mathbb{Z}^3 \rtimes \mathbb{Z}_2$, where the action of $\mathbb{Z}_2$ on $\mathbb{Z}^3$ is given by

$$\begin{pmatrix} -1 & 0 & 0 \\ 0 & 1 & 0 \\ 1 & 0 & 1 \end{pmatrix}.$$

For $E/_Z$ we obtain a presentation

$$E/_Z \; : < a, b, c, \alpha \,|\; [b, a] = [c, a] = [c, b] = 1$$
$$\alpha^2 = 1, \; \alpha a = a^{-1} c \alpha, \; \alpha b = b \alpha, \; \alpha c = c \alpha > .$$

Take φ the non–trivial action of $E/_Z$ on $Z = \text{grp}\{d\}$, which factors through $\mathbb{Z}_2$. Any extension E of $E/_Z$ by Z can be presented as

$$
\begin{aligned}
E \; : < a, b, c, d, \alpha \,| \quad & [b, a] = d^{k_1} & & [d, a] = 1 \quad > \\
& [c, a] = d^{k_2} & & [d, b] = 1 \\
& [c, b] = d^{k_3} & & [d, c] = 1 \\
& \alpha a = a^{-1} c \alpha d^{k_4} & & \alpha d = d^{-1} \alpha \\
& \alpha b = b \alpha d^{k_5} & & \alpha^2 = d^{k_7} \\
& \alpha c = c \alpha d^{k_6} & &
\end{aligned}
$$

for some integers $k_1, k_2, \ldots, k_7$.

However, we can not hope to choose the k_i's independently from each other (we will refer to such a situation by saying that there should be "computational consistency", see chapter 5). It is not too hard to verify the following consistency conditions:

1. $\alpha\alpha^2 = \alpha^2\alpha \Rightarrow \alpha d^{k_7} = d^{k_7}\alpha \Rightarrow d^{-k_7}\alpha = d^{k_7}\alpha \Rightarrow k_7 = 0.$

2. $\alpha a = a^{-1}c\alpha d^{k_4} \Rightarrow b\alpha a = ba^{-1}c\alpha d^{k_4}.$
   The left hand side equals $\alpha b a d^{-k_5} = \alpha a b d^{-k_5 + k_1}$,
   while the right hand side becomes
   $a^{-1}cbd^{-k_1 - k_3}\alpha d^{k_4} = a^{-1}cb\alpha d^{k_4}d^{k_1 + k_3} = a^{-1}c\alpha d^{k_4}bd^{k_1 - k_5}d^{k_3}.$
   Comparing both sides allows us to conclude that $k_3 = 0.$

3. $\alpha a = a^{-1}cd^{-k_4}\alpha \Rightarrow \alpha a^{-1} = ac^{-1}d^{k_4 - k_2}\alpha.$ We use this in:
   $\alpha^2 a = \alpha a^{-1}c\alpha d^{k_4} \Rightarrow a = ac^{-1}d^{k_4 - k_2}\alpha c\alpha d^{k_4} = ad^{2k_4 - k_2 - k_6},$
   implying $k_6 = 2k_4 - k_2.$

As a consequence every extension E can be presented as

$$E \; : \; < a, b, c, d, \alpha \mid \begin{array}{lll} [b, a] = d^{l_1} & [d, a] = 1 & > . \\ [c, a] = d^{l_2} & [d, b] = 1 \\ [c, b] = 1 & [d, c] = 1 \\ \alpha a = a^{-1}c\alpha d^{l_3} & \alpha d = d^{-1}\alpha \\ \alpha b = b\alpha d^{l_4} & \alpha^2 = 1 \\ \alpha c = c\alpha d^{2l_3 - l_2} \end{array} \qquad (2.7)$$

The conditions on the k_i's as given above are seen to be sufficient by realizing that the extensions (2.7) have a faithful representation in $\mathrm{Gl}(5, \mathbb{R})$ as follows:

$$a \mapsto \begin{pmatrix} 1 & -\frac{l_2}{2} & -l_1 & -l_2 & 0 \\ 0 & 1 & 0 & 0 & 1 \\ 0 & 0 & 1 & 0 & 0 \\ 0 & 0 & 0 & 1 & 0 \\ 0 & 0 & 0 & 0 & 1 \end{pmatrix} \qquad b \mapsto \begin{pmatrix} 1 & 0 & 0 & 0 & 0 \\ 0 & 1 & 0 & 0 & 0 \\ 0 & 0 & 1 & 0 & 1 \\ 0 & 0 & 0 & 1 & 0 \\ 0 & 0 & 0 & 0 & 1 \end{pmatrix}$$

$$c \mapsto \begin{pmatrix} 1 & 0 & 0 & 0 & 0 \\ 0 & 1 & 0 & 0 & 0 \\ 0 & 0 & 1 & 0 & 0 \\ 0 & 0 & 0 & 1 & 1 \\ 0 & 0 & 0 & 0 & 1 \end{pmatrix} \qquad d \mapsto \begin{pmatrix} 1 & 0 & 0 & 0 & 1 \\ 0 & 1 & 0 & 0 & 0 \\ 0 & 0 & 1 & 0 & 0 \\ 0 & 0 & 0 & 1 & 0 \\ 0 & 0 & 0 & 0 & 1 \end{pmatrix}$$

$$\text{and } \alpha \mapsto \begin{pmatrix} -1 & \frac{l_2 - 2l_3}{2} & -l_4 & l_2 - 2l_3 & 0 \\ 0 & -1 & 0 & 0 & 0 \\ 0 & 0 & 1 & 0 & 0 \\ 0 & 1 & 0 & 1 & 0 \\ 0 & 0 & 0 & 0 & 1 \end{pmatrix}.$$

The restriction morphism res : $H^2_\varphi(E/_Z, Z) \to H^2(\mathbb{Z}^3, \mathbb{Z})$ maps a class corresponding to the group E (2.7) to a group presented as

$$< a, b, c, d \mid \begin{array}{ll} [b, a] = d^{l_1} & [d, a] = 1 \\ [c, a] = d^{l_2} & [d, b] = 1 \\ [c, b] = 1 & [d, c] = 1 \end{array} >. \qquad (2.8)$$

It is well known that every couple (l_1, l_2) determines exactly one element in $H^2(\mathbb{Z}^3, \mathbb{Z})$. Indeed, recall that $H^2(\mathbb{Z}^3, \mathbb{Z}) \cong \mathbb{Z}^3$, with $(k, l, m) \in H^2(\mathbb{Z}^3, \mathbb{Z})$ representing the group

$$< a, b, c, d \mid \begin{array}{ll} [b, a] = d^k & [d, a] = 1 \\ [c, a] = d^l & [d, b] = 1 \\ [c, b] = d^m & [d, c] = 1 \end{array} >.$$

Let us now make the following observations:

Observation 1: N lies in the image of res $(N \cong (1, 0, 0) \in H^2(\mathbb{Z}^3, \mathbb{Z}))$.

Observation 2: Any couple $(l_1, l_2) = (1, l)$ $(\in \mathrm{Im}(\mathrm{res}))$ determines a group isomorphic to N. Indeed, taking $A = a$, $B = b$, $C = b^l c$, $D = d$ as a new set of generators for N, we obtain as presentation

$$N : < A, B, C, D \mid \begin{array}{ll} [B, A] = D & [D, A] = 1 \\ [C, A] = D^l & [D, B] = 1 \\ [C, B] = 1 & [D, C] = 1 \end{array} >.$$

In fact, remark that one can verify that every 3-tuple $(k, l, m) \in H^2(\mathbb{Z}^3, \mathbb{Z})$ for which $\gcd(k, l, m) = 1$ determines a group isomorphic to N (we will explain this in detail in section 6.2).

Anyhow, we conclude that there are infinitely many cohomologically different groups, all isomorphic to N, in the image of res.

Observation 3: Now take the group

$$E' : < a, b, c, d, \alpha \mid \begin{array}{ll} [b, a] = d & [d, a] = 1 \\ [c, a] = d & [d, b] = 1 \\ [c, b] = 1 & [d, c] = 1 \\ \alpha a = a^{-1} c \alpha & \alpha d = d^{-1} \alpha \\ \alpha b = b \alpha & \alpha^2 = 1 \\ \alpha c = c \alpha d^{-1} & \end{array} >.$$

E' determines an extension of $E/_Z$ by Z, and its restricted extension is isomorphic to N.

Claim 2.4.3 E' cannot be isomorphic to an element in res $^{-1}\{(1,0,0)\}$.

Proof : Consider a new set of generators for E': $A = a$, $B = b$, $C = b^{-1}c$, $D = d$ and $\beta = \alpha$. The presentation of E' becomes:

$$E' : < A, B, C, D, \beta \mid \begin{array}{ll} [B, A] = D & [D, A] = 1 \\ [C, A] = 1 & [D, B] = 1 \\ [C, B] = 1 & [D, C] = 1 \\ \beta A = A^{-1}BC\beta & \beta D = D^{-1}\beta \\ \beta B = B\beta & \beta^2 = 1 \\ \beta C = C\beta D^{-1} & \end{array} > .$$

We claim that this group is not isomorphic to any group with a presentation (2.7) with $l_1 = 1$ and $l_2 = 0$. This means that E' cannot be isomorphic to a group in res $^{-1}\{(1,0,0)\}$. In fact, assume there exists an isomorphism $\psi : E' \to E \in$ res $^{-1}\{(1,0,0)\}$. Taking into account the characteristic subgroups of E and E' we may conclude that ψ has to satisfy the following conditions:

$$\psi(\beta) = a^{x_1} b^{x_2} c^{x_3} d^{x_4} \alpha \quad \text{for } x_1, x_2, x_3, x_4 \in \mathbb{Z}$$
$$\psi(D) = d^{\pm 1} \quad \psi(C) = w(c, d)$$
$$\psi(A) = a^{m_1} b^{m_2} w(c, d) \quad \psi(B) = a^{l_1} b^{l_2} w(c, d)$$

where $\begin{pmatrix} m_1 & m_2 \\ l_1 & l_2 \end{pmatrix} \in$ Aut $\mathbb{Z}^2$ and $w(c, d)$ means a (not further specified) word in c and d. Since ψ is a morphism, we have

$$\psi(\beta)\psi(A) = \psi(A^{-1})\psi(B)\psi(C)\psi(\beta). \qquad (2.9)$$

In E we calculate $\alpha a^{m_1} = a^{-m_1} w(c, d)\alpha$ and $\alpha b^{m_2} = b^{m_2} w(c, d)\alpha$. Using this result and comparing the powers of a and b in both sides of (2.9) allows us to conclude that $l_1 = 0$ and $l_2 = 2m_2$. This implies that

$$\begin{pmatrix} m_1 & m_2 \\ l_1 & l_2 \end{pmatrix} = \begin{pmatrix} m_1 & m_2 \\ 0 & 2m_2 \end{pmatrix} \notin \text{Aut } \mathbb{Z}^2$$

And so ψ does not exist. ∎

This example shows that not all possible AC-groups were counted in the proof of the theorem in [45].

2.5 A new proof for the generalized third Bieberbach theorem

The proof we present here for theorem 2.3.4 is completely different. In fact, it is not clear to us, whether this point of view allows the same "iterative" classification approach to concrete situations, as is achieved on the basis of lemma 2.4.2.

Let us quickly recall some fundamentals of the theory of group extensions with non-abelian kernel (see [50], [38]).

Let N be a group. An extension with kernel N is a short exact sequence $1 \to N \to E \to F \to 1$. This exact sequence induces a homomorphism $\psi : F \to \text{Out}\, N$, which is called an abstract kernel. The extension usually is called compatible with ψ. Given an abstract kernel $\psi : F \to \text{Out}\, N$, the problem of studying all extensions compatible with ψ is well known in literature, and gave rise to the concept of non-abelian 2-cohomology sets . Let us write $\text{Ext}_\psi(F, N)$ for the set of equivalence classes of extensions compatible with ψ. If this set is empty, one says that there is an obstruction to the algebraic realization of the abstract kernel.

Let us denote by ϕ a fixed lifting (most likely not a morphism!) $\phi : F \to \text{Aut}\, N$ for ψ, which is taken to be normalised, i.e. $\phi(1) = 1$.

Now, an extension compatible with ψ is determined by a map $f : F \times F \to N$ (normalised, i.e. $f(g, 1) = f(1, g) = 1$) satisfying a cocycle-type condition

1. $\forall g, h \in F \ \ \phi(g)\phi(h) = \mu(f(g, h))\phi(gh)$

2. $\forall g, h, k \in F \ \ f(g, h)f(gh, k) = \phi(g)\,(f(h, k))\,f(g, hk)$.

The extension E compatible with ψ, determined by f is then obtained as $N \times F$ with multiplication:

$$\forall n, m \in N, \ \forall g, h \in F \ \ (n, g) \cdot_{(\phi, f)} (m, h) = (n\phi(g)(m)f(g, h), gh).$$

Let us denote this particular extension by $E_{(\phi, f)}$. If the set $\text{Ext}_\psi(F, N)$ is not empty it is in bijective correspondence to $H^2(F, Z(N))$, where the F-module structure of $Z(N)$ is induced by ψ.

Let us now notice the following property.

Lemma 2.5.1 *Let ψ_1, $\psi_2 : F \to \text{Out}\, N$ be two abstract kernels which are conjugated by an element $\psi \in \text{Out}\, N$ (i.e. $\forall g \in F$, $\psi_1(g) = \psi^{-1}\psi_2(g)\psi$). Then there is a bijective correspondence ν between the sets $\text{Ext}_{\psi_1}(F, N)$ and $\text{Ext}_{\psi_2}(F, N)$ in such a way that corresponding extensions are isomorphic.*

Proof : Choose a lift $\phi_2 : F \to$ Aut N for ψ_2 (with $\phi_2(1) = 1$) and a lift $\phi \in$ Aut N for ψ. Take $\phi_1 = \phi^{-1}\phi_2\phi$ as a lift for ψ_1. Consider any extension $E_{(\phi_1,f)}$ of $\mathrm{Ext}_{\psi_1}(F, N)$. Take $f' = \phi f$. One checks that f' satisfies the cocycle conditions mentioned above, with respect to ϕ_2. This means that we have an extension $E_{(\phi_2,f')}$. Let $\tilde{\nu}$ be the map sending $E_{(\phi_1,f)}$ to $E_{(\phi_2,f')}$. It is an elementary computation to verify that the map $i : E_{(\phi_1,f)} \to E_{(\phi_2,f')} : (n, g) \to (\phi(n), g)$ is an isomorphism of groups. Finally, remark that $\tilde{\nu}$ maps equivalent extensions onto equivalent extensions and so induces the desired map ν. ∎

Let us now return to the nilpotent case. Assume again that N is torsion free, finitely generated and nilpotent, and that F is a finite group.

Lemma 2.5.2 *If* $1 \to N \to E \to F \to 1$ *is an extension in which* N *is maximal nilpotent and* F *is finite, then the induced abstract kernel* $\psi : F \to$ Out N *is injective.*

Proof : Assume that the induced abstract kernel is not injective. This means that there exists an element in $E \backslash N$ and an element $n_0 \in N$ such that
$$\forall n \in N : xnx^{-1} = n_0 n n_0^{-1}.$$

This implies that the morphism $\varphi_{N,E}$, introduced in section 2.4, is not injective. Hence, by proposition 2.4.1, N is not maximal nilpotent in E. ∎

Remark 2.5.3 *If* N *is abelian, then it is well known that the converse of the statement is also true. However, in general, as the following example shows, this is not the case. I.e.* $\psi : F \to$ Out N *might be injective, without* N *being maximal nilpotent.*

Example 2.5.4

Consider the following torsion free virtually 2–step nilpotent group.

$$E : < a, b, c, \alpha \,\|\; \begin{array}{ll} [b, a] = c^2 & ac = c\alpha \\ [c, a] = 1 & [c, b] = 1 \\ \alpha a = a^{-1}\alpha & \alpha b = b^{-1}\alpha \\ \alpha^2 = c \end{array} > .$$

For $N = \mathrm{grp}\{a, b^2, c\}$, we see that $E/N = \mathbb{Z}_2 \times \mathbb{Z}_2 = \mathrm{grp}\{\bar{b}, \bar{\alpha}\}$. Since conjugation with b, α and $b\alpha$ in E induces automorphisms of N which do not belong to Inn N, the abstract kernel of the extension $1 \to N \to E \to$ $\mathbb{Z}_2 \times \mathbb{Z}_2 \to 1$ is injective. However, as N is contained in grp$\{a, b, c\}$, it is clear that it is not maximal nilpotent.

A correct generalization of remark 2.5.3 towards nilpotent groups, will be given in the following chapter.

A New proof of theorem 2.3.4:

For a fixed abstract kernel $\psi : F \to \text{Out } N$, we know that $\text{Ext}_\psi(F, N)$, if non-empty, is in one–to–one correspondence with $H^2(F, Z(N))$ which is finite. Since we only have to deal with injective abstract kernels (2.5.2) up to conjugation (2.5.1), we should show that there are only finitely many conjugacy classes of finite subgroups in Out N. But Aut N is an arithmetic group ([59]) and Inn $N \cong N/_{Z(N)}$ is a torsion free, finitely generated nilpotent group. This implies that the subgroups K of Aut N such that $[K : \text{Inn } N] < \infty$, lie in finitely many conjugacy classes in Aut N ([59]). By dividing out Inn N, we obtain our result.

∎

Remark 2.5.5 *The use of lemma 2.5.2 in this proof, can be replaced by information obtained in the reduction lemma 2.4.2 showing that we only have to deal with a finite number of finite groups F i.e. those occurring as holonomy group for crystallographic groups.*

Remark 2.5.6 *At the end of this chapter, I would also like to mention the work of F. Grunewald and D. Segal ([37]), who treated generalizations of Bieberbach teorems for affine crystallographic groups. A group $E \subseteq \text{Aff}(\mathbb{R}^n)$ is an affine crystallographic group if E acts properly discontinuously on $\mathbb{R}^n$, with compact quotient. It would however lead us much to far to go into any details here.*

Chapter 3

Algebraic characterizations of almost–crystallographic groups

3.1 Almost–crystallographic groups and essential extensions

We start this section by proving a generalization of remark 2.5.3 in case of nilpotent groups.

Fix a group extension $1 \to N \to E \to F \to 1$ where N is torsion free, finitely generated nilpotent and F is finite as usual. If G denotes the Mal'cev completion of N, then the group extension induces a morphism (analogous and related to the abstract kernal of the extension)

$$\varphi : F \to \mathrm{Out}\,(G)$$

in the following way: For any $\bar{x} \in F$ we choose an element $x \in E$ which maps onto $\bar{x}$. Then conjugation in E by x induces an automorphism of N. This automorphism lifts uniquely to an automorphism $\sigma(\bar{x})$ of G. It is obvious that $\sigma(\bar{x})$ is unique up to inner automorphisms of N (and so of G). Therefore, the map

$$\varphi : F \to \mathrm{Out}\,(G) : \bar{x} \mapsto \sigma(\bar{x})\,\mathrm{Inn}\,(G)$$

is a well defined homomorphism of groups.

Lemma 3.1.1
Let N be a finitely generated, torsion free nilpotent group with Mal'cev

completion G and suppose F is a finite group. Assume that a group extension $1 \to N \to E \to F \to 1$ is given, then

$$N \text{ is maximal nilpotent in } E$$

$$\Updownarrow$$

The induced morphism $\varphi : F \to \mathrm{Out}\,(G)$ is injective.

Proof: First suppose that $\varphi : F \to \mathrm{Out}\,(G)$ is not injective. This means that there exists an element $x \in E$ and an element $g \in G$ such that

$$\forall n \in N : \ xnx^{-1} = gng^{-1},$$

where the left hand side of the above equation is computed in E, while the right hand side has to be evaluated in G. But now, it is easy to see that the homomorphism $\varphi_{N,E}$ of section 2.4 is not injective, and so, by proposition 2.4.1, N is not maximal nilpotent in E.

Conversely, assume that N is not maximal nilpotent in E. Let N' denote a nilpotent group strictly containing N. We distinguish two cases depending on the set of torsion elements $\tau(N')$ of N':

Case 1: $\tau(N') \neq 1$
$\tau(N')$ is a characteristic subgroup of N', and so it is normal in N'. As N is a torsion free normal subgroup of N', we must have that (lemma 1.1.8) that $[\tau(N'), N] = 1$. Thus there exists an element $x \in E \backslash N$ such that

$$\forall n \in N : xnx^{-1} = n.$$

This implies that φ is not injective.

Case 2: $\tau(N') = 1$
In this situation N' is a torsion free nilpotent subgroup, containing N as a subgroup of finite index. Consequently, the Mal'cev completion of N' also equals G (i.e. $N \subseteq N' \subseteq G$). Therefore, conjugating N in E with an element of N' is exactly conjugating with an element of G, from which it follows that φ is not injective.
                                                                        ∎

This lemma offers an alternative way to see that the translational subgroup of an AC–group is maximal nilpotent. Indeed, let us investigate the map $\varphi : F \to \mathrm{Out}\,(G)$ defined above.

Let $\bar{x} \in E/(E \cap G)$, then $x = (g, \alpha)$, for some $g \in G$ and $\alpha \in \mathrm{Aut}\,(G)$. In this case

$$xnx^{-1} = (g, \alpha)(n, 1)(g, \alpha)^{-1} = \mu(g)\alpha(n).$$

Therefore, $\sigma(\bar{x}) = \mu(g)\alpha$ and so

$$\varphi(\bar{x}) = \mu(g)\alpha \operatorname{Inn}(G) = \alpha \operatorname{Inn}(G).$$

As the map $F \to \operatorname{Aut}(G) : \bar{x} \mapsto \alpha$ is injective and as $\operatorname{Inn}(G) \cong G/Z(G)$ is torsion free, the composite map

$$F \to \operatorname{Aut}(G) \to \operatorname{Aut}(G)/\operatorname{Inn}(G)$$

which coincides with φ is injective as well, implying that $N = E \cap G$ is maximal nilpotent in E.

The following lemma will be needed in the proof of the algebraic characterization of AC–groups below.

Lemma 3.1.2 *Let G be a connected and simply connected nilpotent Lie group. Let F be a finite group.*
Any morphism $\psi : F \to \operatorname{Out}(G)$ lifts to a morphism $\varphi : F \to \operatorname{Aut}(G)$ and consequently, any extension of G by F splits.

Proof: We proceed by induction on the nilpotency class c of G. If G is abelian then we have that $\operatorname{Out}(G) = \operatorname{Aut}(G)$ and so we have $\varphi = \psi$. Moreover, given an abstract kernel $\psi : F \to \operatorname{Out}(G) = \operatorname{Aut}(G)$, we know that the set of equivalence classes of extensions stands in one to one correspondence with $H^2_\varphi(F, G \cong \mathbb{R}^n) = 0$ (since F is finite and $\mathbb{R}^n$ is divisible). So, there is at most one class of extensions. Moreover, as $\varphi : F \to \operatorname{Aut}(G)$ is a morphism, we can form the semidirect product group $G \rtimes_\varphi F$, which is a group extension inducing the abstract kernel ψ. So a given extension is equivalent to $G \rtimes_\varphi F$, which implies that such an extension splits.

Suppose now that the nilpotency class of G is $c > 1$. Let $\psi : F \to \operatorname{Out}(G)$ be a morphism. and denote by $\Psi(F) \subseteq \operatorname{Aut}(G)$ the inverse image of $\psi(F)$ under the canonical projection $p : \operatorname{Aut}(G) \to \operatorname{Out}(G)$. There is a short exact sequence

$$1 \to \operatorname{Inn}(G) \to \Psi(F) \to \psi(F) \to 1.$$

As $\operatorname{Inn}(G) \cong G/Z(G)$ is a connected and simply connected nilpotent group of class $c - 1$, the induction hypothesis implies that there is a splitting $\varphi' : \psi(F) \to \Psi(F)$. Of course, by taking $\varphi = \varphi' \circ \psi$, we showed the first claim of the statement. To prove that any extension of G by a finite group splits, we can use the same argumentation as for the abelian case, since $H^2(F, Z(G) \cong \mathbb{R}^n) = 0$ classifies all such extensions and a semidirect product $G \rtimes_\varphi F$ inducing the given abstract kernel exists. ∎

The following theorem is a combination of the results of [48] and [21].

Theorem 3.1.3 Algebraic characterization of AC–groups
The following are equivalent for a polycyclic–by–finite group E:

1. *E is an almost–crystallographic group.*

2. *Fitt (E) is torsion free, maximal nilpotent and of finite index in E.*

3. *E contains a torsion free nilpotent normal subgroup N, which is of finite index in E and such that $C_E(N)$ is torsion free.*

4. *E contains a nilpotent subgroup of finite index and E contains no non trivial finite normal subgroups.*

Proof:
$(1 \Rightarrow 2)$ If E is an AC–group (of a given connected and simply connected nilpotent Lie group G), then we know that $E \cap G$ is a maximal nilpotent and normal subgroup of E. It follows immediately by the first generalized Bieberbach theorem, that $E \cap G = $ Fitt (E) is maximal nilpotent and is of finite index in E.

$(2 \Rightarrow 3)$ We can take $N = $ Fitt (E). For suppose there exists a torsion element x in $C_E($Fitt $(E))$. As Fitt (E) is torsion free, grp(Fitt $(E) \cup \{x\}$) is a nilpotent group, strictly containing Fitt (E) which is a contradiction.

$(3 \Rightarrow 4)$ Let N be a torsion free nilpotent group such that $C_E(N)$ contains no torsion elements. Suppose that E has a finite, non trivial, normal subgroup H. then, by lemma 1.1.8, we know that $[N, H] = 1$, which contradicts the fact that $C_E(N)$ is torsion free.

$(4 \Rightarrow 1)$ Of course, this is the real hard part of the proof.
We are going to show that E can be embedded in Aff(G), in such a way that the image of this embedding is a genuine almost–crystallographic group. In [48], this was done by constructing a pushout in the category of groups. Here we formulate an alternative description of the (same) construction of this embedding.
By restricting to a subgroup of finite index, there is no loss in generality in assuming that N is a normal subgroup of E. Then, the group E fits in a short exact sequence

$$1 \to N \to E \to F \to 1$$

where F is some finite group. The nilpotent group N is torsion free, for if it where not, the set of torsion elements of N would be a non trivial finite normal subgroup of E. In the previous chapter (section 2.5), we explained that E can be described via a "2–cocycle". I.e. there exist functions $\phi : F \to \mathrm{Aut}\,(N)$ and $c : F \times F \to N$, where ϕ is a normalized lift of the abstract kernel $\psi : F \to \mathrm{Out}\,(N)$ induced by the short exact sequence above, and such that the following cocycle conditions are fulfilled:

1. $\forall g, h \in F \;\; \phi(g)\phi(h) = \mu(c(g,h))\phi(gh)$

2. $\forall g, h, k \in F \;\; c(g,h)c(gh,k) = \phi(g)\,(c(h,k))\,c(g,hk).$

Moreover, the extension E can be described as the group with underlying set $N \times F$, and where the multiplication is given by:

$$\forall n, m \in N, \; \forall g, h \in F \;\; (n,g) \cdot_{(\phi,f)} (m,h) = (n\phi(g)(m)c(g,h), gh).$$

Let us denote by ψ' the group homomorphism which is obtained as the composition $F \xrightarrow{\psi} \mathrm{Out}\,(N) \to \mathrm{Out}\,(G)$, where the last arrow is obtained by the unique automorphism extension reults of Mal'cev. Also we use $\phi' : F \xrightarrow{\phi} \mathrm{Aut}\,(N) \to \mathrm{Aut}\,(G)$ as a normalised lift of ψ'. If, moreover, we introduce the map $c' : F \times F \to N \to G$, we see that the pair (ϕ', c') satisfies the cocycle conditions, and so they determine a group extension G' of G by F in such a way that we obtain a commutative diagram

$$
\begin{array}{ccccccccc}
1 & \to & N & \to & E & \to & F & \to & 1 \\
 & & \downarrow i & & \downarrow j & & \| & & \\
1 & \to & G & \to & G' & \to & F & \to & 1
\end{array}
$$

By the previous lemma, we know that the bottom sequence splits, and so, there is an action $\varphi : F \to \mathrm{Aut}\,(G)$ (not necessarily effective) of F on G, such that the group G' is isomorphic with $G \rtimes_\varphi F$. This induces a morphism

$$k : G' \cong G \rtimes_\varphi F \to G \rtimes \mathrm{Aut}\,(G) : (g,f) \mapsto (g, \varphi(f)).$$

We claim that the map $k \circ j : E \to G \rtimes \mathrm{Aut}\,(G) = \mathrm{Aff}(G)$ is the desired embedding. As the restriction of $k \circ j$ to N is the canonical embedding of N into its Mal'cev completion, the only thing left to show is that $k \circ j$ is injective. But as the kernel of the map $k \circ j$ and N only have the neutral element in common, this kernel has to be a finite normal subgroup of E and hence trivial.

At this point we remark that it is not true that F will, in general, be isomorphic to the holonomy group of the almost–crystallographic group $k \circ j(E)$. This follows from the fact that it is not true that the map k itself is injective. The holonomy group of E will be a quotient group of

F.

■

As is seen in this theorem, there are several nice algebraic descriptions of almost–crystallographic groups. These descriptions inspired me to investigate the possibility of generalizing the notion of an almost–crystallographic group to more general groups, in particular the polycyclic–by–finite groups. The following sections are especially devoted to this generalization.

We knew already, by the results of L. Auslander, that any AC–group E induces an essential extension $1 \to N = \text{Fitt}\,(E) \to E \to F \to 1$. Theorem 3.1.3 shows the converse, namely any essential extension determines an almost–crystallographic group. Knowing this, it is obvious that our algebraic formulation of the generalized third Bieberbach theorem is equivalent to the topological one.

3.2   Torsion in the centralizer of a finite index subgroup

One of the statements of theorem 3.1.3 deals with torsion in the centralizer $C_E(N)$. Here N is a normal subgroup of finite index in E. This section is devoted to the investigation of such torsion elements. The basic result is the following:

Lemma 3.2.1
Let $1 \to \mathbb{Z}^k \to E \to F \to 1$ be any central extension, where F is a finite group. Then $\tau(E)$ is a characteristic subgroup of E.

<u>Proof :</u> Once we know that $\tau(E)$ is a group, it follows automatically that it is characteristic. As E is a central extension of $\mathbb{Z}^k$ by F, we may view E as being the set $\mathbb{Z}^k \times F$, where the multiplication $*$ is given by

$$(a, \alpha) * (b, \beta) = (a + b + c(\alpha, \beta), \alpha\beta), \quad \forall a, b \in \mathbb{Z}, \forall \alpha, \beta \in F, \qquad (3.1)$$

for some 2–cocycle $c : F \times F \to \mathbb{Z}^k$. Since $\mathbb{R}^k$ is a vector space, the inclusion map $i : \mathbb{Z}^k \to \mathbb{R}^k$ induces a trivial map $i_* : H^2(F, \mathbb{Z}^k) \to H^2(F, \mathbb{R}^k) = 0$ on the cohomology level. This means that there is a split short exact sequence $1 \to \mathbb{R}^k \to E' \to F \to 1$, where E' denotes the group determined by the cocycle $i(c)$. (I.e. $E' = \mathbb{R}^k \times F$ and multiplication is given by (3.1), where a and b may now belong to $\mathbb{R}^k$.)
So there is a splitting morphism $s : F \to E'$. But it is now easy to see

that s is unique, since for all $f \in F$, $s(f)$ has to be a torsion element and there is only one torsion element in E' mapping to f. This shows that $s(F) = \tau(E')$ and so $\tau(E')$ is a group. The proof finishes, by realizing that $E \subset E'$ and so $\tau(E)$ $(\subseteq \tau(E'))$ has to be a group also.

∎

Remark 3.2.2

In literature, a proof for this lemma is given by means of topological arguments, which can also be used to proof analogous results in more general cases. However, they fail to be useful in the most general case, which we will state below.

Remark 3.2.3

1. The proof of the lemma might suggest that the group E can be decomposed into a direct sum $E = \mathbb{Z}^k \oplus F'$ for some finite group F'. However, this is not true: consider the group $E = (\mathbb{Z} \oplus \mathbb{Z}_2) \rtimes \mathbb{Z}_2$ where $\mathbb{Z}_2 = \{\bar{0}, \bar{1}\}$ and the action of $\mathbb{Z}_2$ on $\mathbb{Z} \oplus \mathbb{Z}_2$ is given by

$$^{\bar{1}}(1, \bar{0}) = (1, \bar{1}) \text{ and } {}^{\bar{1}}(0, \bar{1}) = (0, \bar{1}).$$

   The group $A = 2\mathbb{Z} \subseteq E$ is indeed a free abelian, central subgroup of finite index, so the conditions of the lemma are satisfied, but E cannot be seen as the direct sum of a free abelian group and a finite one.

2. If we examine the conditions of the lemma, we quickly see that the lemma is false if we are not looking at central extensions (E.g. $k = 1$, $F = \mathbb{Z}_2$ and $E = \mathbb{Z} \rtimes \mathbb{Z}_2$, where $\mathbb{Z}_2$ acts non trivially on $\mathbb{Z}$) or at extensions of infinite index (E.g. $k = 0$ and $F = E = \mathbb{Z} \rtimes \mathbb{Z}_2$). However, as the following lemma shows, there is no need for a finitely generated free abelian kernel.

Lemma 3.2.4
Let A be any abelian group. If $1 \to A \to E \to F$ is any central extension, where F is finite, then $\tau(E)$ forms a characteristic subgroup of E.

<u>Proof :</u> $\tau(A)$ is a characteristic subgroup of A, and so there is an induced short exact sequence

$$1 \to A/\tau(A) \to E/\tau(A) \to F \to 1.$$

From this it follows that it will suffice to prove the theorem in case A is torsion free abelian.

Suppose A is torsion free and $x, y \in \tau(E)$. We have to show that $xy \in \tau(E)$. Let E' denote the group generated by x and y, and $A' = A \cap E'$. There is an induced extension

$$1 \to A' \to E' \to F' \to 1$$

where F' is some finite group. Since E' is finitely generated and F' is finite, we may conclude that A' is finitely generated too ([56, p. 117]). Therefore, we are in the situation of lemma 2.1, which implies that $\tau(E')$ is a group, and so $xy \in \tau(E') \subseteq \tau(E)$.

■

Theorem 3.2.5

Let E be any subgroup of finite index in a given group E', then $\tau(C_{E'}E)$ is a subgroup of E'. Moreover, if E is torsion free and normal then $\tau(C_{E'}E)$ is the unique maximal normal torsion subgroup of E'.

Proof : First, let us consider the case where E is normal in E'. There is an exact sequence of subgroups of E':

$$1 \to Z(E) \to C_{E'}E \to F \to 1$$

for some finite group F. It follows immediately from lemma 3.2.4 that $\tau(C_{E'}E)$ is a subgroup of E. It is normal in E, since it is characteristic in another normal subgroup $(C_{E'}E)$. To prove the last statement it is enough to realize that, in case E is torsion free, any normal subgroup T, containing only torsion, commutes with E (see lemma 1.1.8).

If E is not normal in E', we replace E' by the normalizer $N_{E'}E$ of E in E'. Since $C_{E'}E \subseteq N_{E'}E$ we may apply the theorem for normal E to conclude the correctness of the theorem in the general case too.

■

Remark 3.2.6

If $\tau(C_{E'}E)$ is finite, it is the maximal finite normal subgroup of E'. This is e.g. always the case when E' is a polycyclic–by–finite group (see the following section). This observation will be used in the following section.

It is now easy to proof the following lemma due to K.B. Lee ([45]). The original proof of this lemma is based on the Seifert Fiber Space construction with typical fiber a nilmanifold. As it is our intention to avoid the topological arguments as much as possible, we use the theory developed above.

Lemma 3.2.7 *Let* $1 \to N \to E \to F \to 1$ *be any extension of a torsion free nilpotent group* N *by a finite group* F. *Then, the set of torsion elements of* $C_E(N)$ *is a characteristic finite subgroup* H *of* E *and it is the unique finite normal subgroup* H *of* E *such that* E/H *is almost-crystallographic.*

Proof: H is the unique, maximal finite normal subgroup of E. ∎

3.3 Towards a generalization of AC–groups

Lemma 3.3.1 *Let* Γ *be any polycyclic–by–finite group, then* Γ *has a unique maximal finite normal subgroup.*

Proof: Let

$$E_1 \subseteq E_2 \subseteq E_3 \subseteq \cdots E_n \subseteq E_{n+1} \subseteq \cdots$$

be any ascending chain of finite normal subgroups of Γ. This chain is necessarilly finite, since every ascending chain of subgroups of Γ is finite (see [59]). Therefore we can choose a finite normal subgroup H of Γ which is maximal among all finite normal subgroups. Now it is easy to see that H is unique, for if K was another such a normal subgroup, then $H.K$ would contradict the maximality of H. ∎

Remark 3.3.2

For a polycyclic–by–finite group Γ, we will denote its maximal normal finite subgroup by $F(\Gamma)$.

Remark 3.3.3

In general, $F(\Gamma)$ is not maximal among all finite subgroups of Γ. For example, when $\Gamma = \mathbb{Z} \rtimes \mathbb{Z}_2$, where $\mathbb{Z}_2$ acts non trivially on $\mathbb{Z}$, one easily sees that $F(\Gamma) = 1$, while Γ has subgroups of order 2.

Definition 3.3.4 *Let* Γ *be any group. Then* Γ *is said to be almost torsion free if and only if* Γ *has no finite normal subgroups other then the trival one.*

So, a polycyclic–by–finite group Γ is almost torsion free if and only if $F(\Gamma) = 1$. We warn the reader that some authors use the term almost torsion free, when they want to indicate groups which are virtually torsion free. This is a totally different notion from ours, since every polycyclic–by–finite group is virtually torsion free.

We already proved most part of the following theorem:

Theorem 3.3.5 *Let* E *be a finitely generated virtually nilpotent group, then*

$$E \text{ is almost torsion free}$$
$$\Updownarrow$$
$$E \text{ is almost–crystallographic.}$$

Proof: Let E be almost torsion free. Since E is finitely generated, there exists a finitely generated torsion free nilpotent normal subgroup N of finite index in E. This can be seen as follows: There exists a torsion free normal subgroup $\tilde{N}$ of finite index in E, since E is polycyclic–by–finite, and so also (poly-$\mathbb{Z}$)–by–finite. Since E is virtually nilpotent, $N = \mathrm{Fitt}\,(\tilde{N})$ is a characteristic subgroup of $\tilde{N}$ and is of finite index in E.

Therefore we have a short exact sequence

$$1 \to N \to E \to F \to 1$$

with F finite. Following lemma 3.2.7, we may find a finite normal subgroup $H \lhd E$ for which E/H is almost–crystallographic. But the fact that E is almost torsion free implies that $H = 1$ or that E itself is almost–crystallographic.

The converse of the theorem is given by the last statement in theorem 3.1.3

∎

Of course, this theorem also implies that a virtually abelian group Q is crystallographic if and only if Q is almost torsion free.

Knowing this theorem, it is natural, at least as an algebraist, to consider those groups Γ which are almost torsion free as a generalization of the almost–crystallographic groups. We also show how they can be seen as a generalization from the topological point of view. But let us first examine the almost torsion free groups algebraically.

3.4 The closure of the Fitting subgroup

In this section we will define a normal subgroup $\overline{\text{Fitt}}(\Gamma)$ of Γ, which contains the Fitting subgroup of Γ as a normal subgroup of finite index. In fact we take the maximal one with this property. A formal definition looks like:

Definition 3.4.1 *The closure of* $\text{Fitt}(\Gamma)$ *is denoted by* $\overline{\text{Fitt}}(\Gamma)$ *and satisfies*

$$\overline{\text{Fitt}}(\Gamma) = < G \| G \lhd \Gamma \text{ and } [G : \text{Fitt}(G)] < \infty > .$$

We remark that the notion of closure has no topological meaning at all!

It is interesting to note that for any $G \lhd \Gamma$, $G \cap \overline{\text{Fitt}}(\Gamma) = \text{Fitt}(G)$.

Lemma 3.4.2 $\overline{\text{Fitt}}(\Gamma)$ *is the unique maximal subgroup* G *of* Γ *for which*

1. $G \lhd \Gamma$

2. $[G : \text{Fitt}(G)] < \infty$.

Proof: Since any ascending series of subgroups of Γ is finite, we can choose a group G which is maximal amongst those who satisfy 1. and 2. (G contains $\text{Fitt}(\Gamma)$ and so $\text{Fitt}(G) = \text{Fitt}(\Gamma)$). We assert that any other group G_0 combining both conditions stated above is contained in G. To see this, we consider the group $G_0.G$ which is certainly a normal subgroup of Γ. Moreover, since $(G_0.G)/G = G_0/(G_0 \cap G)$ is finite ($G_0 \cap G \supseteq \text{Fitt}(G_0)$),

$$[G_0.G : \text{Fitt}(G_0.G)] = [G_0.G : \text{Fitt}(\Gamma)] = [G_0.G : G][G : \text{Fitt}(\Gamma)] = < \infty.$$

Therefore, we may conclude that $G_0.G = G$ or $G_0 \subseteq G$. This is enough to conclude that $\overline{\text{Fitt}}(\Gamma) = G$. ∎

It is well known that any polycyclic–by–finite group Γ has the property of being (nilpotent–by–abelian)–by–finite. So there is a short exact sequence

$$1 \to X \to \Gamma \to F \to 1$$

where F is finite and X is nilpotent–by–abelian. When dividing out $\text{Fitt}(\Gamma)$, we see that

$$1 \to X/(X \cap \text{Fitt}(\Gamma)) = X/\text{Fitt}(X) \to \Gamma/\text{Fitt}(\Gamma) \to F' \to 1$$

where F' is also finite. Since X is nilpotent–by–abelian, $X/\text{Fitt}(X)$ is abelian. Therefore, we may conclude that $\Gamma/\text{Fitt}(\Gamma)$ is abelian–by–finite, or Γ is nilpotent–by–(abelian–by–finite).

Now we concentrate us on the case of almost torsion free Γ. So, $\text{Fitt}(\Gamma)$ is torsion free and we have a short exact sequence

$$1 \to \overline{\text{Fitt}}(\Gamma) \to \Gamma \to Q \xrightarrow{p} 1$$

where Q is also virtually abelian and almost torsion free. For suppose Q' is a finite normal subgroup of Q, then $G = p^{-1}(Q')$ would be a normal subgroup of Γ satisfying $[G : \text{Fitt}(G)] < \infty$ which implies that $G \subseteq \overline{\text{Fitt}}(\Gamma)$ or $Q' = 1$. So Q is a crystallographic group. Of course, $\overline{\text{Fitt}}(\Gamma)$ is also almost torsion free. Because $F(\overline{\text{Fitt}}(\Gamma))$ is a characteristic subgroup of $\overline{\text{Fitt}}(\Gamma)$ and so a normal subgroup of Γ, which implies that $F(\overline{\text{Fitt}}(\Gamma)) = 1$. We summarize this in the following theorem.

Theorem 3.4.3 *Let Γ be any polycyclic–by–finite group, then*

$$\Gamma \text{ is almost torsion free}$$

$$\Updownarrow$$

$$\Gamma \text{ is (almost–crystallographic)–by–crystallographic.}$$

Moreover, in case Γ is almost torsion free, the intended almost–crystallographic subgroup can be taken as $\overline{\text{Fitt}}(\Gamma)$.

In [59] one can find, as an exercise, the following property: If Γ is a solvable group, then $C_\Gamma(\text{Fitt}(\Gamma)) \subseteq \text{Fitt}(\Gamma)$. We drop the restriction of solvable groups and find the following lemma.

Lemma 3.4.4 *If Γ is any polycyclic–by–finite group, then*

$$C_\Gamma(\text{Fitt}(\Gamma)) \subseteq \overline{\text{Fitt}}(\Gamma).$$

Proof: Γ fits in a short exact sequence

$$1 \to \text{Fitt}(\Gamma) \to \Gamma \xrightarrow{p} Q \to 1$$

where Q is abelian–by–finite. Let A denote an abelian normal subgroup of finite index in Q. Consider the group $\Gamma' = p^{-1}(A)$ which is solvable. Then, since $\text{Fitt}(\Gamma') = \text{Fitt}(\Gamma)$, we see that $C_{\Gamma'}(\text{Fitt}(\Gamma)) \subseteq \text{Fitt}(\Gamma)$ (using the exercise of [59]). Therefore we see that $p(C_\Gamma(\text{Fitt}(\Gamma))) \cap$

$A = \{1\}$. So, $C_\Gamma(\text{Fitt}(\Gamma)).\text{Fitt}(\Gamma)$ is a finite extension of $\text{Fitt}(\Gamma)$, and $C_\Gamma(\text{Fitt}(\Gamma)).\text{Fitt}(\Gamma)$ is a normal subgroup of Γ, it is even characteristic. We may indeed conclude that $C_\Gamma(\text{Fitt}(\Gamma)) \subseteq \overline{\text{Fitt}}(\Gamma)$. ∎

For the following theorem we need some technical facts, which we treat now. Let N be a torsion free, finitely generated nilpotent group of class c. We defined the groups $Z_i(N)$ $(1 \leq i \leq c)$ of the upper central series of N in the first chapter. For any extension $1 \to N \to E \to Q \to 1$ the abstract kernel $\psi : Q \to \text{Out } N$ induces c morphisms $\varphi_i : Q \to \text{Aut } Z_i(N)/Z_{i-1}(N)$ $(1 \leq i \leq c)$. To define the automorphism $\varphi_1(\bar{q})$, for $\bar{q} \in Q$, we consider a lift $q \in E$ of $\bar{q}$ and take as $\varphi_1(\bar{q})$ the automorphism of $Z(N)$ induced by conjugation by q. For $\varphi_2(q)$, we first consider the extension $1 \to N/Z(N) \to E/Z(N) \to Q \to 1$ and then continue as for φ_1. Analogously we build up all φ_i.

Now let e be an element of E such that its projection $\bar{e} \in Q$ has finite order and such that the group N' generated by N and e is nilpotent. We want to show that $p(e) \in \bigcap_{i=1}^{c} \ker(\varphi_i)$. We proceed by induction on the nilpotency class c of N. If N is abelian, the assertion is immediate by lemma 1.1.5. For N of larger nilpotency class, we first consider the extension $1 \to N/Z(N) \to E/Z(N) \to Q \to 1$ and then again, we use lemma 1.1.5. Conversely, if $p(e) \in \bigcap_{i=1}^{c} \ker(\varphi_i)$ ($p(e)$ may be of infinite order) then the group generated by N and e is nilpotent. This is again easy to show by induction on the nilpotency class of N. We are ready now to prove the following theorem.

Theorem 3.4.5 *Let Γ be any polycyclic–by–finite group. Then*

$$\text{Fitt}(\Gamma) \text{ is torsion free and maximal nilpotent in } \Gamma$$

$$\Updownarrow$$

$$\Gamma \text{ is almost torsion free.}$$

Proof: First suppose that $\text{Fitt}(\Gamma)$ is torsion free and maximal nilpotent in Γ. Consider $H \lhd \Gamma$, a finite normal subgroup. Then $H \lhd \text{Fitt}(\Gamma).H$. But $\text{Fitt}(\Gamma).H$ is an almost–crystallographic group (since $\text{Fitt}(\Gamma)$ is maximal nilpotent) and so H has to be trivial.

Conversely, suppose Γ is almost torsion free. It is easy to see that $\text{Fitt}(\Gamma)$ has to be torsion free, since the elements of finite order of $\text{Fitt}(\Gamma)$

form a characteristic subgroup of $\mathrm{Fitt}\,(\Gamma)$. So we can consider the exact sequence

$$1 \to \mathrm{Fitt}\,(\Gamma) \to \Gamma \xrightarrow{p} Q \to 1$$

where $\mathrm{Fitt}\,(\Gamma)$ can play the role of N in the discussion above. First we show that no element $\bar{q}$ of infinite order in Q can be lifted to $q \in E$ in such a way that the group generated by $\mathrm{Fitt}\,(\Gamma)$ and q is nilpotent. Suppose there should exist such a $\bar{q}$. Recall that Q is abelian–by–finite, and so we can find an abelian normal subgroup A of Q of finite index n in Q. Therefore, also $\bar{q}^n \in A$ can be lifted to an element q^n to yield a nilpotent subgroup of E strictly containing $\mathrm{Fitt}\,(\Gamma)$. But when we look at the group $\Gamma' = p^{-1}(A)$ which fits into

$$1 \to \mathrm{Fitt}\,(\Gamma) = \mathrm{Fitt}\,(\Gamma') \to \Gamma' \to A \to 1$$

we see that $\mathrm{Fitt}\,(\Gamma)$ is maximal nilpotent in Γ'. Otherwise, we should find a nilpotent subgroup (normal in Γ') strictly containing $\mathrm{Fitt}\,(\Gamma')$. So the desired $\bar{q}^n \in A$ cannot be found.

Now suppose $\bar{q}$ is of finite order. Then the group generated by q and $\mathrm{Fitt}\,(\Gamma)$ is nilpotent if and only if $p(q) \in \bigcap_{i=1}^{c} \ker(\varphi_i)$, where the φ_i are as above. This intersection is a normal subgroup of Q and is finite, this means that $q \in \overline{\mathrm{Fitt}}\,(\Gamma)$. But now we use the fact that if Γ is almost torsion free, then $\overline{\mathrm{Fitt}}\,(\Gamma)$ is almost–crystallographic. And so, $\mathrm{Fitt}\,(\Gamma)$ is maximal nilpotent in $\overline{\mathrm{Fitt}}\,(\Gamma)$, implying we cannot find q outside $\mathrm{Fitt}\,(\Gamma)$. ∎

We summarize the results found so far in the following theorem. Notice the analogy with theorem 3.1.3.

Theorem 3.4.6 *Let Γ be a polycyclic–by–finite group, then the following are equivalent:*

1. *Γ is an almost torsion free group.*

2. *$\mathrm{Fitt}\,(\Gamma)$ is torsion free and maximal nilpotent in Γ*

3. *Γ contains a torsion free normal subgroup Γ', which is of finite index in Γ and such that $C_\Gamma(\Gamma')$ is torsion free.*

4. *Γ contains no non trivial finite normal subgroups.*

5. *Γ admits an effective and properly discontinuous action on $\mathbb{R}^{h(\Gamma)}$, where $h(\Gamma)$ denotes the Hirsch number of Γ.*

For the proof of number 5. we refer to corollary 3.5.2 (next section) and to theorem 4.2.3 of the following chapter.

We remark that if N is a torsion free nilpotent group of Hirsch number n, then the Mal'cev completion G of N is homeomorphic with $\mathbb{R}^n$. So the action of an almost–crystallographic group E on G can be seen as a properly discontinuous action of E on $\mathbb{R}^n$.

Perhaps it is also worthwhile to notice the following generalization of lemma 3.2.7. The proof of which is easy to deduce from the theory of this chapter.

Lemma 3.4.7 *Let $1 \to \Gamma' \to \Gamma \to F \to 1$ be any extension of a torsion free polycyclic-by-finite group Γ' by a finite group F. Then, the set of torsion elements of $C_\Gamma(\Gamma')$ is a characteristic finite subgroup H of Γ and it is the unique finite normal subgroup H of Γ such that Γ/H is almost torsion free.*

3.5 Almost torsion free groups from the topological point of view

This section is intended to show that, also from a topological standpoint, the almost torsion free groups are interesting groups to investigate. The reader with little topological background should at least try to understand the formulation of corollary 3.5.2.

Theorem 3.5.1 *Let Γ be any group acting on a contractible space $\widetilde{M}$ and containing a normal subgroup Γ' of finite index, which acts freely and properly discontinuous on $\widetilde{M}$, such that the quotient space $M =_{\Gamma'} \backslash \widetilde{M}$ is a closed aspherical manifold. Then, the kernel of the induced action of $F = \Gamma/\Gamma'$ on M is the isomorphic image of the torsion subgroup of $C_\Gamma(\Gamma')$ inside F. In fact, the set of all torsion elements of $C_\Gamma(\Gamma')$ forms a characteristic subgroup of Γ, and equals the kernel of the action of Γ on $\widetilde{M}$.*

<u>Proof:</u> Let t be a torsion element in $C_\Gamma(\Gamma')$. This means that in the commutative diagram

$$
\begin{array}{ccccccccc}
1 & \to & \Gamma' & \to & \Gamma & \to & F & \to & 1 \\
 & & \downarrow & & \downarrow & & \downarrow & & \\
1 & \to & \text{Inn}(\Gamma') & \to & \text{Aut}(\Gamma') & \to & \text{Out}(\Gamma') & \to & 1
\end{array}
$$

the group T generated by t maps trivially into $\text{Aut}(\Gamma')$. (The vertical maps are induced by conjugation in Γ). By [46, Lemma 1] the image of T

in F acts trivially on M. This image is isomorphic to T, as Γ' is torsion free. Conversely, if T' is a subgroup of F acting trivially on M, then T' is the isomorphic image of a subgroup T of Γ (this lifting of T' back to Γ can be done already if the T' action on M has a fixed point). Clearly, then T commutes with Γ' so that $T \subset C_\Gamma(\Gamma')$.

For the last statement, note that a torsion element of Γ acts effectively on $\widetilde{M}$ if and only if the corresponding element of F acts effectively on M.
∎

As an immediate consequence, we find the following reformulation of the above theorem in the situation we're interested in.

Corollary 3.5.2 *Let Γ be a polycyclic-by-finite group with Hirsh number K. Let Γ_c be a torsion free normal subgroup of finite index of Γ, and let F denote the set of torsion elements of $C_\Gamma(\Gamma_c)$. For any properly discontinuous action of $\rho : \Gamma \to \mathbb{R}^K$, F is the kernel of ρ. Moreover, F is the maximal finite normal subgroup of Γ.*

Chapter 4

Canonical type representations

4.1 Introduction

In this chapter we will study special types of actions of a group E (in most cases E will be virtually nilpotent) on some space $\mathbb{R}^n$. We are not only interested in these actions from the topological point of view, but also from a computational point of view. In fact, in order to be able to deal with virtually nilpotent groups, we need some way of computing formally in these kind of groups. Therefore, we want to find nice representations of (virtually) nilpotent groups. Such a "nice" representation is e.g. given by an affine representation $\rho : N \to \mathrm{Aff}(\mathbb{R}^n)$. As is very well known, the group of affine transformations of $\mathbb{R}^n$ can be seen as a subgroup of $\mathrm{Gl}(n+1, \mathbb{R})$. The embedding of $\mathrm{Aff}(\mathbb{R}^n)$ into $\mathrm{Gl}(n+1, \mathbb{R})$ is given by mapping the affine transformation with linear part $A \in \mathrm{Gl}(n, \mathbb{R})$ and translational part $a \in \mathbb{R}^n$ (seen as column vector) onto the element

$$\begin{pmatrix} A & a \\ 0 & 1 \end{pmatrix} \in \mathrm{Gl}(n+1, \mathbb{R}).$$

Moreover, if we identify an element $x \in \mathbb{R}^n$ with the element $\begin{pmatrix} x \\ 1 \end{pmatrix} \in \mathbb{R}^{n+1}$, we can express the image of x under the affine transformation $(A, a) \in \mathrm{Aff}(\mathbb{R}^n)$ by means of a matrix multiplication. I.e.

$$(A, a)x \equiv \begin{pmatrix} A & a \\ 0 & 1 \end{pmatrix} \begin{pmatrix} x \\ 1 \end{pmatrix}.$$

So, an affine representation of a group E can be seen as a matrix representation and already many computer programs can deal very well with (formal) matrices.

From the geometrical point of view, we will describe in this chapter some of the latest developements concerning the conjecture of John Milnor ([52]) stating that any torsion free polycyclic–by–finite group occurs as the fundamental group of a compact, complete affinely flat manifold. We will explain (in section 4.4.5) this notion of a compact, complete affinely flat manifold and see that Milnor's conjecture is false, even in the case of nilpotent groups. However, we will also formulate a conjecture analogous to Milnor's where we replace the word affine by polynomial and prove that this new conjecture holds in case of virtually nilpotent groups. This new conjecture uses the notion of "affine defect" of a group. This is a number which measures up to what extend a given group can serve as a counter example to Milnor's question. We finally compute that the only known counter examples to the conjecture of Milnor are of the smallest possible affine defect, namely one.

4.2   Definition of canonical type structures

It is well known ([59, lemma 6, page 16]) that, if Γ is a polycyclic–by–finite group, then there exists an ascending sequence (or filtration) Γ_* of normal subgroups Γ_i $(0 \leq i \leq c + 1)$ of Γ

$$\Gamma_* : \quad \Gamma_0 = 1 \subset \Gamma_1 \subset \Gamma_2 \subset \cdots \subset \Gamma_{c-1} \subset \Gamma_c \subset \Gamma_{c+1} = \Gamma$$

for which

$$\Gamma_i/\Gamma_{i-1} \cong \mathbb{Z}^{k_i} \text{ for } 1 \leq i \leq c \text{ and some } k_i \in \mathbb{N}_0 \text{ and}$$

$$\Gamma/\Gamma_c \text{ is finite.}$$

Moreover, if desired, each Γ_i can be chosen to be a characteristic subgroup of Γ. Let us call such a filtration (of not necessarily characteristic subgroups) of Γ a **torsion free filtration**. In many cases, we will have that $\Gamma_c = \Gamma_{c+1}$ in which case we do not always write the last term Γ_{c+1} of the torsion free filtration Γ_* of Γ. We use K to denote the Hirsch number (or rank) of Γ. Often, we will also use $K_i = k_i + k_{i+1} + \cdots + k_c$ and $K_{c+1} = 0$. It follows that $K = K_1$.

Definition 4.2.1 *Assume Γ is polycyclic–by–finite with a torsion free filtration Γ_*. A set of generators*

$$A = \{a_{1,1}, a_{1,2}, \ldots, a_{1,k_1}, a_{2,1}, \ldots, a_{c,k_c}, \alpha_1, \ldots, \alpha_r\}$$

of Γ *will be called* compatible with Γ_* *iff*

$$\forall i \in \{1, 2, \ldots, c\} : \ a_{1,1}, a_{1,2}, \ldots, a_{i,k_i} \ generates \ \Gamma_i.$$

It is clear at once that any torsion free filtration Γ_* of Γ admits a compatible set of generators.

In the sequel, we will frequently work with quotient groups of Γ. In order to avoid complicated notations, we will write the same symbols for elements in Γ, as for their coset classes in a quotient group. E.g. we say that Γ_2/Γ_1 is the free abelian group generated by $a_{2,1}, a_{2,2}, \ldots, a_{2,k_2}$.

Now we introduce quickly the basic building blocks for the representations we are interested in (we follow [44] and [14]):

$$\mathcal{M}(\mathbb{R}^K, \mathbb{R}^k) = \{\text{continuous maps } \lambda : \mathbb{R}^K \to \mathbb{R}^k\}$$

$$\mathcal{H}(\mathbb{R}^K) = \{\text{homeomorphisms } h : \mathbb{R}^K \to \mathbb{R}^K\}.$$

$\mathcal{M}(\mathbb{R}^K, \mathbb{R}^k)$ is an abelian group (for the addition of maps) and can be made into a $\mathrm{Gl}(k, \mathbb{R}) \times \mathcal{H}(\mathbb{R}^K)$–module, if we define

$$^{(g,h)}\lambda = g \circ \lambda \circ h^{-1}, \ \forall \lambda \in \mathcal{M}(\mathbb{R}^K, \mathbb{R}^k), \ \forall g \in \mathrm{Gl}(k, \mathbb{R}), \ \forall h \in \mathcal{H}(\mathbb{R}^K).$$

We remark that there is an embedding

$$\mathcal{M}(\mathbb{R}^K, \mathbb{R}^k) \rtimes (\mathrm{Gl}(k, \mathbb{R}) \times \mathcal{H}(\mathbb{R}^K)) \hookrightarrow \mathcal{H}(\mathbb{R}^{k+K})$$

with

$$(\lambda, g, h)(x, y) = (g(x) + \lambda h(y), h(y)), \tag{4.1}$$

$$\forall x \in \mathbb{R}^k, \ \forall y \in \mathbb{R}^K, \ \forall \lambda \in \mathcal{M}(\mathbb{R}^K, \mathbb{R}^k), \ \forall g \in \mathrm{Gl}(k, \mathbb{R}), \ \forall h \in \mathcal{H}(\mathbb{R}^K).$$

Definition 4.2.2 *Assume* Γ *is polycyclic–by–finite with a torsion free filtration* Γ_*. *A representation* $\rho = \rho_0 : \Gamma \to \mathcal{H}(\mathbb{R}^K)$ *will be called* **of canonical type** *with respect to* Γ_* *iff it induces a sequence of representations:*

$$\rho_i : \Gamma/\Gamma_i \to \mathcal{H}(\mathbb{R}^{K_{i+1}}), \ (1 \leq i \leq c)$$

such that for all i the following diagram commutes:

$$
\begin{array}{ccccccccc}
1 & \to & \mathbb{Z}^{k_i} \cong \Gamma_i/\Gamma_{i-1} & \to & \Gamma/\Gamma_{i-1} & \to & \Gamma/\Gamma_i & & \to 1 \\
& & \downarrow j & & \downarrow \rho_{i-1} & & \downarrow B_i \times \rho_i & & \\
1 & \to & \mathcal{M}(\mathbb{R}^{K_{i+1}}, \mathbb{R}^{k_i}) & \to & \mathcal{M} \rtimes (G \times \mathcal{H}) & \to & \mathrm{Gl}(k_i, \mathbb{R}) \times \mathcal{H}(\mathbb{R}^{K_{i+1}}) & \to 1
\end{array}
$$

$$\tag{4.2}$$

where

- $\mathcal{M} \rtimes (G \times \mathcal{H})$ stands for $\mathcal{M}(\mathbb{R}^{K_{i+1}}, \mathbb{R}^{k_i}) \rtimes (\mathrm{Gl}(k_i, \mathbb{R}) \times \mathcal{H}(\mathbb{R}^{K_{i+1}}))$,

- $j(z) : \mathbb{R}^{K_{i+1}} \to \mathbb{R}^{k_i} : x \mapsto z, \forall z \in \mathbb{Z}^{k_i}$ and

- $B_i : \Gamma/\Gamma_i \to \mathrm{Gl}(k_i, \mathbb{Z}) \hookrightarrow \mathrm{Gl}(k_i, \mathbb{R})$ denotes the action of Γ/Γ_i induced on $\mathbb{Z}^{k_i}$ by conjugation in Γ/Γ_{i-1}.

This means that a canonical type representation is nothing else than an iterated Seifert construction with abelian kernel. We explain this in more detail: Let Q be a group acting properly discontinuously on a space W, via $\rho : Q \to \mathcal{H}(W)$. Let $\mathcal{M}(W, \mathbb{R}^k)$ denote the set of continuous mappings of W into $\mathbb{R}^k$. In the same way as above we can make $\mathcal{M}(W, \mathbb{R}^k)$ into a $(\mathrm{Gl}(k, \mathbb{R}) \times \mathcal{H}(W))$–module, and there is an embedding

$$\mathcal{M}(W, \mathbb{R}^k) \rtimes (\mathrm{Gl}(k, \mathbb{R}) \times \mathcal{H}(W)) \hookrightarrow \mathcal{H}(\mathbb{R}^k \times W).$$

Now, consider a group extension $1 \to \mathbb{Z}^k \to E \to Q \to 1$ which induces an automorphism $\varphi : Q \to \mathrm{Aut}\,(\mathbb{Z}^k)$ via conjugation in E. A Seifert construction for the above data is in fact a morphism

$$\psi : E \to \mathcal{M}(W, \mathbb{R}^k) \rtimes (\mathrm{Gl}(k, \mathbb{R}) \times \mathcal{H}(W)) \stackrel{\text{notation}}{=\!=\!=} \mathcal{M} \rtimes (G \times \mathcal{H})$$

such that the following diagram commutes

$$
\begin{array}{ccccccccc}
1 & \to & \mathbb{Z}^k & \to & E & \to & Q & \to & 1 \\
& & \downarrow i & & \downarrow \psi & & \downarrow \varphi \times \rho & & \\
1 & \to & \mathcal{M}(W, \mathbb{R}^k) & \to & \mathcal{M} \rtimes (G \times \mathcal{H}) & \to & \mathrm{Gl}(k, \mathbb{R}) \times \mathcal{H}(W) & \to & 1
\end{array}
$$

The resulting quotient space is also said to be a Seifert Fiber Space with typical fiber a (k–dimensional) torus. This is because the canonical projection

$$p : {}_E\backslash(\mathbb{R}^k \times W) = {}_Q\backslash T^k \times W \longrightarrow {}_Q\backslash W$$

is "something like" a fiber space. Indeed, in general the inverse image $p^{-1}(\bar{w})$ of an element $\bar{w} \in {}_Q\backslash W$ will be diffeomorphic to a torus, but there migth be elements where the "fiber" $p^{-1}(\bar{w})$ is the quotient of the torus by a finite group action. These fibers are called singular, while the first ones are said to be typical.

We are now ready to summarize some important results on (general) representations of canonical type.

For elements $x \in \mathbb{R}^K$ we will label the coordinates as follows:

$$x = (x_{1,1}, x_{2,1}, \ldots, x_{1,k_1}, x_{2,1}, \ldots, x_{2,k_2}, x_{3,1}, \ldots, x_{c,k_c})$$

We also write x_i to indicate all variables $(x_{i,1}, x_{i,2}, \ldots, x_{i,k_i})$. If $\rho = \rho_0 : \Gamma \to \mathcal{H}(\mathbb{R}^K)$ is any representation, then for each $\gamma \in \Gamma$ there exist continuous functions $h_{i,j}^\gamma : \mathbb{R}^K \to \mathbb{R}$ such that

$$\rho(\gamma) : \mathbb{R}^K \to \mathbb{R}^K : x \mapsto (h_{1,1}^\gamma(x), h_{1,2}^\gamma(x), \ldots, h_{c,k_c}^\gamma(x)).$$

In a way similar to the $x_{i,j}$ we use $h_i^\gamma(x)$ to indicate

$$(h_{i,1}^\gamma(x), h_{i,2}^\gamma(x), \ldots, h_{i,k_i}^\gamma(x)).$$

Theorem 4.2.3 *Let Γ be a polycyclic–by–finite group and Γ_* be a torsion free filtration of Γ. Then, there exists a representation $\rho : \Gamma \to \mathcal{H}(\mathbb{R}^K)$ which is of canonical type with respect to Γ_*. Moreover, any two such representations $\rho, \rho' : \Gamma \to \mathcal{H}(\mathbb{R}^K)$ are conjugate to each other inside $\mathcal{H}(\mathbb{R}^K)$. Also, for each canonical type representation $\rho : \Gamma \to \mathcal{H}(\mathbb{R}^K)$ we have:*

1. *Γ acts properly discontinuously on $\mathbb{R}^K$, via ρ and with compact quotient.*

2. *$\ker \rho =$ the maximal finite normal subgroup of Γ, so ρ is effective iff Γ is almost torsion free.*

3. *$\forall \gamma \in \Gamma$:*
   $\rho(\gamma) : \mathbb{R}^K \to \mathbb{R}^K : x \mapsto (h_1^\gamma(x), h_2^\gamma(x), \ldots, h_c^\gamma(x))$ is such that

   $$h_i^\gamma(x) = B_i(\gamma)x_i + g_i^\gamma(x_{i+1}, x_{i+2}, \ldots, x_c),$$

   for some continuous function $g_i^\gamma : \mathbb{R}^{K_{i+1}} \to \mathbb{R}^{k_i}$.

4. *Γ_i acts trivially on $\mathbb{R}^{K_{i+1}}$ and via affine transformations on $\mathbb{R}^{k_i}$.*

<u>Proof:</u> The proof of the existence and uniqueness of a canonical type representation, can be found in [47]. The reader may also consult [47] to see that the action of Γ defined on $\mathbb{R}^K$ is properly discontinuous and that the orbit space is compact. We do not go into detail here, since we will prove the analogous results in more interesting cases later on.

2) follows from corollary 3.5.2.

3) and 4) are immediate consequences of the iterative way of building up ρ
(see also (4.1)).

Remark 4.2.4

Point 3. and 4. from the theorem above can be used to define the concept of a canonical type representation. Such kind of definitions were for instance used in [20].

Having this concept of a (general) canonical type representation, it is clear now, that to get a nicer geometric structure, one needs to use smaller subgroups of $\mathcal{H}(\mathbb{R}^K)$.

In view of the iterative set up, it is however necessary that these subgroups satisfy one crucial condition: assume we restrict $\mathcal{M}(\mathbb{R}^K, \mathbb{R}^k)$ to a subspace $\mathcal{S}(\mathbb{R}^K, \mathbb{R}^k)$ containing the space of constant mappings $\mathbb{R}^k$ and assume we restrict $\mathcal{H}(\mathbb{R}^K)$ to a subgroup $\mathcal{SH}(\mathbb{R}^K)$, then one observes that it is necessary that $\mathcal{S}(\mathbb{R}^K, \mathbb{R}^k)$ is a $(\mathrm{Gl}(k, \mathbb{R}) \times \mathcal{SH}(\mathbb{R}^K))$-submodule such that there is a monomorphism,

$$\mathcal{S}(\mathbb{R}^K, \mathbb{R}^k) \rtimes (\mathrm{Gl}(k, \mathbb{R}) \times \mathcal{SH}(\mathbb{R}^K)) \hookrightarrow \mathcal{SH}(\mathbb{R}^{k+K}).$$

Possible examples of such situations are

Smooth representations: Let $C^\infty(\mathbb{R}^K, \mathbb{R}^k)$ denote the vector space of functions $f : \mathbb{R}^K \to \mathbb{R}^k$, which are infinitely many times differentiable and denote by $C^\infty(\mathbb{R}^K)$ the group, under composition of maps, of smooth diffeomorphisms of $\mathbb{R}^K$. Then by considering $C^\infty(\mathbb{R}^K, \mathbb{R}^k)$ instead of $\mathcal{M}(\mathbb{R}^K, \mathbb{R}^k)$ and replacing $\mathcal{H}(\mathbb{R}^K)$ by $C^\infty(\mathbb{R}^K)$, we find the so-called canonical type smooth representations. Here, results analogous to theorem 4.2.3 can be formulated (see [14] and [47]).

Affine representations: Restrict $\mathcal{M}(\mathbb{R}^K, \mathbb{R}^k)$ to $\mathrm{Aff}(\mathbb{R}^K, \mathbb{R}^k)$ (the vector space of affine mappings) and $\mathcal{H}(\mathbb{R}^K)$ to $\mathrm{Aff}(\mathbb{R}^K)$ (the affine group of $\mathbb{R}^K$). We will treat this kind of representations in a few moments.

Polynomial representations: Write $P(\mathbb{R}^K, \mathbb{R}^k)$ to refer to the vector space of polynomial mappings $p : \mathbb{R}^K \to \mathbb{R}^k$. So p is given by k polynomials in K variables. $P(\mathbb{R}^K)$ will be used to indicate the set of all polynomial diffeomorphisms of $\mathbb{R}^K$, with an inverse which is also a polynomial mapping. This is a group where the multiplication is given by composition. It is not hard to verify that $P(\mathbb{R}^K, \mathbb{R}^k)$ is an $\mathrm{Aut}(\mathbb{Z}^k) \times P(\mathbb{R}^K)$-module and that the resulting semidirect product group

$$P(\mathbb{R}^K, \mathbb{R}^k) \rtimes (\mathrm{Aut}\, \mathbb{Z}^k \times P(\mathbb{R}^K)) \subseteq P(\mathbb{R}^{K+k})$$

since by definition

$$\forall\, (p,g,h) \in P(\mathbb{R}^K, \mathbb{R}^k) \rtimes (\text{Aut } \mathbb{Z}^k \times P(\mathbb{R}^K)),\ \forall\, (x,y) \in \mathbb{R}^{k+K}:$$

$$^{(p,g,h)}(x,y) = (gx + p(h(y)), h(y)).$$

Restrict $\mathcal{M}(\mathbb{R}^K, \mathbb{R}^k)$ to $P(\mathbb{R}^K, \mathbb{R}^k)$ and $\mathcal{H}(\mathbb{R}^K)$ to $P(\mathbb{R}^K)$, and we will speak of canonical type polynomial representations. We will also deal with these kind of representations later in this chapter.

In each of these restricted situations one now faces existence and uniqueness questions. In order to be able to solve these problems, we will first give a general treatment of the Seifert Fiber Space construction in the following section.

4.3 An algebraic description of the Seifert Fiber Space construction

In this section we will first prove a general algebraic lemma and then apply this lemma to the Seifert Construction situation. This theory was developed by K.B. Lee in [44].

Let Q and Q_1 be groups and suppose that there are two abelian groups Z and S such that Z is a Q–module, while S is a Q_1–module. Assume moreover that two morphisms $i : Z \hookrightarrow S$ (i is an embedding) and $j : Q \to Q_1$ (j is not necessarily injective) are given which are compatible with the module structures. I.e.

$$\forall \alpha \in Q, \forall z \in Z : i(^{\alpha}z) =^{j(\alpha)} i(z).$$

The aim of this section is to describe all pairs (E, f), where E is a group extension of Q by Z and $f : E \to S \rtimes Q_1$ is a morphism of groups such that the following diagram is commutative:

$$
\begin{array}{ccccccccc}
0 & \to & Z & \to & E & \to & Q & \to & 1 \\
  &   & \downarrow i & & \downarrow f & & \downarrow j & & \\
0 & \to & S & \to & S \rtimes Q_1 & \to & Q_1 & \to & 1
\end{array}
$$

Of course, before we can start the description of all such pairs, we must know when to consider two such pairs as equal or as different.

Definition 4.3.1 *Two pairs (E, f) and (E', f') as above are said to be equivalent if and only if E and E' are equivalent as group extensions of Z by Q via an isomorphism θ (inducing the identity on Z and on Q) such that there exists an element $s \in S$ with $f'\theta = \mu(s)f$.*

So if $(E, f) \sim (E', f')$ via an isomorphism θ, then there exists an element $s \in S$ for which the following diagram commutes:

$$
\begin{array}{ccc}
E & \overset{\theta}{\longrightarrow} & E' \\
f \downarrow & & \downarrow f' \\
S \rtimes Q_1 & \overset{\mu(s)}{\longrightarrow} & S \rtimes Q_1
\end{array}
$$

Let us denote the set of equivalence classes of pairs (E, F) by $H(Q; Z, S)$. The following lemma and its proof are crucial to understand most of the remaining chapter. This is because we will show a one to one correspondence between $H(Q; Z, S)$ and a 1–cohomology group $H^1(Q, \frac{S}{Z})$, where the exact correspondence will be explained during the proof.

Lemma 4.3.2 *There is a bijection between $H(Q; Z, S)$ and $H^1(Q, \frac{S}{Z})$*

Proof: Consider the short exact sequence of Q modules

$$
1 \to Z \overset{i}{\longrightarrow} S \overset{p}{\longrightarrow} \frac{S}{Z} \to 1,
$$

where the Q–module structure of S is, of course, given by

$$
\forall q \in Q, \forall s \in S : {}^q s = {}^{j(q)} s.
$$

We will define a map $\Omega : H^1(Q, \frac{S}{Z}) \to H(Q; Z, S)$ and prove that this map is indeed a bijection.

Construction of Ω:
Let $\lambda : Q \to \frac{S}{Z}$ be any crossed homomorphism (i.e. take $\lambda \in Z^1(Q, \frac{S}{Z})$). Choose any lift $\tilde{\lambda} : Q \to S$ of λ (so $p\tilde{\lambda} = \lambda$) with $\tilde{\lambda}(1) = 0$. This $\tilde{\lambda}$ induces a pair $(E(\tilde{\lambda}), f_{\tilde{\lambda}})$ as follows:

1. Let $E(\tilde{\lambda})$ be the group with underlying set $Z \times Q$ and with multiplication

$$
\forall z, y \in Z, \forall \alpha, \beta \in Q : (z, \alpha)(y, \beta) = (z + {}^\alpha y + \delta\tilde{\lambda}(\alpha, \beta), \alpha\beta)
$$

where $\delta\tilde{\lambda}(\alpha, \beta) = {}^\alpha \tilde{\lambda}(\beta) - \tilde{\lambda}(\alpha\beta) + \tilde{\lambda}(\alpha)$. This means in fact that $E(\tilde{\lambda})$ is the extension representing the cohomology class $\delta\langle\lambda\rangle \in H^2(Q, Z)$, where

$$
\delta : H^1(Q, \frac{S}{Z}) \to H^2(Q, Z)
$$

is the connecting homomorphism in the long exact cohomology sequence

$$\cdots \to H^1(Q,S) \xrightarrow{p_*} H^1(Q,\frac{S}{Z}) \xrightarrow{\delta} H^2(Q,Z) \xrightarrow{i_*} H^2(Q,S) \to \cdots$$

induced by the short exact sequence $0 \to Z \to S \to S/Z \to 0$ of Q–modules.

2. Define $f_{\tilde{\lambda}} : E(\tilde{\lambda}) \to S \rtimes Q_1 : (z,\alpha) \mapsto (i(z) + \tilde{\lambda}(\alpha), j(\alpha))$.
   In the sequel, we will write z in stead of $i(z)$, which is justified by the fact that i is injective. So, $f_{\tilde{\lambda}}(z,\alpha) = (z + \tilde{\lambda}(\alpha), j(\alpha))$. It is easily verified that $f_{\tilde{\lambda}} : E(\tilde{\lambda}) \to S \rtimes Q_1$ is a morphism of groups.

The first thing to show is that the choice of lift $\tilde{\lambda}$ is unimportant. So consider another lift of λ, say $\tilde{\tilde{\lambda}}$. Then we define

$$g : Q \to Z : \alpha \mapsto \tilde{\tilde{\lambda}}(\alpha) - \tilde{\lambda}(\alpha).$$

The fact that $g(\alpha)$ takes images in Z follows from the fact that $\tilde{\tilde{\lambda}}$ and $\tilde{\lambda}$ are both lifts of the same λ. Using this g, we introduce a map

$$\theta : E(\tilde{\lambda}) \to E(\tilde{\tilde{\lambda}}) : (z,\alpha) \mapsto (z - g(\alpha), \alpha).$$

Some elementary computations show that θ is an isomorphism of groups inducing the identity on both Z and Q and such that

$$f_{\tilde{\lambda}} = f_{\tilde{\tilde{\lambda}}} \circ \theta.$$

This shows that the pairs $(E(\tilde{\lambda}), f_{\tilde{\lambda}})$ and $(E(\tilde{\tilde{\lambda}}), f_{\tilde{\tilde{\lambda}}})$ are equivalent. This implies that the choice of lift does not play an essential role.

So far, we only defined Ω on $Z^1(Q,\frac{S}{Z})$. Therefore, we investigate what happens in case we consider two cohomologous 1–cocycles λ and λ'. There exists an element $\bar{s} \in \frac{S}{Z}$ such that $\lambda'(\alpha) - \lambda(\alpha) = \delta\bar{s}(\alpha) = \bar{s} - ^\alpha\bar{s}$. Let $s \in S$ be a lift of $\bar{s}$ and suppose that $\tilde{\lambda}$ is a lift of λ, then we can take $\tilde{\lambda}' = \tilde{\lambda} + \delta s$ as a lift for λ'. Remark that $E(\tilde{\lambda}) = E(\tilde{\lambda}')$, since $\delta\tilde{\lambda}' = \delta\tilde{\lambda} + \delta\delta s = \delta\tilde{\lambda}$. Moreover, $f_{\tilde{\lambda}'} = \mu(s) \circ f_{\tilde{\lambda}}$. This shows that λ and λ' determine two equivalent pairs.

Conclusion: the map

$$\Omega : H^1(Q,\frac{S}{Z}) \to H(Q;Z,S) : \langle\lambda\rangle \mapsto (E(\tilde{\lambda}), f_{\tilde{\lambda}})$$

is well defined.

Ω is injective:

Suppose that $\Omega\langle\lambda\rangle = \Omega\langle\lambda'\rangle$. So there exists an equivalence $\theta : E(\tilde{\lambda}) \to E(\tilde{\lambda}')$ of group extensions and an element $s \in S$ such that $\mu(s) \circ f_{\tilde{\lambda}} = f_{\tilde{\lambda}'} \circ \theta$. Define the map $g : Q \to Z$ by the condition

$$\theta(0, \alpha) = (-g(\alpha), \alpha).$$

Remark that

$$\mu(s) \circ f_{\tilde{\lambda}}(0, \alpha) = (s + \tilde{\lambda}(\alpha) -^\alpha s, j(\alpha))$$

while

$$f_{\tilde{\lambda}'} \circ \theta = (-g(\alpha) + \tilde{\lambda}'(\alpha), j(\alpha))$$

which implies that

$$s + \tilde{\lambda}(\alpha) -^\alpha s = -g(\alpha) + \tilde{\lambda}'(\alpha) \Rightarrow \lambda' - \lambda = \delta\bar{s} \text{ with } \bar{s} = p(s).$$

Therefore, seen as elements of $H^1(Q, \frac{S}{Z})$, $\langle\lambda\rangle = \langle\lambda'\rangle$, which was to be shown.

Ω is surjective:

Suppose (E, f) is a pair satisfying the necessary conditions. E is an extension of Q by Z and so there exists a 2–cocycle $c : Q \times Q \to Z$ such that E can be seen as the set $Z \times Q$ and where the multiplication is given by

$$\forall z, y \in Z; \forall \alpha, \beta \in Q : (z, \alpha)(y, \beta) = (z +^\alpha y + c(\alpha, \beta), \alpha\beta).$$

Determine a map $q : E \to S$ by $f(z, \alpha) = (z + q(z, \alpha), j(\alpha))$. However, as $(z, \alpha) = (z, 1)(0, \alpha)$ and $f(z, 1) = (z, 1)$, it follows that

$$q(z, \alpha) = q(0, \alpha)$$

and so, from now on, we consider q as being a map from Q to Z. Evaluating f on both sides of

$$(z, \alpha)(y, \beta) = (z +^\alpha y + c(\alpha, \beta), \alpha\beta)$$

shows that

$$(z + q(\alpha) +^\alpha y +^\alpha q(\beta), j(\alpha)j(\beta)) = (z +^\alpha y + c(\alpha, \beta) + q(\alpha\beta), j(\alpha\beta))$$

or $c(\alpha, \beta) = \delta q(\alpha\beta)$.

This shows that $(E, f) = \Omega(\langle p(q) \rangle)$, which finishes the last part of the proof of the lemma. ∎

We will apply this lemma in the following concrete situtation. Let Q be a group equiped with two morphisms:

$$\varphi : Q \to \text{Gl}(k, \mathbb{Z}) \text{ and } \rho : Q \to \mathcal{H}(W), W \text{ is a topological space.}$$

For Q_1 we choose a subgroup of $\text{Gl}(k, \mathbb{R}) \times \mathcal{H}(W)$, which contains the image of $j = \varphi \times \rho$. For S we shall consider a Q_1–submodule $S(W, \mathbb{R}^k)$ of $\mathcal{M}(W, \mathbb{R}^k)$, containing the constant maps (the Q_1–module structure is decribed in the previous section). Then, the above lemma gives a lot of information on the pairs (E, f), where E is an extension of $\mathbb{Z}^k$ by Q, compatible with φ and where $f : E \to S(W, \mathbb{R}^k) \rtimes Q_1$ is a morphism making the diagram

$$
\begin{array}{ccccccccc}
0 & \to & \mathbb{Z}^k & \to & E & \to & Q & \to & 1 \\
 & & \downarrow i & & \downarrow f & & \downarrow j & & \\
0 & \to & S(W, \mathbb{R}^k) & \to & S(W, \mathbb{R}^k) \rtimes Q_1 & \to & Q_1 & \to & 1
\end{array}
$$

commutative. Here $i(z) : W \to \mathbb{R}^k : w \mapsto z$.

The information the lemma provides is given by the connecting homomorphism

$$\delta : H^1(Q, \frac{S(W, \mathbb{R}^k)}{\mathbb{Z}^k}) \to H^2(Q, \mathbb{Z}^k).$$

For a given extension E, there exists a morphism $f : E \to S(W, \mathbb{R}^k) \rtimes Q_1$ as above if and only if the extension E corresponds to a cohomology class in the image of δ. So, the eventual surjectivity of the connecting homomorphism δ guarantees the existence of a morphism f for any extension. Moreover, the number of such morphisms f, up to conjugation with an element of $S(W, \mathbb{R}^k)$, is measured by the kernel of the δ. So an injective connecting homomorphism implies the uniqueness of the morphism f.

4.4 Canonical type affine representations

We will first of all concentrate on the most useful type of canonical type representations namely the affine representations. These are interesting since they can be seen as faithful matrix representations, and thus they allow a formal computational approach via very simple computer algorithms.

However, as we will see, such a representation does not always exists, but fortunately, there is no trouble in case one considers only finite extensions of groups of low nilpotency classes. We will deal with the uniqueness problem of these representation in the following section.

Let N be a finitely generated, torsion free nilpotent group and consider any **central series** with torsion free factors

$$N_* : 1 = N_1 \subseteq N_2 \subseteq N_3 \supseteq \cdots \supseteq N_c = N(= N_{c+1}).$$

We will refer to such a filtration as a **torsion free central series**. Examples of such central series are given by the upper central series and the adapted lower central series of N. As in the general case we also denote by k_i $(1 \leq i \leq c)$ the rank of N_i/N_{i+1}. So, $N_i/N_{i+1} \cong \mathbb{Z}^{k_i}$. We rewrite the definition of a canonical type affine representation with respect to the torsion free central series N_*. The reader should convince himself that, in this restricted setting, this new definition is really an alternative one to the general definition 4.2.2.

Definition 4.4.1 *A faithful representation* $\rho : N \to \mathrm{Aff}(\mathbb{R}^K)$ *will be called "of canonical type" with respect to* N_* *if and only if*

1. *the matrix parts of* ρ *are blocked upper-triangular with the identity matrices of size* $k_1, k_2, \cdots, k_c$ *as diagonal entries,*

2. *the subgroup* N_i *of* N *acts on the ith block* $(\mathbb{R}^{k_i})$ *as translations* $(= \mathbb{Z}^{k_i})$ *and trivially on* $\mathbb{R}^{K_{i+1}}$.

4.4.1   Iterating canonical type affine representations

Let us describe the Seifert construction in this context:
Consider a group N with a canonical type affine representation $\rho : N \to \mathrm{Aff}(\mathbb{R}^K) : n \mapsto \rho(n) = (A(n), a(n))$, where $A(n)$ denotes the linear $(K \times K$–matrix) and $a(n) \in \mathbb{R}^K$ the translational part of $\rho(n)$. Analogously, elements of the additive group $\mathrm{Aff}(\mathbb{R}^K, \mathbb{R}^k)$ of affine mappings from $\mathbb{R}^K$ to $\mathbb{R}^k$ are given by a couple (D, d) where D denotes the $(k \times K)$ linear part and d $(\in \mathbb{R}^k)$ denotes the constant part. We regard $\mathbb{Z}^k$ as a subgroup of constant mappings in $\mathrm{Aff}(\mathbb{R}^K, \mathbb{R}^k)$ i.e. z $(\in \mathbb{Z}^k) \mapsto (0, z)$ $(\in \mathrm{Aff}(\mathbb{R}^K, \mathbb{R}^k))$.

There is an action of $\mathrm{Aff}(\mathbb{R}^K)$ (and so of N) on $\mathrm{Aff}(\mathbb{R}^K, \mathbb{R}^k)$ as follows: if $h \in \mathrm{Aff}(\mathbb{R}^K)$ and $\lambda \in \mathrm{Aff}(\mathbb{R}^K, \mathbb{R}^k)$ then $^h\lambda = \lambda \circ h^{-1}$. Clearly $\mathbb{Z}^k$ becomes a trivial N–module. The semidirect product $\mathrm{Aff}(\mathbb{R}^K, \mathbb{R}^k) \rtimes \mathrm{Aff}(\mathbb{R}^K)$ acts on $\mathbb{R}^k \times \mathbb{R}^K = \mathbb{R}^{k+K}$ if we define for $(x, y) \in \mathbb{R}^k \times \mathbb{R}^K$, $^{(\lambda, h)}(x, y) =$

$(x + \lambda(h(y)), h(y))$. It follows immediately that if $h = (A, a)$ and $\lambda = (D, d)$, this action is given by

$$\begin{pmatrix} I & DA \\ 0 & A \end{pmatrix} \cdot \begin{pmatrix} x \\ y \end{pmatrix} + \begin{pmatrix} Da + d \\ a \end{pmatrix} \tag{4.3}$$

and so is clearly affine.

The iteration problem. Given the previous set up, it is natural to consider the following problem:
given a central extension $1 \to \mathbb{Z}^k \to E \to N \to 1$, can we extend ρ to a representation $\rho': E \to \mathrm{Aff}(\mathbb{R}^K, \mathbb{R}^k) \rtimes \mathrm{Aff}(\mathbb{R}^K)$ such that the following diagram is commutative:

$$
\begin{array}{ccccccccc}
1 & \to & \mathbb{Z}^k & \to & E & \to & N & \to & 1 \\
& & \downarrow & & \downarrow \rho' & & \downarrow \rho & & \\
1 & \to & \mathrm{Aff}(\mathbb{R}^K, \mathbb{R}^k) & \to & \mathrm{Aff}(\mathbb{R}^K, \mathbb{R}^k) \rtimes \mathrm{Aff}(\mathbb{R}^K) & \to & \mathrm{Aff}(\mathbb{R}^K) & \to & 1
\end{array}
$$

If yes, clearly ρ' will be again canonical.

As this iteration problem can be seen as a Seifert Fiber Space construction, crucial information about this iteration problem is contained in the connecting homomorphism δ of the long exact cohomology sequence

$$\to H^1(N, \mathrm{Aff}(\mathbb{R}^K, \mathbb{R}^k)/\mathbb{Z}^k) \xrightarrow{\delta} H^2(N, \mathbb{Z}^k) \to H^2(N, \mathrm{Aff}(\mathbb{R}^K, \mathbb{R}^k)) \to \cdots \tag{4.4}$$

according to the exact sequence of N-modules

$$0 \to \mathbb{Z}^k \to \mathrm{Aff}(\mathbb{R}^K, \mathbb{R}^k) \to \mathrm{Aff}(\mathbb{R}^K, \mathbb{R}^k)/\mathbb{Z}^k \to 0 .$$

If E can be represented by a 2-cocycle f, with $\langle f \rangle \in H^2(N, \mathbb{Z}^k)$ lying in the image of δ then the existence of an extended ρ' is true.

In [54], Nisse announced a very general proposition stating that δ is surjective, even in the case of non central extensions and more generally for polycyclic groups N. However, as shown in [49], a (solvable, not nilpotent) counter-example in the case of non central extensions casts doubt on this formulation. In the next section (see example 4.4.13) we will show that Nisse's formulation is incorrect, even in the case of central extensions and for nilpotent groups N.

Assume we are given a cohomology class $\langle f \rangle \in H^2(N, \mathbb{Z}^k)$, representing an extension $E = \mathbb{Z}^k \times N$ where the multiplication in E is given by

$$\forall z, z_1 \in \mathbb{Z}^k, \forall n, n_1 \in N : (z, n)(z_1, n_1) = (z + z_1 + f(n, n_1), nn_1).$$

If we can compute explicitly a 1–cochain, say $\gamma : N \to \mathrm{Aff}(\mathbb{R}^K, \mathbb{R}^k)$ killing the class $\langle f \rangle$ in $H^2(N, \mathrm{Aff}(\mathbb{R}^K, \mathbb{R}^k))$, then $\rho'(z, n)$ is given by $\rho'(z, n) = (z + \gamma(n), \rho(n))$. As (4.3) shows, this extended representation will again be of canonical type.

This problem, in principle, now becomes a computational one. Indeed, we should find $\gamma : N \to \mathrm{Aff}(\mathbb{R}^K, \mathbb{R}^k) : x \mapsto \gamma(x) = (D(x), d(x))$, such that $\delta\gamma(x, y) = f(x, y)$ $(\forall x, y \in N)$. More explicitly, this means finding a matrix part $D(x)$ and a translational part $d(x)$, satisfying

$$^x(D(y), d(y)) - (D(xy), d(xy)) + (D(x), d(x)) = (0, f(x, y))$$

or equivalently

$$D(y)A(x^{-1}) - D(xy) + D(x) \;=\; 0 \qquad\qquad (4.5)$$
$$D(y)(a(x^{-1})) + d(y) - d(xy) + d(x) \;=\; f(x, y). \qquad (4.6)$$

Since $\mathbb{Z}^k$ is a trivial N–module, this problem can be treated componentwise.

Now what looks like a 2–condition problem, surprisingly is a 1–condition problem, as we point out in the following proposition:

Proposition 4.4.2 *Assume $\rho : N \to \mathrm{Aff}(\mathbb{R}^K)$ is a representation of canonical type, and $1 \to \mathbb{Z} \to E \to N \to 1$ is a central extension, determined by $\langle f \rangle \in H^2(N, \mathbb{Z})$.*
Then, $\langle f \rangle$ lies in the image of $\delta : H^1(N, \mathrm{Aff}(\mathbb{R}^K, \mathbb{R})/\mathbb{Z}) \to H^2(N, \mathbb{Z})$ iff one can find $(D, d) : N \to \mathrm{Aff}(\mathbb{R}^K, \mathbb{R})$ satisfying condition (4.6).

Proof: We will show that condition (4.5) is automatically satisfied, once condition (4.6) is fulfilled. So, assume (4.6) is satisfied. Since ρ is of canonical type, we know that the translational parts $a(x)$ (for $x \in N$) are spanning the whole vector space $\mathbb{R}^K$. Thus, it will be enough to show that

$$(D(y)A(x^{-1}) - D(xy) + D(x))a(z) = 0, \quad \forall z \in N. \qquad (4.7)$$

Now, from $\rho(x^{-1}z) = \rho(x^{-1})\rho(z)$ it follows at once that $a(x^{-1}z) = a(x^{-1}) + A(x^{-1}).a(z)$. This and the assumed relation (4.6) allow us to write (4.7) as

$$D(y)(a(x^{-1}z) - a(x^{-1})) - D(xy)a(z) + D(x)a(z)$$
$$= \; f(z^{-1}x, y) - f(x, y) - f(z^{-1}, xy) + f(z^{-1}, x)$$
$$= \; -\delta f(z^{-1}, x, y)$$
$$= \; 0.$$

Note that we used both the fact that f is a 2–cocycle and that $\mathbb{Z}^k$ is considered as a trivial N–module.

4.4.2 Canonical type representations and matrices over polynomial rings

The examples we'll study later on, inspired us to detect an interesting property for canonical type affine representations. Basically, what we saw in all examples, were upper triangular matrices with polynomial entries and degrees going up towards the right upper corner of the matrix. We now prove this is what should happen.

For any commutative ring R with identity, we write $\mathbf{UT}_K(R)$ for the (multiplicative) group of upper-triangular $(K \times K)$-matrices with entries in R and 1's on the diagonal. A matrix A in $\mathbf{UT}_K(R)$ is called blocked upper triangular of type $(k_1, \ldots, k_c)$ (with $\sum_{i=1}^{c} k_i = K$) if and only if A has identity matrix blocks of size $k_1, \ldots, k_c$ on its diagonal. The subgroup of matrices in $\mathbf{UT}_K(R)$ which are of this type is denoted $\mathbf{BUT}_{\sum k_i}(R)$. From now on, we will speak of unitriangular and blocked unitriangular matrices.

The only eventual nonzero entries (resp. blocks) in a matrix A of $\mathbf{UT}_K(R)$ (resp. $\mathbf{BUT}_{\sum k_i}(R)$) are the entries (resp. blocks) $a_{i,j}$ (resp. $A_{i,j} = (k_i \times k_j)$-block) with $j \geq i$. For these entries $a_{i,j}$ (resp. $A_{i,j}$) we call $(j - i)$ their distance from the diagonal.

From now on we take R to be the ring $F[X_1, \ldots, X_m]$ of polynomials in m variables over a field F. We use this in the following definition:

Definition 4.4.3 *A matrix A in $\mathbf{UT}_K(R)$ (resp. $\mathbf{BUT}_{\sum k_i}(R)$) is said to have the Diagonal Distance Degree property (DDD-property) if and only if each $a_{i,j}$ $(j > i)$ (resp. each entry in $A_{i,j} =$ the $(k_i \times k_j)$-block in A) is a polynomial of total degree $\leq (j - i)$. Such a matrix will be called a DDD-matrix (resp. a blocked DDD-matrix of type $\sum k_i$).*

Lemma 4.4.4
The set of all (blocked) DDD-matrices in $\mathbf{UT}_K(R)$ (resp. $\mathbf{BUT}_{\sum k_i}(R)$) forms a subgroup of $\mathbf{UT}_K(R)$ (resp. $\mathbf{BUT}_{\sum k_i}(R)$).

<u>Proof:</u> If A and B are DDD-matrices (resp. blocked DDD-matrices of type $\sum k_i$), then

$$(A \cdot B)_{i,j} = \sum_{t=1}^{K} A_{i,t} \cdot B_{t,j}.$$

and so, it is easily seen that the degree of $(A \cdot B)_{i,j}$ is bounded above by $(j - i)$.

To prove that A^{-1} has also the required DDD-property, one can proceed by induction on K (resp. c). For $K = 1$ (resp. $c = 1$) the claim is evident. Assume now that $K > 1$ (resp. $c > 1$). Then A can be viewed as a matrix

$$A = \begin{pmatrix} A' & a \\ 0 & 1 \end{pmatrix} \quad (\text{resp.} \quad \begin{pmatrix} A' & a \\ 0 & I_{k_c} \end{pmatrix})$$

where A' is in $\mathbf{UT}_{K-1}(R)$ (resp. $\mathbf{BUT}_{k_1 + \cdots + k_{c-1}}(R)$) and has the DDD-property. It is well known that

$$A^{-1} = \begin{pmatrix} A'^{-1} & -A'^{-1} \cdot a \\ 0 & 1 \end{pmatrix} \quad (\text{resp.} \quad \begin{pmatrix} A'^{-1} & -A'^{-1} \cdot a \\ 0 & I_{k_c} \end{pmatrix}).$$

By assumption A'^{-1} has the DDD-property. One verifies easily that the entries in $A'^{-1} \cdot a$ have also the required property.

∎

The following lemma is very important from the computational point of view! It shows that it is possible to compute A^l for a formal parameter l, for all unitriangular matrices A.

Lemma 4.4.5
Assume A is a fixed blocked unitriangular matrix of type $(k_1, \ldots, k_c)$ with entries in F. If $R = F[X]$, then there exists a DDD-matrix $B(X) \in \mathbf{BUT}_{\sum k_i}(R)$ such that $\forall l \in \mathbb{Z}\ A^l = B(l)$.

Proof: If $l \in \mathbb{Z}$, then clearly $A^l = ((A - I) + I)^l = \displaystyle\sum_{t=0}^{c} \binom{l}{t}(A - I)^t$.

Evidently, A^l will again be unitriangular. If $j > i$, then

$$(A^l)_{i,j} = \left(\sum_{t=0}^{c} \binom{l}{t}(A - I)^t \right)_{i,j} = \sum_{t=0}^{j-i} \binom{l}{t}((A - I)^t)_{i,j}$$

which is clearly seen to be of degree at most $(j - i)$ in l. So, it is sufficient to take $B(X) \in \mathbf{BUT}_{\sum k_i}(R)$ with

$$B(X)_{i,j} = \sum_{t=0}^{j-i} \binom{X}{t}((A - I)^t)_{i,j}.$$

∎

Remark 4.4.6 *The upper-bound on the degree of the polynomials in $B(X)$ depends also on the nilpotency degree of the matrix $(A - I)$, as is seen directly in the proof of the lemma. E.g. if A is blocked unitriangular of type $\sum k_i$, and $(A - I)$ has m bottom rows of blocks which are zero then $(A - I)^{c+1-m} = 0$ and so the degree of the blocks $B(X)_{i,j}$ in $B(X)$ will be less than or equal to $Min\{j - i, c - m\}$.*

Let us return to nilpotent groups. Consider a finitely generated torsion free nilpotent group N of rank $K = \sum_{i=1}^{c} k_i$ and a torsion free central series N_*. Moreover, we fix a set of generators

$$\{a_{1,1}, a_{1,2}, \ldots, a_{1,k_1}, a_{2,1}, \ldots, a_{c,k_c}\}$$

of N, which is compatible with N_*. These generators are labeled with two indices and we propose the following (somewhat bizarre) way of ordering these labels: label (i, j) is said to be less than or equal to label (m, n) if and only if $m < i$ or $((m = i)$ and $(j \leq n))$. Then the commutator presentation can be written as

$$< a_{c,1}, a_{c,2}, \ldots, a_{c,k_c}, a_{c-1,1}, \ldots, a_{c-1,k_{c-1}}, \ldots, a_{1,1}, \ldots, a_{1,k_1} \|$$

$$[a_{i,t_j}, a_{m,t_n}] = \text{word in } a_{l,t_p}\text{'s } (l, t_p) > (i, t_j) > (m, t_n) > . \quad (4.8)$$

Regarding $\mathrm{Aff}(\mathbb{R}^K)$ as embedded in $\mathrm{Gl}(K + 1, \mathbb{R})$ as usual, we are ready for the following theorem:

Theorem 4.4.7 *Assume N is a group of type $(k_1, \ldots, k_c)$ with a commutator presentation as in (4.8). Denote R for $\mathbb{R}[x_{c,1}, \ldots, x_{1,k_1}]$. A representation $\rho : N \to \mathrm{Aff}(\mathbb{R}^K) \hookrightarrow \mathrm{Gl}(K + 1, \mathbb{R})$ of N is of canonical type if and only if for $n = a_{c,1}^{x_{c,1}} \ldots a_{1,k_1}^{x_{1,k_1}} \in N$, $\rho(n)$ is a DDD-matrix in $\mathbf{BUT}_{(\sum k_i)+1}(R)$ combining the following properties:*

1. *the total degree in the variables $(x_{i,1}, \ldots, x_{i,k_i})$ of the entries of $\rho(n)$ is less than or equal to i, more precisely: polynomial entries containing the variables $(x_{i,1}, \ldots, x_{i,k_i})$ occur*

   - *in the linear part only in the blocks of the r-th row, for $r \leq i-1$, in terms of total degree at most $i - r$,*

   - *or, in the translational part in the blocks of the r-th row, for $r \leq i$ in terms of total degree at most $i + 1 - r$. Moreover, the i-th block of the images of the generators $a_{i,t}$ $(1 \leq t \leq k_i)$ spans $\mathbb{R}^{k_i}$;*

2. if an entry $a_{i,j}$ $(j > i)$ in $\rho(n)$ is not zero, then it is a polynomial without constant term.

<u>Proof:</u> The basic fact in the proof is given by the definition itself of a representation of canonical type. It follows immediately, that, for each i $(1 \leq i \leq c)$ the generators $a_{i,j}$ $(1 \leq j \leq k_i)$ are mapped by ρ to a matrix of the type

$$
\begin{pmatrix}
I_{k_1} & \star & \star & \star & \cdots & \star & \star \\
0 & I_{k_2} & \star & \star & \cdots & \star & \star \\
\vdots & \vdots & \vdots & \vdots & & \vdots & \vdots \\
0 & 0 & I_{k_i} & 0 & \cdots & 0 & B_{i,j} \\
0 & 0 & 0 & I_{k_{i+1}} & \cdots & 0 & 0 \\
\vdots & \vdots & \vdots & \vdots & & \vdots & \vdots \\
0 & 0 & 0 & 0 & \cdots & I_{k_c} & 0 \\
0 & 0 & 0 & 0 & \cdots & 0 & 1
\end{pmatrix} .
$$

Here $B_{i,j} \in \mathbb{R}^{k_i}$ and $\mathbb{R}^{k_i}$ is spanned by $\{B_{i,1}, B_{i,2}, \ldots B_{i,k_i}\}$.

To finish the proof it is sufficient to realize that ρ is a homomorphism and to use the lemmas given above together with remark (4.4.6). Then one proves successively that the following matrices satisfy the conditions (1) and (2) listed in the theorem:

1. $\rho(a_{i,j}^{x_{i,j}})$
2. $\rho(a_{i,1}^{x_{i,1}} a_{i,2}^{x_{i,2}} \ldots a_{i,k_i}^{x_{i,k_i}})$
3. $\rho(a_{c,1}^{x_{i,1}} \ldots a_{m,j}^{x_{m,j}} \ldots a_{1,k_1}^{x_{1,k_1}})$ (by induction on m).

The sufficiency of the conditions listed is also easily verified.   ∎

Let G be the Mal'cev completion of N. It is clear that the representation obtained in the theorem above, is also a Lie group representation for G. Indeed, for $\rho : N \rightarrow \mathrm{Aff}(\mathbb{R}^K)$ as before, we get a representation of G in $\mathrm{Aff}(\mathbb{R}^K)$ by allowing also reals to be substituted for the variables $x = (x_{1,1}, \ldots, x_{c,k_c})$.

Via exp and log this Lie group G is in one-to-one correspondence with its Lie algebra $\mathfrak{g}$. It becomes natural to ask for the meaning of canonical type on the Lie algebra level. Therefore let us define the concept of a canonical representation of a nilpotent Lie algebra into $\mathfrak{aff}(\mathbb{R}^L)$, the semidirect product $\mathbb{R}^L \times \mathfrak{gl}(\mathbb{R}^L)$.

We first introduce the concept of a central series of a Lie algebra, which is analogous to a torsion free central series of a nilpotent group.

Definition 4.4.8 *Let $\mathfrak{g}$ be a Lie algebra. By a central series of $\mathfrak{g}$, we mean a series*

$$\mathfrak{g}_* : 0 = \mathfrak{g}_0 \subseteq \mathfrak{g}_1 \subseteq \mathfrak{g}_2 \subseteq \cdots$$

of subalgebras of $\mathfrak{g}$ which satisfies $[\mathfrak{g}, \mathfrak{g}_{i+1}] \subseteq \mathfrak{g}_i$.
The central series $\mathfrak{g}_*$ is said to be of length c iff $\mathfrak{g}_{c-1} \neq \mathfrak{g}$ and $\mathfrak{g}_c = \mathfrak{g}_{c+1} = \cdots = \mathfrak{g}$.

For a given nilpotent Lie algebra $\mathfrak{g}$, there are two well known central series of finite length, namely the lower central series and the upper central series. Given a central series $\mathfrak{g}_*$ of length c we denote the dimension of $\mathfrak{g}_i/\mathfrak{g}_{i-1}$ by l_i (for $1 \leq i \leq c$). We use $L = l_1 + l_2 + \cdots + l_c$ to refer to the dimension of $\mathfrak{g}$. So, we can choose a basis

$$A_{1,1}, A_{1,2}, \ldots, A_{1,l_1}, A_{2,1}, \ldots, A_{c,l_c} \tag{4.9}$$

of $\mathfrak{g}$ in such a way that $\mathfrak{g}_i$ is spanned by all vectors

$$A_{1,1}, A_{1,2}, \ldots, A_{1,l_1}, A_{2,1}, \ldots, A_{i,l_i}$$

We will refer to such a basis as a basis which is **compatible** with the given central series $\mathfrak{g}_*$.

Definition 4.4.9 *An embedding $\rho : \mathfrak{g} \to \mathfrak{aff}(\mathbb{R}^L)$ is of canonical type with respect to a central series $\mathfrak{g}_*$, if and only if for an appropriate choice of a compatible basis (4.9)*

$$\rho(A_{i,j}) = \begin{pmatrix} O_{l_1} & * & * & * & \cdots & * & * \\ 0 & O_{l_2} & * & * & \cdots & * & * \\ \vdots & \vdots & \vdots & \vdots & & \vdots & \vdots \\ 0 & 0 & O_{l_i} & 0 & \cdots & 0 & E_{i,j} \\ 0 & 0 & 0 & O_{l_{i+1}} & \cdots & 0 & 0 \\ \vdots & \vdots & \vdots & \vdots & & \vdots & \vdots \\ 0 & 0 & 0 & 0 & \cdots & O_{l_c} & 0 \\ 0 & 0 & 0 & 0 & \cdots & 0 & 0 \end{pmatrix}$$

where $E_{i,j} = (0, 0, \ldots, 0, 1, 0, \ldots, 0)^{tr}$ ($(l_i \times 1)$–matrix with 1 on the j-th spot).

Using again $a_{1,1}, a_{1,2}, \ldots, a_{c,k_c}$ to refer to the generators of a torsion free nilpotent group N (4.8) (compatible with a given torsion free central series), we take $A_{i,j} = \log(a_{i,j})$. As was explained in the first chapter, we see that the sequences of subspaces $\mathfrak{g}_0 \subseteq \mathfrak{g}_1 \subseteq \cdots \subseteq \mathfrak{g}_c$, where $\mathfrak{g}_i$ is the subspace spanned by $A_{1,1} \ldots A_{i,k_i}$, forms a central series $\mathfrak{g}_*$ of $\mathfrak{g}$. In particular $\mathfrak{g}$ is spanned by all $A_{i,j}$'s. Remark that in this situation $l_i = k_i$ $(1 \leq i \leq c)$ and $L = K$.

Assume that $\rho : N \rightarrow \mathrm{Aff}(\mathbb{R}^K)$ is a canonical type affine representation and write $A(x)$ for $\rho(n)$ where $n = a_{c,1}^{x_{c,1}} \ldots a_{1,k_1}^{x_{1,k_1}}$.

Theorem 4.4.10 *The map $\tilde{\rho} = \log \rho \exp$ is a linear representation of $\mathfrak{g}$ into $\mathfrak{aff}(\mathbb{R}^K)$, which is of canonical type and $\tilde{\rho}(x_{1,1}A_{1,1}+\cdots+x_{c,k_c}A_{c,k_c}) = B(x)$. Here, $B(x)$ is obtained from $A(x)$ by replacing all the diagonal identity-blocks by zero-blocks as well as by replacing all degree ≥ 2 parts of $A(x)$ by zero.*

Proof:

$\tilde{\rho}$ ($= d\rho$, the differential of ρ) is a Lie algebra morphism, and so $\tilde{\rho}(x_{i,j}A_{i,j}) = x_{i,j}\tilde{\rho}(A_{i,j})$. This implies that the entries of this matrix, will be of degree 1 in the variable $x_{i,j}$.

On the other hand we see that for all $x_{i,j} \in \mathbb{Z}$, $\tilde{\rho}(x_{i,j}A_{i,j}) = \log(\rho(a_{i,j}^{x_{i,j}}))$. This means that $\log(\rho(a_{i,j}^{x_{i,j}}))$, as matrix, has degree–1 entries in the variable $x_{i,j}$.

We determine these entries by observing that

$$\log(\rho(a_{i,j}^{x_{i,j}})) = (\rho(a_{i,j}^{x_{i,j}}) - I) + \underbrace{\sum_{k=2}^{c} \frac{(-1)^{k+1}}{k}(\rho(a_{i,j}^{x_{i,j}}) - I)^k}_{\text{Containing only terms of degree} \geq 2} .$$

So the degree–1 terms come from $\rho(a_{i,j}^{x_{i,j}}) = A(x)_{|x=(0,\ldots,0,x_{i,j},0,\ldots,0)}$. In turn, this implies that

$$\tilde{\rho}(x_{1,1}A_{1,1} + \cdots + x_{c,k_c}A_{c,k_c}) = x_{1,1}\log(\rho(a_{1,1})) + \cdots + x_{c,k_c}\log(\rho(a_{c,k_c}))$$
$$= B(x).$$

Now, by looking at the matrices of $\tilde{\rho}(A_{i,j})$, one concludes that $\tilde{\rho}$ is of canonical type.

Conversely, if one considers a canonical embedding $\tilde{\rho}$ of $\mathfrak{g}$ into $\mathfrak{aff}(\mathbb{R}^K)$, then one can easily see that $\rho = \exp \tilde{\rho} \log : G \rightarrow \mathrm{Aff}(\mathbb{R}^K)$ induces a representation of N, which is of canonical type.

∎

Now that we have a better picture of canonical type affine represent-ations, we want to come back to the iteration problem as stated in the previous section. Also we can point out here that there has already been some interest in the literature for similar-looking iterative work concern-ing complete normal Koszul-Vinberg (KV) structures on nilpotent Lie algebras ([9]). Remark that although a canonical Lie algebra represent-ation determines a complete KV–structure, this KV–structure will not necessarily be normal.

The following example, communicated to us on the Lie algebra level by Dan Segal and Fritz Grunewald (to whom we express our gratitude), shows however that a great amount of care is necessary with respect to the "universal" nature of both iterative approaches.

For clarity, we prefer to present the example twice: once on the Lie algebra level and once on the group level. As a consequence it will follow that

1. the final theorem, called the "Lifting theorem", in [9], is incorrect as stated there.

2. the iteration problem as stated previously does generally not have a positive answer; a fortiori the announcement of a positive answer to a much more general version in [54] is incorrect.

3. one should pay attention to have a well understanding of the 3-step nilpotent case in [44].

Example 4.4.11 (Lie algebra level)

To permit the reader an easier comparison with the situation in [9], we use the notations and terminology adopted there. Consider the 4-dimensional 3-step nilpotent Lie algebra $\mathfrak{g} = < A_1, A_2, A_3, A_4 >$ where the brackets are defined by

$$[A_1, A_2] = A_3, \ [A_1, A_3] = A_4, \ [A_2, A_3] = 0 = [\mathfrak{g}, A_4].$$

It is easily seen that $\bar{\mathfrak{g}} = \mathfrak{g}/ < A_4 >=< \bar{A}_1, \bar{A}_2, \bar{A}_3 >=$ is the Heisenberg algebra. In $\bar{\mathfrak{g}}$, let us consider the flag of ideals $F(\bar{\mathfrak{g}})$, given by

$$F(\bar{\mathfrak{g}}) : \bar{\mathfrak{g}} = \bar{\mathfrak{g}}_3 \supset \bar{\mathfrak{g}}_2 =< \bar{A}_1, \bar{A}_3 > \supset \bar{\mathfrak{g}}_1 =< \bar{A}_3 > \supset 0.$$

Remark that this flag is finer then the lower central series of $\bar{\mathfrak{g}}$. It is not hard to verify that the following linear representation $\tilde{\rho}$ of $\bar{\mathfrak{g}}$ is a complete, normal Koszul-Vinberg structure (KV-structure); $\tilde{\rho}$ is defined by

$$\tilde{\rho}(\bar{A}_2)\bar{A}_1 = \bar{A}_3, \quad \tilde{\rho}(\bar{A}_1)\bar{A}_2 = 2\bar{A}_3, \quad \tilde{\rho}(\bar{A}_i)\bar{A}_j = 0 \text{ in all other cases.}$$

We now proceed to show that this KV-structure does not lift to a normal KV-structure ρ on $\mathfrak{g}$. Taking into account the lower central series of $\mathfrak{g}$, one verifies that a normal KV-structure ρ on $\mathfrak{g}$ must satisfy

$$\rho(A_i)A_4 = 0, \quad \rho(A_3)A_3 = 0, \quad \rho(\mathfrak{g})\mathfrak{g} =< A_3, A_4 > .$$

Furthermore, a lifting of $\tilde{\rho}$ must have at least the following properties:

$$\rho(A_2)A_1 = A_3 + \alpha A_4, \quad \rho(A_1)A_1 = \gamma A_4,$$

$$\rho(A_1)A_3 = \beta A_4, \quad \rho(A_3)A_1 = (\beta - 1)A_4.$$

However, from the definition of KV-structure in [9] it follows that we should also have

$$\rho(A_2)\rho(A_1)A_1 - \rho(A_1)\rho(A_2)A_1 = \rho([A_2, A_1])A_1 \Rightarrow -\beta A_4 = (1 - \beta)A_4$$

and this is clearly a contradiction.

Remark 4.4.12
This situation is "exceptional", because every complete, normal KV-struc-ture $(k \neq 0)$

$$\tilde{\rho}(\bar{A}_2)\bar{A}_1 = k\bar{A}_3, \quad \tilde{\rho}(\bar{A}_1)\bar{A}_2 = (k + 1)\bar{A}_3, \quad \tilde{\rho}(\bar{A}_i)\bar{A}_j = 0 \text{ in all other cases,}$$

on $\bar{\mathfrak{g}}$, with $k \neq 1$ extends to a normal KV-structure on $\mathfrak{g}$.

Example 4.4.13 (group level)

We now reconsider this example on the group level. Take the 3-step nilpotent group:

$$N_3 :< a_1, a_2, a_3, a_4 \parallel [a_2, a_1] = a_3^{-1}, [a_3, a_1] = a_4 \qquad > .$$
$$[a_3, a_2] = [a_4, a_1] = [a_4, a_2] = [a_4, a_3] = 1$$

$$(4.10)$$

Although this group can be given a canonical type representation, as will be described in a following section, we show that it can serve as a critical example with respect to the iteration problem.

N_3 can be seen as a central extension of

$$N_2 :< a_1, a_2, a_3 \parallel [a_2, a_1] = a_3^{-1}, [a_3, a_1] = [a_3, a_2] = 1 > .$$

Consider the following canonical type affine representation ($\tilde{\rho}$) of N_2:

$$\tilde{\rho}(a_1^{x_1} a_2^{x_2} a_3^{x_3}) = \begin{pmatrix} A(x) & a(x) \\ 0 & 1 \end{pmatrix} = \begin{pmatrix} 1 & x_2 & 2x_1 & 2x_1 x_2 + x_3 \\ 0 & 1 & 0 & x_1 \\ 0 & 0 & 1 & x_2 \\ 0 & 0 & 0 & 1 \end{pmatrix}.$$

We show that $\tilde{\rho}$ can not be lifted to a canonical type affine representation $\rho : N_3 \rightarrow \mathrm{Aff}(\mathbb{R}^n)$ (remark that N_3 admits only one torsion free central series of length 3). For suppose that this ρ exists, then it must be of the form

$$\rho(a_1) = \begin{pmatrix} 1 & A_1 & A_2 & A_3 & A_4 \\ 0 & 1 & 0 & 2 & 0 \\ 0 & 0 & 1 & 0 & 1 \\ 0 & 0 & 0 & 1 & 0 \\ 0 & 0 & 0 & 0 & 1 \end{pmatrix} \quad \rho(a_2) = \begin{pmatrix} 1 & B_1 & B_2 & B_3 & B_4 \\ 0 & 1 & 1 & 0 & 0 \\ 0 & 0 & 1 & 0 & 0 \\ 0 & 0 & 0 & 1 & 1 \\ 0 & 0 & 0 & 0 & 1 \end{pmatrix}$$

$$\rho(a_3) = \begin{pmatrix} 1 & C_1 & C_2 & C_3 & C_4 \\ 0 & 1 & 0 & 0 & 1 \\ 0 & 0 & 1 & 0 & 0 \\ 0 & 0 & 0 & 1 & 0 \\ 0 & 0 & 0 & 0 & 1 \end{pmatrix} \quad \rho(a_4) = \begin{pmatrix} 1 & 0 & 0 & 0 & 1 \\ 0 & 1 & 0 & 0 & 0 \\ 0 & 0 & 1 & 0 & 0 \\ 0 & 0 & 0 & 1 & 0 \\ 0 & 0 & 0 & 0 & 1 \end{pmatrix}$$

for some real numbers $A_1, A_2, \ldots, C_4$. If ρ is a homomorphism of groups, the relations appearing in (4.10) must be satisfied if we replace a_1 by $\rho(a_1)$, a_2 by $\rho(a_2)$, a_3 by $\rho(a_3)$ and a_4 by $\rho(a_4)$. This leads to a system of linear equations in the parameters $A_1, A_2, \ldots, C_4$:

$$\begin{cases} C_1 = 0 \\ -A_1 + C_2 = 0 \\ 2B_1 + C_3 = 0 \\ A_1 - A_3 + B_1 + B_2 - C_1 + C_4 = 0 \\ -1 - A_1 + C_2 = 0 \\ -B_1 + C_3 = 0, \end{cases}$$

which is easily seen to be inconsistent. Therefore, ρ cannot exist.

These examples of course, do not contradict the conjecture of Milnor. But, recently Y. Benoist ([4]) and D. Burde & F. Grunewald ([12]) proved that there exist nilmanifolds (of dimension 11) which does not admit a complete affinely flat structure. We will come back to these examples later on.

4.4.3   Virtually 2-step nilpotent groups

In this section we will prove that for any connected and simply connected
2–step nilpotent Lie group G, there exists a faithful affine representation
of $G \rtimes \mathrm{Aut}\,(G)$ letting G act simply transitively on some space $\mathbb{R}^n$. It
will follow that any AC–group, with a 2–step nilpotent Fitting subgroup
allows a canonical type affine representation. We use $\mathfrak{g}$ to denote the Lie
algebra of G.

In the 2–step nilpotent case, the group commutators and Lie brackets
are very nicely related.

Lemma 4.4.14 $\forall a, b \in G : \ \log[a, b] = [\log a, \log b].$

<u>Proof:</u> Let us denote $A = \log a$ and $B = \log b$. Using the Campbell–
Baker–Hausdorff formula we find that:

$$
\begin{aligned}
[a, b] = a^{-1}b^{-1}ab &= \exp(-A)\exp(-B)\exp(A)\exp(B) \\
&= \exp(-A - B + \frac{1}{2}[-A, -B])\exp(A + B + \frac{1}{2}[A, B]) \\
&= \exp([A, B] + \\
&= \quad \frac{1}{2}[-A - B + \frac{1}{2}[A, B], A + B + \frac{1}{2}[A, B]]) \\
&= \exp([A, B]).
\end{aligned}
$$

∎

The 2–step nilpotent group G fits in a short exact sequence

$$
1 \to [G, G] \to G \to G/[G, G] \to 1.
$$

Both $[G, G]$ and $G/[G, G]$ are real, finite dimensional vectorspaces (group
operation = addition of vectors). We choose a basis $\{b_1, b_2, \ldots, b_m\}$ of
$[G, G]$ and a basis $\{\bar{a}_1, \bar{a}_2, \ldots, \bar{a}_n\}$ of $G/[G, G]$. We also fix a lift $a_i \in G$
for $\bar{a}_i$ ($\forall i,\ 1 \le i \le n$). Now, any element g of G can be written, in a
unique way, in the form

$$
g = a_1^{x_1} a_2^{x_2} \ldots a_n^{x_n} b_1^{y_1} b_2^{y_2} \ldots b_m^{y_m}, \quad x_i \in \mathbb{R}, y_j \in \mathbb{R}. \tag{4.11}
$$

We define an alternating, bilinear map $L : \mathbb{R}^n \times \mathbb{R}^n \to \mathbb{R}^m$ as follows.
Let $u, v \in \mathbb{R}^n = G/[G, G]$ and consider any lifts $\tilde{u}, \tilde{v}$ of u and v. Let
$L(u, v) = [\tilde{u}, \tilde{v}] \in [G, G] = \mathbb{R}^m$. We remark that this definition does
not depend upon the chosen lifts. Suppose that L is determined by the
parameters

$$
l_{i,p,j}, \ 1 \le i, j \le n, \ 1 \le p \le m,
$$

where
$$[a_i, a_j] = b_1^{l_{i,1,j}} b_2^{l_{i,2,j}} \ldots b_m^{l_{i,m,j}}.$$
So, $l_{i,p,j} = -l_{j,p,i}$ and $l_{i,p,i} = 0$.

Lemma 4.4.15 *With the notations above, $\mathfrak{g}$ has a presentation of the form*

$$\mathfrak{g}:< A_1, A_2, \ldots, A_n, B_1, \ldots, B_m \| [A_i, A_j] = \sum_{p=1}^{m} l_{i,p,j} B_p \ (1 \leq j < i \leq n) >$$
$$[B_i, A_j] = 0 \ (1 \leq i \leq m, \ 1 \leq j \leq n)$$
$$[B_i, B_j] = 0 \ (1 \leq j < i \leq m).$$

<u>Proof:</u> The proof is immediate, since we know that $\mathfrak{g}$ has a basis

$$\{A_1 = \log(a_1), \ldots, A_n = \log(a_n), B_1 = \log(b_1), \ldots, B_m = \log(b_m)\}.$$

The Lie brackets may be computed via lemma 4.4.14. ∎

We introduce the following notation:

1. For all $i \in \{1, 2, \ldots, n\}$, L_i denotes the $m \times n$–matrix, with $(L_i)_{p,q} = l_{i,p,q}$.

2. For all $i \in \{1, 2, \ldots, n\}$, u_i denotes the n–column with a 1 on the i–th place and a 0 elsewhere.

3. For all $i \in \{1, 2, \ldots, m\}$, v_i denotes the m–column with a 1 on the i–th place and a 0 elsewhere.

With these symbols in mind, it is well known (and easy to check) that $\mathfrak{g}$ can be faithfully represented as a Lie algebra of $(m+n+1)$–matrices. This is obtained as follows:

$$A_i \mapsto \begin{pmatrix} 0_m & \frac{1}{2}L_i & 0 \\ 0 & 0_n & u_i \\ 0 & 0 & 0_1 \end{pmatrix} \qquad B_i \mapsto \begin{pmatrix} 0_m & 0 & v_i \\ 0 & 0_n & 0 \\ 0 & 0 & 0_1 \end{pmatrix}.$$

Here, 0_j stands for a $j \times j$ zero matrix.

From this representation, one may obtain a faithful representation φ of G into $\mathrm{Aff}(\mathbb{R}^{m+n}) \subset \mathrm{Gl}(m+n+1, \mathbb{R})$, just by exponentiating the matrices above. So, this representation is determined completely by:

$$b_i = \exp B_i \mapsto \exp \begin{pmatrix} 0_m & 0 & v_i \\ 0 & 0_n & 0 \\ 0 & 0 & 0_1 \end{pmatrix} =$$

$$I_{m+n+1} + \begin{pmatrix} 0_m & 0 & v_i \\ 0 & 0_n & 0 \\ 0 & 0 & 0_1 \end{pmatrix} = \begin{pmatrix} I_m & 0 & v_i \\ 0 & I_n & 0 \\ 0 & 0 & I_1 \end{pmatrix}, \tag{4.12}$$

$$a_i = \exp A_i \mapsto \exp \begin{pmatrix} 0_m & \frac{1}{2}L_i & 0 \\ 0 & 0_n & u_i \\ 0 & 0 & 0_1 \end{pmatrix} =$$

$$I_{m+n+1} + \begin{pmatrix} 0_m & \frac{1}{2}L_i & 0 \\ 0 & 0_n & u_i \\ 0 & 0 & 0_1 \end{pmatrix} + \begin{pmatrix} 0_m & 0 & \frac{1}{2}L_i u_i = 0 \\ 0 & 0_n & 0 \\ 0 & 0 & 0_1 \end{pmatrix} =$$

$$\begin{pmatrix} I_m & \frac{1}{2}L_i & 0 \\ 0 & I_n & u_i \\ 0 & 0 & I_1 \end{pmatrix}. \tag{4.13}$$

Here, I_j denotes the $j \times j$ identity matrix.

This affine representation for G can be interpreted in a very nice way. We introduce the following coordinate system on G:

Definition 4.4.16 *The coordinate system of G (associated to the basis $B_1, \ldots, B_m, A_1, \ldots, A_n$) is the map $m : G \to \mathbb{R}^{m+n}$ which maps each element $g \in G$ to the coordinate of $\log(g)$ (with respect to the basis $B_1, \ldots, B_m, A_1, \ldots, A_n$).*

It follows that the affine representation φ is exactly the coordinate expression for the multiplication in G. I.e.

Lemma 4.4.17 *Let G be a 2–step nilpotent Lie group, with a coordinate system m as above, then we have that*

$$\forall g, h \in G : m(gh) = \varphi(g)m(h).$$

Proof: By the Campbell–Baker–Hausdorff formula we find that

$$\log(gh) = \log g + \log h + \frac{1}{2}[\log g, \log h].$$

This implies that for a fixed g, the coordinate of gh is an affine function of $m(h)$. In other words, there exists a map

$$\psi : G \to \mathrm{Aff}(\mathbb{R}^{m+n}) : g \mapsto \psi(g) \text{ such that } \forall g, h \in G : m(gh) = \psi(g)m(h).$$

It is easy to see that ψ is a morphism of groups for which $\varphi(a_i) = \psi(a_i)$ $(1 \leq i \leq n)$ and $\varphi(b_i) = \psi(b_i)$ $(1 \leq i \leq m)$. Therefore, φ and ψ coincide on the whole of G.

                                                            ■

Let us return to the discrete case now and assume that N is any torsion free finitely generated 2–step nilpotent group. N fits in a short exact sequence

$$1 \to \sqrt[N]{[N,N]} \cong \mathbb{Z}^m \to N \to N/\sqrt[N]{[N,N]} \cong \mathbb{Z}^n \to 1$$

for some $m, n \in \mathbb{N}$. Choose a set of generators $b_1, \ldots, b_m, a_1, \ldots a_n$ of N in such a way that $b_1, \ldots, b_m$ generate $\sqrt[N]{[N,N]}$. This means that the set $b_1, \ldots, b_m, a_1, \ldots a_n$ is a compatible set of generators for the torsion free central series

$$N_* : 1 = N_0 \subseteq N_1 = \sqrt[N]{[N,N]} \subseteq N_2 = N(= N_3).$$

As explained in the first chapter, the elements

$$B_1 = \log(b_1), \ldots, B_m = \log(b_m), A_1 = \log(a_1), \ldots, A_n = \log(a_n)$$

form a basis for the Lie algebra $\mathfrak{g}$ of the Mal'cev completion G of N. It follows that the a_i and b_i play the same role as in the continuous case. In particular, it follows that the restriction of the affine representation $\varphi : G \to \text{Aff}(\mathbb{R}^{m+n})$ to N induces a faithful affine representation of N. Moreover, by looking at the matrices corresponding to the generators of N, we see that this representation is of canonical type with respect to the series N_*. We summarize these observations in a theorem.

Theorem 4.4.18 *Let N be a 2–step nilpotent group with a presentation*

$$N :< a_1, a_2, \ldots, a_n, b_1, \ldots, b_m \parallel [a_i, a_j] = \prod_{p=1}^{m} b_p^{l_{i,p,j}} \ (1 \leq j < i \leq n) \ > .$$

$$[b_i, a_j] = 0 \ (1 \leq i \leq m, \ 1 \leq j \leq n)$$
$$[b_i, b_j] = 0 \ (1 \leq j < i \leq m)$$

Then there exists a canonical type affine representation $\varphi : N \to \text{Aff}(\mathbb{R}^{m+n})$ with respect to the torsion free central series

$$N_* : \ 1 = N_0 \subseteq N_1 = \text{grp}\{b_1, b_2, \ldots, b_m\} \subseteq N_2 = N(= N_3)$$

which is completely determined by the matrices given in (4.12) and (4.13).

Proof: Everything is already shown above except for one thing: From the presentation of N above, we are not allowed to conclude that the group generated by $b_1, b_2, \ldots, b_m$ is really the subgroup $\sqrt[N]{[N, N]}$ of N. Indeed, we can only assume that $\sqrt[N]{[N, N]} \subseteq \mathrm{grp}\{b_1, b_2, \ldots, b_m\}$. However, this does not play an essential role in the theory developped above and everything works well in this more general case too. (One just has to use $G' =$ the Mal'cev completion of $\mathrm{grp}\{b_1, b_2, \ldots, b_m\}$ in stead of $[G, G]$). ∎

Definition 4.4.19 *We call a canonical type affine representation of a 2-step nilpotent group N as obtained in the theorem above a* **stable** *affine representation.*

A 2–step nilpotent group N can have more than one stable affine representation, since such a representation is determined by a particular choice of generators of N. Nevertheless, all these stable representations have one thing in common: they extend to a canonical type affine representation of any AC–group having this nilpotent group as Fitting subgroup. To prove this, we look at the continuous case again.

Proposition 4.4.20 *Let G be a connected and simply connected $m + n$-dimensional 2-step nilpotent Lie group. Then there exists a faithful affine representation $\tilde{\varphi} : G \rtimes \mathrm{Aut}\,(G) \to \mathrm{Aff}(\mathbb{R}^{m+n})$ which restricts to φ (given by (4.12) and (4.13)) on G.*

Proof: By expressing the natural action of $G \rtimes \mathrm{Aut}\,(G)$ on G in terms of the coordinate system m introduced before, we can define a homomorphism $\psi : G \rtimes \mathrm{Aut}\,(G) \to \mathcal{H}(\mathbb{R}^{m+n})$, where $\mathcal{H}(\mathbb{R}^{m+n})$ denotes the group of homeomorphisms of $\mathbb{R}^{m+n}$. To be precise, this homomorphism is given by

$$\forall g, h \in G, \forall \alpha \in \mathrm{Aut}\,(G) : \psi(g, \alpha)m(h) = m^{((g,\alpha)}h) = m(g\alpha(h)).$$

We already know that $\psi(g, 1) = \varphi(g)$ (see lemma 4.4.17). This implies that we will finish the proof, provided we can show that $\psi(1, \alpha)$ is an affine map. But $\psi(1, \alpha)m(h) = m(\alpha(h)) =$ the coordinate of $\log(\alpha(h))$ with respect to the Lie algebra basis introduced above. However, by the commutative diagram 1.2, we know that $\log(\alpha(h)) = d\alpha(\log h)$, where $d\alpha$, the differential of α is a linear map. It follows that the coordinate expression for $d\alpha(\log h)$ is linear in $m(h)$. This implies that the homeomorphism $\psi(1, \alpha)$ of $\mathbb{R}^{m+n}$ is linear and so a fortiori affine. ∎

We return to the discrete case again and reformulate the previous result for AC–groups.

Theorem 4.4.21 *Let E be any AC–group with a 2–step nilpotent Fitting subgroup N with a presentation*

$$N :< a_1, a_2, \ldots, a_n, b_1, \ldots, b_m \parallel [a_i, a_j] = \prod_{p=1}^{m} b_p^{l_{i,p,j}} \ (1 \le j < i \le n) \quad >.$$
$$[b_i, a_j] = 0 \ (1 \le i \le m, \ 1 \le j \le n)$$
$$[b_i, b_j] = 0 \ (1 \le j < i \le m)$$

If $\mathrm{grp}\{b_1, b_2, \ldots, b_m\}$ *is a normal subgroup of E, then there exists a canonical type affine representation* $\varphi : E \to \mathrm{Aff}(\mathbb{R}^{m+n})$ *with respect to the torsion free filtration*

$$E_* : \quad 1 = E_0 \subseteq E_1 = \mathrm{grp}\{b_1, b_2, \ldots, b_m\} \subseteq E_2 = N \subseteq E_3 = E$$

and where the induced representation of the Fitting subgroup is completely determined by the matrices given in (4.12) and (4.13).

<u>Proof:</u> The existence of a faithful affine representation is immediate now, since we know that any AC–group E embeds into $G \rtimes \mathrm{Aut}\,(G)$ in such a way that N is mapped identically onto itself. The reader in invited to see that this representation is of canonical type. (One can also use lemma 4.5.10)   ∎

Remark 4.4.22

For any AC–group E, we can always find a subgroup $\mathrm{grp}\{b_1, b_2, \ldots, b_m\}$ which is normal in E. It suffices to choose for it a characteristic subgroup of N e.g. $Z(N)$ or $\sqrt[N]{[N,N]}$. So, any AC–group with a 2–step nilpotent Fitting subgroup admits a canonical type affine representation which restricts to a stable representation of the Fitting subgroup.

Definition 4.4.23 *Let E be an AC–group with a 2–step nilpotent Fitting subgroup N. An affine representation of E is said to be a stable representation iff the restricted representation of N is stable.*

4.4.4   Virtually 3-step nilpotent groups

The reason why we are so interested in stable affine representations is that they can be used to build up affine representations for AC–groups with a 3-step nilpotent Fitting subgroup. We make this clear in the following theorem, which is a more concrete version of theorem 2.5 of [44]:

Theorem 4.4.24 *Let Q be any AC-group of a 2-step nilpotent Lie group G, and let $\lambda : Q \to \mathrm{Aff}(\mathbb{R}^n)$ be a stable representation of Q. Then for any extension*

$$1 \to \mathbb{Z}^m \to E \xrightarrow{p} Q \to 1$$

inducing a morphism $\varphi : Q \to \mathrm{Aut}\,\mathbb{Z}^m$ with $\varphi(\mathrm{Fitt}(Q)) = 1$, there exists an embedding $\tilde\lambda : E \to \mathrm{Aff}(\mathbb{R}^{m+n})$ with

$$\forall e \in E : \tilde\lambda(e) = \begin{pmatrix} \varphi(p(e)) & * \\ 0 & \lambda(p(e)) \end{pmatrix},$$

moreover $\forall z \in \mathbb{Z}^m : \tilde\lambda(z) = \begin{pmatrix} I_k & 0 & z \\ 0 & I_n & 0 \\ 0 & 0 & I_1 \end{pmatrix}.$

<u>Proof:</u> Following sections 4.3 and 4.4.1, the statement is equivalent to proving that the connecting homomorphism

$$\delta : H^1(Q, \mathrm{Aff}(\mathbb{R}^n, \mathbb{R}^m)/\mathbb{Z}^m) \to H^2(Q, \mathbb{Z}^m)$$

with respect to the short exact sequence of Q–modules

$$1 \to \mathbb{Z}^m \to \mathrm{Aff}(\mathbb{R}^n, \mathbb{R}^m) \to \mathrm{Aff}(\mathbb{R}^n, \mathbb{R}^m)/\mathbb{Z}^m \to 1$$

is surjective. We recall that the Q–module structure on the abelian group of affine mappings of n–space into m-space is given by $\forall q \in Q, \forall h \in \mathrm{Aff}(\mathbb{R}^n, \mathbb{R}^m) : {}^q h = \varphi(q)h\lambda(q)^{-1}$. The group $\mathbb{Z}^m$ is to be considered as a submodule of $\mathrm{Aff}(\mathbb{R}^n, \mathbb{R}^m)$ by identifying $z \in \mathbb{Z}^m$ with the constant map $\mathbb{R}^n \to \mathbb{R}^m : x \mapsto z$.

We prove the theorem in two steps. First we show it for torsion free nilpotent groups of class 2, thereafter we look at the general case.

Case 1: $Q = N$ is 2-step nilpotent

So, let N be a uniform lattice of a 2-step nilpotent connected and simply connected Lie group G. We have to show that any central extension $1 \to \mathbb{Z}^m \to E \to N \to 1$ can be given an affine representation into $\mathrm{Aff}(\mathbb{R}^{m+n})$ build up from the stable representation of N. Such an extension E can be seen as the uniform lattice of a connected, simply connected nilpotent Lie group G', which itself is a central extension of G by $\mathbb{R}^m$. The stable embedding of N, is the restriction of an embedding of G. By the work of Scheuneman [58] the embedding of G (to be exact of its Lie algebra $\mathfrak{g}$) is of such a kind that it can be lifted to an embedding of G' into $\mathrm{Aff}(\mathbb{R}^{m+n})$.

The restriction of this embedding to E gives the desired result in the nilpotent case.

Case 2: General Q

To see what happens in the virtual nilpotent case, we look at this problem from the cohomological point of view (in fact, this is the first place where we really need the cohomological approach). Consider an AC–group Q with a Fitting subgroup $\mathrm{Fitt}(Q) = N$ of class 2. Note the following commutative diagram with exact columns:

$$
\begin{array}{ccc}
\downarrow & & \downarrow \\
H^1(Q, \mathrm{Aff}(\mathbb{R}^n, \mathbb{R}^m)/\mathbb{Z}^m) & \xrightarrow{\text{res}} & H^1(N, \mathrm{Aff}(\mathbb{R}^n, \mathbb{R}^m)/\mathbb{Z}^m) \\
\delta_1 \downarrow & & \delta_2 \downarrow \\
H^2(Q, \mathbb{Z}^m) & \xrightarrow{\text{res}} & H^2(N, \mathbb{Z}^m) \\
i_1 \downarrow & & i_2 \downarrow \\
H^2(Q, \mathrm{Aff}(\mathbb{R}^n, \mathbb{R}^m)) & \xrightarrow{\text{res}} & H^2(N, \mathrm{Aff}(\mathbb{R}^n, \mathbb{R}^m)) \\
\downarrow & & \downarrow
\end{array}
$$

We have to show that δ_1 is surjective or that $i_1 = 0$. By the first case we know that δ_2 is surjective or that $i_2 = 0$. By the fact that $\mathrm{Aff}(\mathbb{R}^n, \mathbb{R}^m)$ is a vectorspace, we know that the restriction map at the bottom of the diagram is injective. Now, it follows easily that also $i_1 = 0$, which was to be shown. ∎

Remark 4.4.25

The previous theorem is a fortiori also valid in case Q is crystallographic, i.e. Q has an abelian Fitting subgroup $\mathbb{Z}^n$. In this case stable means that any element $z \in \mathbb{Z}^n$ of the Fitting subgroup acts on $\mathbb{R}^n$ exactly as translation by z

Corollary 4.4.26 *Let E be an AC-group, with a 3-step nilpotent Fitting subgroup, then E admits a canonical type affine representation.*

Proof: Let N denote the Fitting subgroup of E. By, lemma 2.4.2, there is a short exact sequence

$$1 \to Z \to E \to E/Z = Q \to 1,$$

where $Z = \sqrt[N]{[N, [N, N]]}$ and $Q = E/Z$ is an AC-group with a 2-step nilpotent Fitting subgroup. By the previous section we know that E/Z

admits a stable affine representation, which can be lifted, due to the theorem above, to an affine representation of E. Moreover, a close look to the matrix form of this representation shows that we can assume that this representation is of canonical type with respect to the following torsion free filtration E_*:

$$E/\ E_0 = 1 \subseteq E_1 = Z \subseteq E_2 = \sqrt[N]{[N,N]} \subseteq E_3 = N \subseteq E_4 = E.$$

∎

4.4.5   What about the general case?

Our study of canonical type affine representations is related to the so called affinely flat manifolds. An affinely flat manifold of dimension n is a (smooth) manifold M provided with an atlas $\mathcal{A} = \{\mu_\alpha : U_\alpha \to \mathbb{R}^n \ \| \ \alpha \in I\}$ of coordinate homeomorphisms, such that the transition functions

$$t_{\alpha,\beta} : \mu_\beta \circ \mu_\alpha^{-1} : \mu_\alpha(U_\alpha \cap U_\beta) \to \mu_\beta(U_\alpha \cap U_\beta), \text{ for } \alpha, \beta \in I, \ U_\alpha \cap U_\beta \neq \phi$$

extend (uniquely) to an affine transformation: $\tilde{t}_{\alpha,\beta} : \mathbb{R}^n \to \mathbb{R}^n : x \mapsto A_{\alpha,\beta} x + a_{\alpha,\beta}$, where $A_{\alpha,\beta}$ is an invertible $(n \times n)$–matrix and $a_{\alpha,\beta} \in \mathbb{R}^n$.

If one imposes even more requirements on the affine transformations $\tilde{t}_{\alpha,\beta}$ one finds special classes of affinely flat manifolds. For example, if all $A_{\alpha,\beta}$ are orthogonal matrices, so if all $\tilde{t}_{\alpha,\beta}$ are rigid motions of Euclidian n–space, one obtains the flat Riemannian manifolds M. Another interesting subclass of these affinely flat manifolds are the Lorentz–flat manifolds, i.e. those affinely flat manifolds for which each $A_{\alpha,\beta}$ belongs to the Lorentz group $O(n-1,1)$.

A geodesic in an affinely flat manifold M is a curve $\gamma : [a,b] \to M$ which is locally a straight line, traversed at constant speed. The manifold M is said to be complete if any geodesic can be defined on $\mathbb{R}$. This means that every "partial" geodesic $\gamma : [a,b] \to M$ can be extended to a "full" geodesic $\tilde{\gamma} : \mathbb{R} \to M$. It is known that any n–dimensional connected complete affinely flat manifold can be constructed as a quotient space $M = {}_E\backslash\mathbb{R}^n$ where E acts freely and properly discontinuous, via affine transformations on $\mathbb{R}^n$ ([3]). We are especially interested in the case where the manifold M is compact. Here, we should mention the famous Auslander conjecture stating that if a complete affinely flat manifold $M = {}_E\backslash\mathbb{R}^n$ is compact, then E necessarily has to be a polycyclic–by–finite group. In fact, Auslander formulated this as a theorem ([2]), but unfortunately his proof contained a gap. So, the problem is still open and is therefore nowadays referred to as the Auslander conjecture. Nevertheless, it follows that the class of polycyclic–by–finite groups is an interesting class of groups to study. At this point we also want to mention the paper of F. Grunewald and G. Margulis ([36]) in which a general classification scheme for fundamental groups of compact, complete Lorentz–flat manifolds is given.

In 1977 ([52]), John Milnor showed that every torsion free polycyclic–by–finite group Γ can be realized as the fundamental group of a (not necessarily compact) complete affinely flat manifold M. Moreover, it is known that any such Γ appears as the fundamental group of a compact manifold (e.g. consider the quotient space $\rho(\Gamma)\backslash\mathbb{R}^{h(\Gamma)}$, where $h(\Gamma)$ denotes the Hirsch number of Γ and $\rho : \Gamma \to \mathcal{H}(\mathbb{R}^{h(\Gamma)})$ is any canonical type representation, see theorem 4.2.3). Therefore, J. Milnor wondered whether these two results could be joined and formulated the following question (often referred to as Milnor's conjecture):

> Given a torsion free polycyclic–by–finite group Γ, is it possible to construct a manifold M which is both compact and complete affinely flat with fundamental group equal to Γ?

For a long time only a few special cases (e.g. Γ is a virtually 3–step nilpotent group), giving positive answer to this question in these special cases, were known, but nevertheless many people started to believe that the answer to Milnor's question had to be yes. However, in 1992 Y. Benoist ([4]) constructed an example of a 10–step nilpotent group of rank 11, which contradicted the question of Milnor. This example was generalized to a family of examples by D. Burde & F. Grunewald ([12]).

In fact Benoist and Burde & Grunewald do not work with groups but rather on the Lie algebra level. It is possible to understand this change of language by the work of D. Fried, W.M. Goldman and M. Hirsch ([31]) and [30]. Indeed, in case we restrict ourselves to nilpotent groups we have the following results, where we use N to denote a finitely generated, torsion free nilpotent group of rank K, G is meant to be the Mal'cev completion of N and $\mathfrak{g}$ is the Lie algebra of G.

Suppose N acts properly discontinuously on $\mathbb{R}^K$ via affine motions. Refer to this action by $\varphi : N \to \text{Aff}(\mathbb{R}^K)$. It follows from the proof of theorem 7.1 of [30], that φ extends uniquely to a simply transitive and affine action $\tilde{\varphi} : G \to \text{Aff}(\mathbb{R}^K)$ of G on $\mathbb{R}^K$. Conversely, given a simply transitive action $\tilde{\varphi}$ of G on $\mathbb{R}^K$, one obtains a properly discontinuous action of N on $\mathbb{R}^K$ by considering the restriction φ of $\tilde{\varphi}$ to N. Denote $d\tilde{\varphi} : \mathfrak{g} \to \mathfrak{aff}(\mathbb{R}^K)$ is the differential of $\tilde{\varphi}$. As mentioned before, any element of $\mathfrak{aff}(\mathbb{R}^K)$ consists of a linear part and a translational part, determined by the two maps:

$$\mathrm{lin} : \mathfrak{aff}(\mathbb{R}^K) \to \mathfrak{gl}(K, \mathbb{R}) : \begin{pmatrix} A & a \\ 0 & 0 \end{pmatrix} \mapsto A$$

and

$$\mathrm{tr} : \mathfrak{aff}(\mathbb{R}^K) \to \mathbb{R}^K : \begin{pmatrix} A & a \\ 0 & 0 \end{pmatrix} \mapsto a.$$

It follows from [31], that

1. the linear parts of $d\tilde{\varphi}$ consist of nilpotent matrices and

2. the translational parts of $d\tilde{\varphi}(\mathfrak{g})$ form all of n–space. More precise, $\mathrm{tr} \circ d\tilde{\varphi} : \mathfrak{g} \to \mathbb{R}^n$ is an isomorphism of vector spaces.

Conversely, given a Lie algebra homomorphism $\rho : \mathfrak{g} \to \mathfrak{aff}(\mathbb{R}^K)$, such that $\mathrm{tr} \circ \rho : \mathfrak{g} \to \mathbb{R}^K$ is an isomorphism of vectorspaces and such that the linear parts of ρ consist of nilpotent matrices, there exists a unique simply transitive action $\tilde{\varphi} : G \to \mathrm{Aff}(\mathbb{R}^K)$, with $d\tilde{\varphi} = \rho$. We call such a ρ a **complete affine structure** on $\mathfrak{g}$. We remark that one also speaks of an affine structure of a polycyclic–by–finite group Γ, by which one means a representation $\rho : \Gamma \to \mathrm{Aff}(\mathbb{R}^K)$, which lets ρ act properly discontinuously on $\mathbb{R}^K$ and with compact quotient.

We conclude that there are the following one to one correspondences (See also [43])

Affine structures of N.

$\updownarrow$ 1-1

Simply transitive affine actions of G.

$\updownarrow$ 1-1

Complete affine structures on $\mathfrak{g}$.

This shows that it is indeed possible that, at least in case of nilpotent groups, one approaches Milnor's problem at the Lie algebra level as was done by Y. Benoist (and later by D. Burde and F. Grunewald). The Lie algebra of Y. Benoist which does not admit a complete affine structure is given in the last section of this chapter. Finally, we want to mention that, we were able to construct a complete affine structure on a broad class of 4–step nilpotent Lie algebras ([19]).

4.5   Canonical type polynomial representations

Now, that we know that not all finitely generated nilpotent (and so a fortiori not all polycyclic–by–finite) groups admit an affine structure the following notions make sense:

Let us call a quotient manifold $_E\backslash\mathbb{R}^K$, where E acts freely, properly discontinuously, via polynomial diffeomorphisms on $\mathbb{R}^K$ a *polynomial manifold*. Moreover, if all diffeomorphisms involved are of degree $\leq s$, we say that the (polynomial) manifold is of degree $\leq s$.

Definition 4.5.1 *Let E be a polycyclic–by–finite group with Hirsch rank $h(E)$. The **affine defect** of E, denoted by $d(E)$ is defined as*

$$Min\left\{ s \in \mathbb{N} \| \begin{array}{l} E \text{ acts properly discontinuous and via polynomial} \\ \text{diffeomorphisms of degree} \leq s+1 \text{ on } \mathbb{R}^{h(E)} \end{array} \right\},$$

if there exists such an action for E; $d(E) = \infty$ if E does not allow any properly discontinuous action via polynomial diffeomorphisms of bounded degree on $\mathbb{R}^{h(E)}$.

As is easily seen, a torsion free, polycyclic–by–finite group E has affine defect zero iff E occurs as the fundamental group of a compact, complete affinely flat manifold. Therefore, this affine defect number somehow measures the obstruction for E to be realized as the fundamental group of a compact, complete affinely flat manifold. Analogously to the notion of an affine structure, we introduce the concept of a polynomial structure of a polycyclic–by–finite group Γ, which is nothing else than a representation $\rho : \Gamma \to P(\mathbb{R}^K)$, letting Γ act properly discontinuous and with compact quotient.

It is our intention to show in this book that the affine defect of a virtually c–step nilpotent group is always finite. In fact there exists an upperbound depending (in a linear way) only on the nilpotency class c.

More generally and very recently P. Igodt and myself obtained some nice existence results of polynomial structures for polycyclic–by–finite groups. It would lead us to far away from our subject to discuss these results here in detail but let me at least state the two main facts:

1. As a first result ([22]) we obtained that any polycyclic group Γ, for which $\Gamma/\text{Fitt}(\Gamma)$ is free abelian, has an affine defect satisfying: $d(\Gamma) < h(\Gamma)$. This is an importatnt result as we know that any polycyclic–by–finite group Γ' has a subgroup Γ of finite index satisfying the condition above. (The way to prove this first result was analogous to "the first approach" we follow below.)

2. As a second result ([23]), we very recently obtained that any polycyclic–by–finite group Γ admits a polynomial structure of finite degree.
   However the way of proving this was not of that kind that we could
   deduce any information on how small/large this finite degree really
   is. Remark that this result implies that there is a sentence to much
   now in de definition of affine defect. Indeed, it cannot happen that
   a polycyclic–by–finite group has an infinite affine defect. (The way
   to prove this second result was analogous to "the second approach"
   we follow below.)

These two results together lead to the following conjecture:

> **Conjecture** There exists a linear function $\nu : \mathbb{N} \to \mathbb{N}$, such
> that for any polycyclic–by–finite group Γ we have that the
> affine defect $d(\Gamma)$ satisfies $d(\Gamma) < \nu(h(\Gamma))$, where $h(\Gamma)$ as usual
> denotes the Hirsch length of Γ.

Stated otherwise, it is reasonable to believe that any polycyclic–by–finite
group Γ has a polynomial structure of degree $\leq \nu(h(\Gamma))$, for some linear
function ν. It even seems to be possible to assume that we can take
$\nu(n) = n - 1$ (at least if we ignore the case $n = 1$).

4.5.1   The first approach

As always, N denotes a torsion free, finitely generated nilpotent group of
rank K and class c.

Although, everything we will say in the sequel works for general torsion free central series of a group N, we will restrict ourselves to central
series which are as short as possible. So, the torsion free filtrations we
are looking at are central series of length c:

$$N_* : \quad N_0 = 1 \subset N_1 \subset N_2 \subset \cdots \subset N_{c-1} \subset N_c = N.$$

From now on, if we speak of a polynomial representation $\rho : N \to$
$P(\mathbb{R}^K)$ which is of canonical type, we mean of canonical type with respect
to some torsion free central series of length c.

Referring to theorem 4.2.3 and to the fact that we are dealing with a
central series, we can state the following corollary, which provides us an
alternative definition of a canonical type polynomial representation.

Corollary 4.5.2
*Assume that N is as above and that N_* is some torsion free central series.
A representation $\rho : N \to P(\mathbb{R}^K)$ is of canonical type with respect to N_**

$$\Updownarrow$$

$\forall n \in N : \rho(n) : \mathbb{R}^K \to \mathbb{R}^K : x \mapsto (p_1^\gamma(x), p_2^\gamma(x), \ldots, p_c^\gamma(x))$ *is such that*

$$p_i^\gamma(x) = x_i + Q_i^n(x_{i+1}, x_{i+2}, \ldots, x_c),$$

for some polynomial mapping $Q_i^n : \mathbb{R}^{K_{i+1}} \to \mathbb{R}^{k_i}$. *Moreover, N_i acts trivially on $\mathbb{R}^{K_{i+1}}$ and as translations on the i-th block $\mathbb{R}^{k_i}$, in such a way that these translations form the whole subgroup $\mathbb{Z}^{k_i}$.*

Example 4.5.3

Let $N :< a_{1,1}, a_{1,2}, a_{2,1}, a_{2,2}, a_{3,1}, a_{3,2}$ ‖

$$
\begin{aligned}
[a_{3,2}, a_{3,1}] &= a_{1,1}^{-1} a_{2,1} a_{2,3}^2 \\
[a_{2,1}, a_{3,1}] &= a_{1,1}^{-2} \\
[a_{2,2}, a_{3,2}] &= a_{1,1} \\
[a_{2,3}, a_{2,2}] &= a_{1,1} \\
[a_{2,2}, a_{2,1}] &= a_{1,1}^2
\end{aligned}
$$
>

all other commutators trivial.

One can check that there is a canonical type representation $\rho : N \longrightarrow P(\mathbb{R}^6)$ of this N given in terms of the images of the generators (the central series used, is suggested by the labeling of the generators):

$$
\begin{aligned}
\rho(a_{1,1})(x) &= (1 + x_{1,1}, x_{2,1}, x_{2,2}, x_{2,3}, x_{3,1}, x_{3,2}) \\
\rho(a_{2,1})(x) &= (x_{1,1} - 2x_{3,1}, 1 + x_{2,1}, x_{2,2}, x_{2,3}, x_{3,1}, x_{3,2}) \\
\rho(a_{2,2})(x) &= (x_{1,1} + 2x_{2,1} + x_{3,2}, x_{2,1}, 1 + x_{2,2}, x_{2,3}, x_{3,1}, x_{3,2}) \\
\rho(a_{2,3})(x) &= (x_{1,1} + x_{2,2}, x_{2,1}, x_{2,2}, 1 + x_{2,3}, x_{3,1}, x_{3,2}) \\
\rho(a_{3,1})(x) &= (x_{1,1}, x_{2,1}, x_{2,2}, x_{2,3}, 1 + x_{3,1}, x_{3,2}) \\
\rho(a_{3,2})(x) &= (x_{1,1} + 2x_{2,2}x_{3,1} - x_{3,1}^2, x_{2,1} + x_{3,1}, x_{2,2}, x_{2,3} + 2x_{3,1}, \\
&\qquad x_{3,1}, 1 + x_{3,2}).
\end{aligned}
$$

The following theorem, intrinsically due to Mal'cev ([51]), will imply the existence of canonical type polynomial representations for all AC–groups.

Theorem 4.5.4 *Let G be any connected, simply connected c–step nilpotent Lie group G of dimension K. Then $G \rtimes \text{Aut}(G)$ embeds into $P(\mathbb{R}^K)$ in such a way that the image of $G \rtimes \text{Aut}(G)$ consists of polynomials of degree $\leq s$ for $s = \max(1, c - 1)$ and so that the action of G on $\mathbb{R}^K$ is simply transitive.*

<u>Proof:</u>
Write g for the Lie algebra of G. We know that there is a one to one

correspondence between $\mathfrak{g}$ and G via the exponential mapping. We follow Mal'cev's ideas to describe the group structure of G.

Consider any central series of length c of $\mathfrak{g}$:

$$\mathfrak{g}_* : 0 = \mathfrak{g}_0 \subseteq \mathfrak{g}_1 \subseteq \mathfrak{g}_2 \subseteq \cdots \subseteq \mathfrak{g}_c$$

and choose a basis

$$A_{1,1}, A_{1,2}, \ldots, A_{1,k_1}, A_{2,1}, \ldots, A_{2,k_2}, A_{3,1} \ldots, A_{c,k_c}$$

of $\mathfrak{g}$, which is compatible with $\mathfrak{g}_*$. Now, introduce a system of coordinates on G by defining for all g of G its coordinate $m(g)$ to be the coordinate of $\log(g)$ with respect to the basis $A_{1,1}, \ldots, A_{c,k_c}$.

Consider x (resp. y) $\in G$, with $\log x = \sum x_{i,j} A_{i,j}$ (resp. $\log y = \sum y_{i,j} A_{i,j}$). The Campbell–Baker–Hausdorff formula implies that

$$
\begin{aligned}
xy &= \exp(\sum x_{i,j} A_{i,j}) \exp(\sum y_{i,j} A_{i,j}) \\
&= \exp\left(\sum (x_{i,j} + y_{i,j}) A_{i,j} + \frac{1}{2} \sum x_{p,q} y_{r,s} [A_{p,q}, A_{r,s}] + \cdots \right).
\end{aligned}
$$

A close examination of the previous expression now shows that the (i, j)-th coordinate of the product xy satisfies:

$$m(xy)_{i,j} = x_{i,j} + y_{i,j} + P_{i,j}(x_{i+1,1}, \ldots, x_{c,k_c}, y_{i+1,1}, \ldots, y_{c,k_c})$$

where $P_{i,j}$ is a polynomial of total degree $\leq c - i + 1$ in all the variables $x_{p,q}$ and $y_{r,s}$ (use the nilpotency of $\mathfrak{g}$). Moreover, when regarded as a polynomial in the $x_{p,q}$-variables (resp. $y_{r,s}$-variables) alone, $P_{i,j}$ has degree $\leq c - i$.

So, in terms of the coordinates introduced for G, the product is expressed by polynomial functions. We use these functions to define the embedding of G in $P(\mathbb{R}^K)$ as follows:

$$\rho : G \rightarrow P(\mathbb{R}^K) : g \mapsto \rho(g) \text{ with } \forall y \in \mathbb{R}^K :$$

$$(\rho(g)(y))_{i,j} = m(g)_{i,j} + y_{i,j} + P_{i,j}(m(g)_{i+1,1}, \ldots, m(g)_{c,k_c}, y_{i+1,1}, \ldots, y_{c,k_c})$$

It is now easy to see that this ρ indeed defines an action of G on $\mathbb{R}^K$ (in fact, it is just left translation in G!), which is simply transitive.

We can extend this embedding of G to an embedding of $G \rtimes \mathrm{Aut}\,(G) \rightarrow P(\mathbb{R}^K)$ by defining:

$$\rho : G \rtimes \mathrm{Aut}\,(G) \rightarrow P(\mathbb{R}^K) : (g, \alpha) \mapsto \rho(g, \alpha) \text{ with}$$

$$\forall y \in \mathbb{R}^K : \rho(g, \alpha)(y) = m(g\alpha(m^{-1}(y))).$$

(This is just the coordinate expression for the usual action of $G \rtimes \mathrm{Aut}\,(G)$ on itself.) In order to show that the image of ρ consists indeed of polynomials of degree $\leq s$ it is enough to show that $\rho(1, \alpha)$ is polynomial of degree ≤ 1 forall $\alpha \in \mathrm{Aut}\,(G)$. So, consider $\alpha \in \mathrm{Aut}\,(G)$ and assume that $y = (y_{1,1}, \ldots, y_{c,k_c})$. Let $h \in G$ be determined by $h = m^{-1}(y)$, then there is a commutative diagram

$$
\begin{array}{ccccc}
h & G & \stackrel{\alpha}{\longrightarrow} & G & \alpha(h) \\
\uparrow \exp & \uparrow \exp & & \uparrow \exp & \uparrow \exp \\
y_{1,1}A_{1,1} + \cdots + y_{c,k_c}A_{c,k_c} & \mathfrak{g} & \stackrel{d\alpha}{\longrightarrow} & \mathfrak{g} & d\alpha(y_{1,1}A_{1,1} + \cdots + y_{c,k_c}A_{c,k_c})
\end{array}
$$

where $d\alpha$ denotes the differential of α. As $d\alpha$ is a linear map, the coordinate expression for $\alpha(h) = \alpha(m^{-1}(y))$ is linear in the components of y as well.

∎

Corollary 4.5.5 *If G admits a lattice N, then we consider any torsion free central series N_* of characteristic subgroups of N. If $a_{1,1}, a_{1,2}, \ldots, a_{c,k_c}$ denotes a set of generators of N which is compatible with N_*, we may choose the basis used in the theorem above to be determined by*

$$A_{1,1} = \log a_{1,1}, A_{1,2} = \log a_{1,2}, \ldots, A_{c,k_c} = \log a_{c,k_c}.$$

Restricting the representation $\rho : G \to P(\mathbb{R}^K)$ (based on the chosen basis of $\mathfrak{g}$) to N, we see that we obtain a canonical type representation of N into $P(\mathbb{R}^K)$. Moreover, if E is any AC–group containing N as its Fitting subgroup, we see that the restriction of ρ to E is a canonical type polynomial representation with respect to the torsion free filtration E_* (see also lemma 4.5.10):

$$E_* : E_0 = N_0 = 1 \subset E_1 = N_1 \subset \cdots \subset E_{c-1} = N_{c-1} \subset E_c = N_c \subset E_{c+1} = E.$$

Corollary 4.5.6 *Let E be any AC–group with a c–step nilpotent Fitting subgroup, then the affine defect of E is $\leq \mathrm{Max}(0, c - 2)$.*

4.5.2 The second approach

For technical reasons we need the following (rather well known) lemma.

Lemma 4.5.7 *Let* $\Gamma \subseteq \mathbb{R}^k$ *be any set of elements which span* $\mathbb{R}^k$ *as a vector space, for some* $k \in \mathbb{N}_0$. *Let* $p(x_1, x_2, \ldots, x_k, x_{k+1}, \ldots, x_n)$ *be a polynomial in* n *variables for* $n \geq k$, *with real coefficients. Suppose*

$$p(x_1+z_1, x_2+z_2, \ldots, x_k+z_k, x_{k+1}, \ldots, x_n) = p(x_1, x_2, \ldots, x_k, x_{k+1}, \ldots, x_n)$$

for all $(z_1, z_2, \ldots, z_k) \in \Gamma$ *and all* $(x_1, x_2, \ldots, x_n) \in \mathbb{R}^n$.

Then, the polynomial $p(x_1, x_2, \ldots, x_k, x_{k+1}, \ldots, x_n)$ *does not depend on the variables* $x_1, x_2, \ldots, x_k$.

Proof: First of all we remark that we may suppose that

$$\{(1, 0, \ldots, 0), (0, 1, \ldots, 0), \ldots, (0, 0, \ldots, 1)\} \subseteq \Gamma.$$

We prove this lemma by induction on the number k.

- Suppose $k = 1$.

    –We start with the investigation of the case $n = 1$. So $p(x)$ is a polynomial in one variable for which $p(x) = p(x + 1)$. We proceed by induction on the degree d of the polynomial $p(x)$. For $d = 0$ there is nothing to show. So suppose $d > 0$ and the lemma holds for lower degrees. Since $\frac{dp}{dx}$ is a polynomial of degree $d - 1$ for which

$$\frac{dp}{dx}(x) = \frac{dp}{dx}(x + 1),$$

    we conclude that $\frac{dp}{dx} = r$ for some $r \in \mathbb{R}$. Therefore $p(x) = \int r\,dx = rx + s$ for some $r, s \in \mathbb{R}$. The relation $p(x) = p(x+1)$ now easily implies that $r = 0$ or that $p(x)$ is independent of x.

    –For $n > 1$, we have to consider polynomials of the form $p(x_1, x_2, \ldots, x_n)$ for which

$$p(x_1, x_2, \ldots, x_n) = p(x_1 + 1, x_2, \ldots, x_n) \quad \forall(x_1, x_2, \ldots, x_n).$$

We arrange the polynomial $p(x_1, \ldots, x_n)$ in the following form:

$$p(x_1, \ldots, x_n) = q_0(x_2, \ldots, x_n) + x_1 q_1(x_2, \ldots, x_n) + \cdots + x_1^l q_l(x_2, \ldots, x_n)$$

where the $q_i(x_2, \ldots, x_n)$ denote some polynomials in $n - 1$ variables. When we fix real numbers $x_2^0, x_3^0, \ldots, x_n^0$, we see that $p(x_1, x_2^0, \ldots, x_n^0)$ is a polynomial in one variable x_1 satisfying the conditions of the lemma with $k = 1$. Therefore we may conclude that

$$q_i(x_2^0, x_3^0, \ldots, x_n^0) = 0, \ \forall i > 0.$$

Since this happens for all choices of $x_2^0, x_3^0, \ldots, x_n^0$, it follows that

$$q_i(x_2, x_3, \ldots, x_n) = 0, \ \forall i > 0,$$

implying that $p(x_1, x_2, \ldots, x_n)$ is independent of x_1.

• Next, we suppose that $k > 1$ and the lemma holds for smaller values of k. By first considering the fact that

$$p(x_1, x_2, \ldots, x_n) = p(x_1 + 1, x_2, \ldots, x_n) \ \forall x_1, x_2, \ldots, x_n,$$

we may conclude that $p(x_1, \ldots, x_n)$ is independent of x_1. Now, we use the induction hypothesis to say that $p(x_1, x_2, \ldots, x_n)$ is independent of $x_1, x_2, \ldots, x_k$. ∎

Lemma 4.5.8 *Suppose N and N_* are as before. For any canonical type polynomial representation of N with respect to N_* we have*

$$P(\mathbb{R}^{K_i}, \mathbb{R}^k)^{N_i/N_{i-1}} \cong P(\mathbb{R}^{K_{i+1}}, \mathbb{R}^k)$$

for all i, $1 \leq i < c$, as N/N_i–module.

Proof:
Since $P(\mathbb{R}^{K_i}, \mathbb{R}^k)$ is an N/N_{i-1}–module via ${}^h p = p \circ \rho_{i-1}(h)^{-1}$, where $\rho_{i-1} : N/N_{i-1} \to P(\mathbb{R}^{K_i})$, it makes sense to speak of $P(\mathbb{R}^{K_i}, \mathbb{R}^k)^{N_i/N_{i-1}}$. Let $p(x) \in P(\mathbb{R}^{K_i}, \mathbb{R}^k)$. Then it is easy to see that ${}^z p(x) = p(x)$ for all $z \in N_i/N_{i-1}$ and $x \in \mathbb{R}$ if and only if $p(x)$ is independent of the variables $x_{i,1}, x_{i,2}, \ldots x_{i,k_i}$ (see lemma 4.5.7). Therefore, we may identify $p(x)$ with an element of $P(\mathbb{R}^{K_{i+1}}, \mathbb{R}^k)$. This identification is of course compatible with the N/N_i–action. ∎

Theorem 4.5.9 *Let E be a group containing a finitely generated torsion free nilpotent group N of rank K as a subgroup of finite index. Let φ :*

$E \longrightarrow \text{Aut}(\mathbb{Z}^k)$ *be any morphism with* $\varphi(N) = 1$ *and let* $\rho : E \longrightarrow$ $P(\mathbb{R}^K)$ *be any representation restricting to a canonical type polynomial representation of* N, *with respect to some torsion free central series. With a* E-*module structure on* $P(\mathbb{R}^K, \mathbb{R}^k)$ *via*

$$^h p = \varphi(h) \circ p \circ \rho(h)^{-1}, \quad p \in P(\mathbb{R}^K, \mathbb{R}^k), \quad h \in E,$$

we have that $H^i(E, P(\mathbb{R}^K, \mathbb{R}^k)) = 0$ *for all* $i \geq 1$.

<u>Proof:</u> First we consider any torsion free abelian group $\mathbb{Z}^l$ with a canonical type polynomial representation into $P(\mathbb{R}^l)$ and we assume $l \leq K$. We consider $P(\mathbb{R}^l) \subseteq P(\mathbb{R}^K)$, by lettting $P(\mathbb{R}^l)$ act on the l first components of $\mathbb{R}^K$ and by leaving the other $K - l$ components fixed. If $l = 1$ then $H^1(\mathbb{Z}, P(\mathbb{R}^K, \mathbb{R}^k)) = 0$ by theorem 2.5 of [39] (notice that for an abelian group $\mathbb{Z}^l$, the concept of a canonical type polynomial representation is the same as a "special affine" representation in [39].) For $i > 1$ $H^i(\mathbb{Z}, P(\mathbb{R}^K, \mathbb{R}^k)) = 0$ because $\mathbb{Z}$ has cohomological dimension one. Suppose $H^i(\mathbb{Z}^{l-1}, P(\mathbb{R}^K, \mathbb{R}^k)) = 0$ for all $i \geq 1$. Without loss of generality we assume that $\mathbb{Z}^l$ has a normal subgroup $\mathbb{Z}$, with torsion free quotient $\mathbb{Z}^l/\mathbb{Z}$ and acting only on the first component of $\mathbb{R}^K$. As in lemma 4.5.8 we find that $P(\mathbb{R}^K, \mathbb{R}^k)^{\mathbb{Z}} \cong P(\mathbb{R}^{K-1}, \mathbb{R}^k)$ as $\mathbb{Z}^l/\mathbb{Z} = \mathbb{Z}^{l-1}$ modules. Since $H^i(\mathbb{Z}, P(\mathbb{R}^K, \mathbb{R}^k)) = 0$ for all i, the restriction–inflation exact sequence yields isomorphisms:

$$0 = H^i(\mathbb{Z}^{l-1}, P(\mathbb{R}^{K-1}, \mathbb{R}^k)) \cong H^i(\mathbb{Z}^l, P(\mathbb{R}^K, \mathbb{R}^k)).$$

This proves (something more than) the theorem for torsion free abelian groups.

Now we consider any finitely generated nilpotent group N of nilpotency class $c > 1$ and we assume the theorem holds for groups of lower nilpotency class. By the above we know that $H^i(N_1, P(\mathbb{R}^K, \mathbb{R}^k)) = 0$ for $i \geq 1$. Therefore, the restriction–inflation exact sequence now provides us isomorphisms:

$$0 = H^i(N/N_1, P(\mathbb{R}^{K_2}, \mathbb{R}^k)) \cong H^i(N, P(\mathbb{R}^K, \mathbb{R}^k)).$$

The generalization for virtually nilpotent groups can be obtained as follows: Because $P(\mathbb{R}^K, \mathbb{R}^k)$ is a vector space and N is of finite index in E, we know that the restriction map

$$H^i(E, P(\mathbb{R}^K, \mathbb{R}^k)) \overset{\text{res}}{\Rightarrow} H^i(N, P(\mathbb{R}^K, \mathbb{R}^k)) = 0$$

is injective, which ends the proof.

■

Lemma 4.5.10 *Let E be a group containing a torsion free, finitely generated nilpotent group N as a normal subgroup of finite index. If $\rho : E \to P(\mathbb{R}^K)$ is a representation of E, restricting to a canonical type polynomial representation of N, with respect to a torsion free central series N_* consisting of normal subgroups of E, then ρ itself is canonical, with respect to E_*, where E_* is given by:*

$$E_* : E_0 = N_0 = 1 \subset E_1 = N_1 \subset \cdots \subset E_{c-1} = N_{c-1} \subset E_c = N_c \subset E_{c+1} = E.$$

<u>Proof:</u> It suffices to show that for $e \in E$, $\rho(e)$ is such that

$$\rho(e) : \mathbb{R}^K \longrightarrow \mathbb{R}^K$$
$$: x = (x_{1,1}, x_{1,2}, \ldots, x_{1,k_1}, x_{2,1}, \ldots, x_{i,j}, \ldots, x_{c,k_c}) \mapsto \rho(e)(x),$$

with $\rho(e)(x) = (P_{1,1}^e(x), \ldots, P_{i,j}^e(x), \ldots, P_{c,k_c}^e(x))$, and $P_{i,j}^e(x)$ is a polynomial depending only on the variables $x_{p,q}$ with $p \geq i$.

First we show that $P_{i,j}^e(x)$ does not depend on $x_{1,q}$, if $i > 1$. Let z be any element of N_1, then there exists a $z' \in N_1$ such that

$$ez = z'e. \tag{4.14}$$

We will look at the effect of both sides of (4.14) on a general element x with coordinates $(x_{1,1}, x_{1,2}, \ldots, x_{c,k_c})$.
Remember that z (resp. z') acts on $\mathbb{R}^K$ by some translation on the first k_1 components, say by $(z_{1,1}, \ldots, z_{1,k_1})$ (resp. $(z'_{1,1}, \ldots, z'_{1,k_1})$). On the one hand we have that

$$
\begin{aligned}
^{ez}x &= {}^e({}^zx) \\
&= {}^e(z_{1,1} + x_{1,1}, \ldots, z_{1,k_1} + x_{1,k_1}, x_{2,1}, x_{2,2}, \ldots, x_{c,k_c}) \\
&= (P_{1,1}^e(z_{1,1} + x_{1,1}, \ldots, x_{c,k_c}), \ldots, P_{c,k_c}^e(z_{1,1} + x_{1,1}, \ldots, x_{c,k_c})).
\end{aligned}
$$

While on the other hand, we see that

$$
\begin{aligned}
^{ez}x = {}^{z'e}x &= {}^{z'}({}^ex) \\
&= (z'_{1,1}, \ldots, z'_{1,k_1}, 0, \ldots, 0) + (P_{1,1}^e(x), \ldots, P_{c,k_c}^e(x)) \\
&= (P_{1,1}^e(x) + z'_{1,1}, \ldots, P_{1,k_1}^e + z'_{1,k_1}, P_{2,1}^e(x), \ldots, P_{c,k_c}^e(x)).
\end{aligned}
$$

These two computations result in $\forall i > 1$, $\forall z \in N_1$:

$$P_{i,j}^e(z_{1,1} + x_{1,1}, \ldots, z_{1,k_1} + x_{1,k_1}, x_{2,1}, \ldots, x_{c,k_c}) =$$

$$P^e_{i,j}(x_{1,1}, \ldots, x_{1,k_1}, x_{2,1}, \ldots, x_{c,k_c}).$$

Since the set of all $(z_{1,1}, \ldots, z_{1,k_1})$ forms a lattice of $\mathbb{R}^{k_1}$, we conclude that the polynomial $P^e_{i,j}(x_{1,1}, \ldots, x_{c,k_c})$ has to be independent of the co-ordinates $x_{1,q}$ $(1 \leq q \leq k_1)$ (use lemma 4.5.7).

The same technique can now be used to proceed this proof. One shows by induction on p that $P^e_{i,j}$ is independent of $x_{p,q}$ if $i > p$. ∎

Remark 4.5.11 We will call a filtration E_* as in the lemma above, a torsion free pseudo–central series of E. If E is virtually nilpotent as above, then, if we consider a canonical type representation we will always mean "canonical with respect to a torsion free pseudo–central series of E".

We finish this section with the following lemma:

Lemma 4.5.12 *Let E be a virtually abelian group containing a group $\mathbb{Z}^k$ as a normal subgroup of finite index. Then a canonical type polynomial representation is an affine representation.*

<u>Proof:</u> Let $\rho : E \to P(\mathbb{R}^k)$ be the canonical type polynomial representation. We know that the abelian normal subgroup acts on $\mathbb{R}^k$ via translations:

$$^z x = z + x \quad \forall z \in \mathbb{Z}^k \text{ and } \forall x \in \mathbb{R}^k.$$

There is a homomorphism $\varphi : E \to \text{Aut}(\mathbb{Z}^k)$ such that $eze^{-1} = \varphi(e)z$. Fix an element e and assume $\varphi(e)$ is given by a matrix $(\alpha_{i,j})_{1 \leq i,j \leq k}$. We denote $\rho(e) = (P_1(x), P_2(x), \ldots, P_k(x))$, where the $P_i(x)$ are some polynomials in k variables. For any element $x = (x_1, x_2, \ldots, x_k)$ and any $z = (z_1, z_2, \ldots, z_k) \in \mathbb{Z}^k$ we have that

$$^{ez}x = ^{(\varphi(e)z)e}x$$
$$^e(x_1 + z_1, \ldots, x_k + z_k) = \varphi(e)z + (P_1(x), \ldots, P_k(x))$$
$$\Downarrow$$
$$P_i(x_1 + z_1, \ldots, x_k + z_k) = \alpha_{i,1}z_1 + \cdots + \alpha_{i,k}z_k + P_i(x_1, \ldots, x_k).$$
$$(1 \leq i \leq k)$$

Now, consider the polynomial

$$Q(x) = P_i(x) - \alpha_{i,1}x_1 - \cdots - \alpha_{i,k}x_k - P(0).$$

The previous computation shows that $Q(z) = 0 \; \forall z \in \mathbb{Z}^k$. This implies that $Q(x) = 0$ or that $P_i(x)$ is a polynomial of degree 1.

■

4.5.3 Existence and uniqueness of Polynomial Manifolds

In this section we make intensive use of the basic facts of the theory of Seifert Fiber Space constructions.

Theorem 4.5.13 *Let E be a group containing a normal subgroup N, which is torsion free, finitely generated of rank K, nilpotent of class c and of finite index in E and fix a pseudo–central series E_* of E. Then there exists a canonical type polynomial representation of E with respect to E_*.*

Moreover, if ρ_1, ρ_2 are two canonical type polynomial representations of E into $P(\mathbb{R}^K)$, with respect to E_*, then there exists a polynomial map $p \in P(\mathbb{R}^K)$ such that $\rho_2 = p^{-1} \circ \rho_1 \circ p$.

In case E is torsion free this means that the manifolds $\rho_1(E)\backslash\mathbb{R}^K$ and $\rho_2(E)\backslash\mathbb{R}^K$ are "polynomially diffeomorphic".

<u>Proof:</u>
<u>Existence:</u>
We will proceed by induction on the nilpotency class c of N. If N is abelian (or 2- or 3-step nilpotent), then the existence is known, since we already obtained even a canonical type affine representation for these groups. Now, suppose that N is of class $c > 1$ and the existence is guaranteed for lower nilpotency classes. Using the induction hypothesis, the group E/N_1 can be furnished with a canonical type polynomial representation

$$\bar{\rho} : E/N_1 \to P(\mathbb{R}^{K_2}).$$

We obtain an embedding $i : N_1 \cong \mathbb{Z}^{k_1} \to P(\mathbb{R}^{K_2}, \mathbb{R}^{k_1})$ if we define $i(z) :$ $\mathbb{R}^{K_2} \to \mathbb{R}^{k_1} : x \mapsto z$. We are looking for a map ρ making the following diagram commutative: $(P \rtimes (A \times P) \overset{\text{not}}{=} P(\mathbb{R}^{K_2}, \mathbb{R}^{k_1}) \rtimes (\text{Aut } \mathbb{Z}^{k_1} \times P(\mathbb{R}^{K_2})))$

$$1 \to \quad N_1 \quad \to \quad E \quad \to \quad E/N_1 \quad \to 1$$

$$\downarrow i \qquad\qquad \downarrow \rho \qquad\qquad \downarrow \varphi \times \bar{\rho}$$

$$1 \to P(\mathbb{R}^{K_2}, \mathbb{R}^{k_1}) \to P \rtimes (A \times P) \to \text{Aut } \mathbb{Z}^{k_1} \times P(\mathbb{R}^{K_2}) \to 1$$

$$\cap$$

$$P(\mathbb{R}^K)$$

where $\varphi : E/N_1 \to \operatorname{Aut} \mathbb{Z}^{k_1} = \operatorname{Aut} N_1$ denotes the morphism induced by the extension $1 \to N_1 \to E \to E/N_1 \to 1$.

The existence of such a map ρ is now guaranteed by the surjectiveness of δ in the long exact cohomology sequence

$$\cdots \to 0 \to H^1(E/N_1, P(\mathbb{R}^{K_2}, \mathbb{R}^{k_1})/\mathbb{Z}^{k_1}) \xrightarrow{\delta} H^2(E/N_1, \mathbb{Z}^{k_1}) \to 0 \to \cdots$$
$$\|\hspace{6.5cm}\|$$
$$H^1(E/N_1, P(\mathbb{R}^{K_2}, \mathbb{R}^{k_1})) \qquad (\text{theorem 4.5.9}) \qquad H^2(E/N_1, P(\mathbb{R}^{K_2}, \mathbb{R}^{k_1}))$$

$$(4.15)$$

Uniqueness:

Again we proceed by induction on the nilpotency class c of N. For $c = 1$ the result is well known. Indeed, two canonical type representations of a virtually abelian group are even known to be affinely conjugated. (To see this, we can use the generalized second Bieberbach theorem).

So we suppose that N is of class $c > 1$ and that the theorem holds for smaller nilpotency classes. The representations ρ_1, ρ_2 induce two canonical type polynomial representations

$$\bar{\rho}_1, \bar{\rho}_2 : E/N_1 \to \mathbb{R}^{K_2}.$$

By the induction hypothesis, there exists a polynomial diffeomorphism $\bar{q} : \mathbb{R}^{K_2} \to \mathbb{R}^{K_2}$ such that

$$\bar{\rho}_2 = \bar{q}^{-1} \circ \bar{\rho}_1 \circ \bar{q}.$$

Lift this $\bar{q}$ to a polynomial diffeomorphism q of $\mathbb{R}^{K_1}$ as follows:

$$\forall x \in \mathbb{R}^{k_1}, \forall y \in \mathbb{R}^{K_2} : q : \mathbb{R}^{K_1} \to \mathbb{R}^{K_1} : (x, y) \mapsto q(x, y) = (x, \bar{q}(y)).$$

Let us denote $\psi = q^{-1} \circ \rho_1 \circ q$. Then we see that ψ and ρ_2 are two canonical type polynomial representations of E, which induce the same representation of E/N_1. This means that ψ and ρ_2 can be seen as the result of a Seifert construction with respect to the same data (cfr. the commutative diagram above).

Now we use the injectiveness of δ in (4.15), which implies that ψ and ρ_2 are conjugated to each other by an element $r \in P(\mathbb{R}^{K_2}, \mathbb{R}^{k_1})$ (seen as an element of $P(\mathbb{R}^K)$!). So we may conclude that

$$\rho_2 = r^{-1} \circ \psi \circ r = r^{-1} \circ q^{-1} \circ \rho_1 \circ q \circ r = p^{-1} \circ \rho_1 \circ p$$

if we take $p = q \circ r$.

Corollary 4.5.14 *Since any canonical type polynomial representation ρ of E is polynomially conjugated to a polynomial representation of finite degree by the previous section, ρ itself has to be of finite degree.*

Remark 4.5.15 Theorem 4.5.13 is a generalization of theorem 1.20 in [31] and of theorem 4.1 in [39]. In this second paper, the authors are concerned with canonical type *affine* representations (called "special" there), with respect to the upper central series. They claim that we can take the polynomial to conjugate with, of degree $\leq c!$. Certainly there is a polynomial conjugation, but the degree does not seem to be right. We indicate a gap in the proof of theorem 4.1 of [39].

This gap appears in the induction argument in [39]. Indeed, an induction argument assumes that, after conjugation by a polynomial on the $N/N_1 = N/Z(N)$-level, the result on the N-level is again a canonical type *affine* representation. Of course, the conjugation does not touch the canonical type character of the representation. However, as the subsequent example shows, one might obtain a polynomial representation which stays outside the affine group.

Example 4.5.16

Let N be the following finitely generated torsion free 3–step nilpotent group:

$$N :< a_1, a_2, b, c \parallel \begin{matrix} [a_2, a_1] = b, \ [b, a_1] = c, \\ [b, a_2] = [c, a_1] = [c, a_2] = [c, b] = 1 \end{matrix} \quad > .$$

We propose the following two canonical type affine representations of N (upper central series), where we indicate only the images of the generators:

$$\rho_1(a_1) = \left(\begin{pmatrix} 1 & -1 & 0 & 1 \\ 0 & 1 & 0 & -1 \\ 0 & 0 & 1 & 0 \\ 0 & 0 & 0 & 1 \end{pmatrix}, \begin{pmatrix} 0 \\ 0 \\ 1 \\ 0 \end{pmatrix} \right) \qquad \rho_1(a_2) = \left(\begin{pmatrix} 1 & 0 & 0 & 0 \\ 0 & 1 & 0 & 0 \\ 0 & 0 & 1 & 0 \\ 0 & 0 & 0 & 1 \end{pmatrix}, \begin{pmatrix} 0 \\ 0 \\ 0 \\ 1 \end{pmatrix} \right)$$

$$\rho_1(b) = \left(\begin{pmatrix} 1 & 0 & 0 & 0 \\ 0 & 1 & 0 & 0 \\ 0 & 0 & 1 & 0 \\ 0 & 0 & 0 & 1 \end{pmatrix}, \begin{pmatrix} 0 \\ 1 \\ 0 \\ 0 \end{pmatrix} \right) \qquad \rho_1(c) = \left(\begin{pmatrix} 1 & 0 & 0 & 0 \\ 0 & 1 & 0 & 0 \\ 0 & 0 & 1 & 0 \\ 0 & 0 & 0 & 1 \end{pmatrix}, \begin{pmatrix} 1 \\ 0 \\ 0 \\ 0 \end{pmatrix} \right)$$

and

$$\rho_2(a_1) = \left(\begin{pmatrix} 1 & -\frac{2}{3} & 0 & 0 \\ 0 & 1 & 0 & -\frac{1}{2} \\ 0 & 0 & 1 & 0 \\ 0 & 0 & 0 & 1 \end{pmatrix}, \begin{pmatrix} 0 \\ 0 \\ 1 \\ 0 \end{pmatrix} \right) \qquad \rho_2(a_2) = \left(\begin{pmatrix} 1 & 0 & 0 & 0 \\ 0 & 1 & \frac{1}{2} & 0 \\ 0 & 0 & 1 & 0 \\ 0 & 0 & 0 & 1 \end{pmatrix}, \begin{pmatrix} 0 \\ 0 \\ 0 \\ 1 \end{pmatrix} \right)$$

$$\rho_2(b) = \left(\begin{pmatrix} 1 & 0 & \frac{1}{3} & 0 \\ 0 & 1 & 0 & 0 \\ 0 & 0 & 1 & 0 \\ 0 & 0 & 0 & 1 \end{pmatrix}, \begin{pmatrix} \frac{2}{3} \\ 1 \\ 0 \\ 0 \end{pmatrix} \right) \quad \rho_2(c) = \left(\begin{pmatrix} 1 & 0 & 0 & 0 \\ 0 & 1 & 0 & 0 \\ 0 & 0 & 1 & 0 \\ 0 & 0 & 0 & 1 \end{pmatrix}, \begin{pmatrix} 1 \\ 0 \\ 0 \\ 0 \end{pmatrix} \right).$$

The presentation for $N/Z(N)$ can be read from the presentation of N, by simply replacing all c's by 1. The induced canonical type affine representations $\bar{\rho}_1, \bar{\rho}_2 : N/Z(N) \to \mathrm{Aff}(\mathbb{R}^3)$ are also easy to write down:

- In the linear part of ρ_i ($i = 1, 2$), we discard the first column and the first row.

- In the translational part, we omit the first term.

Following the general theory of Seifert Fiber Spaces and the proof of theorem 4.1 in [39] or of theorem 4.5.13 we find that there must exist a polynomial mapping $p(x)$ in $P(\mathbb{R}^2, \mathbb{R})$ such that

$$\bar{\rho}_2 = p^{-1} \circ \bar{\rho}_1 \circ p$$

when $p(x)$ is seen as an element of $P(\mathbb{R}^2, \mathbb{R}) \rtimes P(\mathbb{R}^2)$ or as an element of $P(\mathbb{R}^3)$. Some elementary computations show that the only possible p's are of the form:

$$p : \mathbb{R}^3 \to \mathbb{R}^3 : \begin{pmatrix} z \\ y \\ x \end{pmatrix} \mapsto \begin{pmatrix} z - \frac{xy}{2} + r \\ y \\ x \end{pmatrix}$$

where r can be any real number. If we now lift this p to a polynomial mapping of $\mathbb{R}^4$, say

$$q : \mathbb{R}^4 \to \mathbb{R}^4 : \begin{pmatrix} t \\ z \\ y \\ x \end{pmatrix} \mapsto \begin{pmatrix} t \\ z - \frac{xy}{2} + r \\ y \\ x \end{pmatrix}$$

we find that

$$q^{-1} \rho_1(a_1) q : \mathbb{R}^4 \to \mathbb{R}^4 : \begin{pmatrix} t \\ z \\ y \\ x \end{pmatrix} \mapsto \begin{pmatrix} t + x - z + \frac{xy}{2} - r \\ z - \frac{x}{2} \\ y + 1 \\ x \end{pmatrix}$$

which is clearly not affine. Moreover, the same can be said about any lift of p of the form

$$q : \mathbb{R}^4 \to \mathbb{R}^4 : \begin{pmatrix} t \\ z \\ y \\ x \end{pmatrix} \mapsto \begin{pmatrix} m(x, y, z, t) \\ z - \frac{xy}{2} + r \\ y \\ x \end{pmatrix}$$

where $m(x, y, z, t)$ denotes some polynomial of degree ≤ 2 in the four variables.

Finally, we indicate how ρ_1 and ρ_2 can be seen to be conjugated by a polynomial. Consider

$$
p : \mathbb{R}^4 \to \mathbb{R}^4 : \begin{pmatrix} t \\ z \\ y \\ x \end{pmatrix} \mapsto \begin{pmatrix} t + \frac{xy}{3} + \frac{xy^2}{6} - \frac{2z}{3} - \frac{yz}{3} \\ z - \frac{xy}{2} \\ y \\ x \end{pmatrix} .
$$

Then one can see that

$$
p^{-1} : \mathbb{R}^4 \to \mathbb{R}^4 : \begin{pmatrix} t \\ z \\ y \\ x \end{pmatrix} \mapsto \begin{pmatrix} t + \frac{2z}{3} + \frac{yz}{3} \\ z + \frac{xy}{2} \\ y \\ x \end{pmatrix}
$$

from which some easy calculations result in

$$
\rho_2(n) = p^{-1} \circ \rho_1(n) \circ p \quad \forall n \in N.
$$

We remark that, in spite of this example, we were able, using totally different techniques, to determine an upper bound for the degree of p, again in terms of the nilpotency class c ([16]).

4.5.4 Groups of affine defect one

The only known connected, simply connected nilpotent Lie groups not acting simply transitively and affinely on Euclidean n-space, are those constructed by Benoist ([4]) and Burde and Grunewald ([12]). As we told already, it is the situation on the Lie algebra level which is investigated in both papers.

The Lie algebras in question are 11-dimensional; an interesting family of them is denoted by $\mathfrak{a}(-2, s, t)$ and has a basis $e_1, e_2, \ldots, e_{11}$. Moreover, the brackets of these Lie algebras are completely determined by

$$
\begin{aligned}
&[e_1, e_i] = e_{i+1} \ 2 \leq i \leq 10 \\
&[e_1, e_{11}] = 0 \\
&[e_2, e_3] = e_5 \text{ and} \\
&[e_2, e_5] = -2e_7 + se_8 + te_9.
\end{aligned}
$$

The results of Benoist, Burde and Grunewald imply that, in case $s \neq 0$, $\mathfrak{a}(-2, s, t)$ has no faithful 12-dimensional linear representation. Therefore, no lattice of the corresponding Lie group $\mathcal{A}(-2, s, t)$, can act properly discontinuously on $\mathbb{R}^{11}$, via affine motions and with compact quotient.

In this section we indicate, how one can construct a uniform lattice Γ of the Lie group $\mathcal{A}(-2, s, t)$, acting properly discontinuous on $\mathbb{R}^{11}$ via polynomial diffeomorphisms, in such a way that the polynomial mappings used are of degree ≤ 2. It will follow that these lattices Γ have affine defect 1.

To do so, we will embed this Γ into a connected, simply connected Lie group P (of finite dimension) of polynomial diffeomorphisms of $\mathbb{R}^{11}$. In fact, let P be the group consisting of all mappings

$$f : \mathbb{R}^{11} \to \mathbb{R}^{11} : \begin{pmatrix} y_{11} \\ y_{10} \\ \vdots \\ y_2 \\ y_1 \end{pmatrix} \mapsto \begin{pmatrix} y_{11} + p_{11}(y_1, y_2, \ldots, y_{10}) \\ y_{10} + p_{10}(y_1, y_2, \ldots, y_9) \\ \vdots \\ y_2 + p_2(y_1) \\ y_1 + p_1 \end{pmatrix}$$

where $p_i(y_1, y_2, \ldots, y_{i-1})$ is a polynomial of degree ≤ 1 in the variables $y_1, \ldots y_{i-1}$, except p_{11} which is of degree ≤ 2.

In view of the results mentioned above, an embedding $\Gamma \hookrightarrow P$ is really the best (non affine) thing we could expect.

The first step to clear this job, was to construct a unitriangular matrix representation of $\mathcal{A}(-2, s, t)$. This was not difficult since Burde and Grunewald ([12]) give an explicit way to construct a 34–dimensional (and with some more trouble, a 22–dimensional) matrix representation for the Lie algebra $\mathfrak{a}(-2, s, t)$, using its universal enveloping algebra. These matrices are upper triangular, with 0's on the diagonal, and so by exponentiating we find the desired representation on the Lie group level.

For the sake of simplicity we will restrict ourselves to the cases where $s, t \in \mathbb{Z}$, however, we remark here that things work out for rational s and t too.

Let Γ denote the subgroup of $\mathcal{A}(-2, s, t)$ generated by $A_1, A_2, \ldots, A_{11}$, where

$$
\begin{array}{lll}
A_1 = \exp(e_1) & A_2 = \exp(e_2) & A_3 = \exp(e_3) \\
A_4 = \exp(\tfrac{1}{2}e_4) & A_5 = \exp(\tfrac{1}{6}e_5) & A_6 = \exp(\tfrac{1}{24}e_6) \\
A_7 = \exp(\tfrac{1}{120}e_7) & A_8 = \exp(\tfrac{1}{720}e_8) & A_9 = \exp(\tfrac{1}{25200}e_9) \\
A_{10} = \exp(\tfrac{1}{1008000}e_{10}) & A_{11} = \exp(\tfrac{1}{90720000}e_{11})
\end{array}
$$

Making use of the matrix representation of $\mathcal{A}(-2, s, t)$, and so of Γ, mentioned above, one checks that Γ is a uniform lattice of $\mathcal{A}(-2, s, t)$ with

$$\Gamma = \langle A_1, \ldots, A_{11} \| \qquad \rangle$$

$$[A_2, A_1] = A_3^{-1} A_4 A_5^2 A_6^{-5} A_7^{109} A_8^{4(79-60s)} A_9^{6(-1108-525s-1400t)} \cdot$$
$$A_{10}^{-2267555+151088s-126000t} \cdot$$
$$A_{11}^{2(203230225-12930435s+677376s^2+4372200t)}$$

$$[A_3, A_1] = A_4^{-2} A_5^3 A_6^{-4} A_7^{185} A_8^{6(-91-60s)} A_9^{35(829+180s-360t)} \cdot$$
$$A_{10}^{80(45515-3066s+3150t)} \cdot$$
$$A_{11}^{180(-5202055+135870s+18816s^2-190050t)}$$

$$[A_3, A_2] = A_5^{-6} A_7^{-120} A_8^{360(3+s)} A_9^{840(-26-15s+15t)} \cdot$$
$$A_{10}^{6720(-585-28s-75t)} A_{11}^{3780(-41975-13035s-2688s^2+5150t)}$$

$$[A_4, A_1] = A_5^{-3} A_6^6 A_7^{-10} A_8^{15} A_9^{16905} A_{10}^{140(-2425-864s)} \cdot$$
$$A_{11}^{90(836225+60480s+28224s^2-222075t)}$$

$$[A_4, A_2] = A_6^{-12} A_8^{-900} A_9^{12600s} A_{10}^{8400(341+60t)} A_{11}^{13494600s}$$

$$[A_4, A_3] = A_7^{-180} A_8^{360s} A_9^{12600t} A_{10}^{-5896800} A_{11}^{-85050(4725+334s)}$$

$$[A_5, A_1] = A_6^{-4} A_7^{10} A_8^{-20} A_9^{175} A_{10}^{-1400} A_{11}^{21708750}$$

$$[A_5, A_2] = A_7^{40} A_8^{-120s} A_9^{840(13-5t)} A_{10}^{94080s} A_{11}^{1260(-12350+4032s^2-7725t)}$$

$$[A_5, A_3] = A_8^{-360} A_9^{4200s} A_{10}^{168000t} A_{11}^{91003500}$$

$$[A_5, A_4] = A_9^{-11340} A_{10}^{80640s} A_{11}^{3780(-448s^2+3525t)}$$

$$[A_6, A_1] = A_7^{-5} A_8^{15} A_9^{-175} A_{10}^{1750} A_{11}^{-31500}$$

$$[A_6, A_2] = A_8^{150} A_9^{-2100s} A_{10}^{42000(-13-2t)} A_{11}^{-6095250s}$$

$$[A_6, A_3] = A_9^{2520} A_{10}^{1680s} A_{11}^{1890(448s^2-1525t)}$$

$$[A_6, A_4] = A_{10}^{-113400} A_{11}^{1814400s}$$

$$[A_6, A_5] = A_{11}^{-10843875}$$

$$[A_7, A_1] = A_8^{-6} A_9^{105} A_{10}^{-1400} A_{11}^{31500}$$

$$[A_7, A_2] = A_9^{546} A_{10}^{-17136s} A_{11}^{189(-6175-896s^2-4950t)}$$

$$[A_7, A_3] = A_{10}^{65520} A_{11}^{-695520s}$$

$$[A_7, A_4] = A_{11}^{4465125}$$

$$[A_8, A_1] = A_9^{-35} A_{10}^{700} A_{11}^{-21000}$$

$$[A_8, A_2] = A_{10}^{-7280} A_{11}^{-141120s}$$

$$[A_8, A_3] = A_{11}^{-505575}$$

$$[A_9, A_1] = A_{10}^{-40} A_{11}^{1800}$$

$$[A_9, A_2] = A_{11}^{-4275}$$

$$[A_{10}, A_1] = A_{11}^{-90}$$

$$[A_i, A_j] = 0 \text{ for } 1 \le j < i \le 11 \text{ and } i+j > 11$$

In fact, after having taken $A_1 = \exp(e_1)$ and $A_2 = \exp\{e_2\}$, the others have been chosen carefully so that they generate a lattice. It is obvious that this choice is not unique.

Now that we have enough information about the group structure, we are going to build up an embedding $\Gamma \to P$. In order to do so we will first realize the rank 10 group $\Gamma/Z(\Gamma) = \Gamma/\langle A_{11}\rangle$ as a properly discontinuous group of affine motions of $\mathbb{R}^{10}$. The idea behind this is the fact that the adjoint representation $\mathrm{ad} : \mathfrak{a}(-2,s,t) \to \mathfrak{gl}(\mathfrak{a}(-2,s,t))$ is an 11-dimensional representation factoring through $\mathfrak{a}(-2,s,t)/Z(\mathfrak{a}(-2,s,t)) = \mathfrak{a}(-2,s,t)/\langle e_{11}\rangle$. By exponentiating this representation, we may realize $\mathcal{A}(-2,s,t)/Z(\mathcal{A}(-2,s,t))$ as a group of affine motions of $\mathbb{R}^{10}$. However, in this way the group $\mathcal{A}(-2,s,t)/Z(\mathcal{A}(-2,s,t))$ does not act simply transitive on $\mathbb{R}^{10}$ as we would like. Therefore we slightly modify the adjoint representation in order to obtain, on the Lie group level, a simply transitive action. This altered adjoint representation, denoted by ϕ is determined completely by the images of e_1 and e_2:

$$\phi(e_1) = \begin{pmatrix}
0 & 1 & 0 & 0 & 0 & 0 & 0 & 0 & 0 & 0 & 0 \\
0 & 0 & 1 & 0 & 0 & 0 & 0 & 0 & 0 & 0 & 0 \\
0 & 0 & 0 & 1 & 0 & 0 & 0 & 0 & 0 & 0 & 0 \\
0 & 0 & 0 & 0 & 1 & 0 & 0 & 0 & 0 & 0 & 0 \\
0 & 0 & 0 & 0 & 0 & 1 & 0 & 0 & 0 & 0 & 0 \\
0 & 0 & 0 & 0 & 0 & 0 & 1 & 0 & 0 & 0 & 0 \\
0 & 0 & 0 & 0 & 0 & 0 & 0 & 1 & 0 & 0 & 0 \\
0 & 0 & 0 & 0 & 0 & 0 & 0 & 0 & 1 & 0 & 0 \\
0 & 0 & 0 & 0 & 0 & 0 & 0 & 0 & 0 & 0 & 0 \\
0 & 0 & 0 & 0 & 0 & 0 & 0 & 0 & 0 & 0 & -1 \\
0 & 0 & 0 & 0 & 0 & 0 & 0 & 0 & 0 & 0 & 0
\end{pmatrix}$$

and

$$\phi(e_2) = \begin{pmatrix}
0 & 0 & \frac{19}{16} & \frac{28\,s}{25} & \frac{448\,s^2+2475\,t}{2000} & 0 & 0 & 0 & 0 & 0 & 0 \\
0 & 0 & 0 & \frac{26}{5} & \frac{51\,s}{25} & 2t & 0 & 0 & 0 & 0 & 0 \\
0 & 0 & 0 & 0 & -\frac{13}{5} & 2s & t & 0 & 0 & 0 & 0 \\
0 & 0 & 0 & 0 & 0 & -5 & s & 0 & 0 & 0 & 0 \\
0 & 0 & 0 & 0 & 0 & 0 & -2 & 0 & 0 & 0 & 0 \\
0 & 0 & 0 & 0 & 0 & 0 & 0 & 1 & 0 & 0 & 0 \\
0 & 0 & 0 & 0 & 0 & 0 & 0 & 0 & 1 & 0 & 0 \\
0 & 0 & 0 & 0 & 0 & 0 & 0 & 0 & 0 & 0 & 0 \\
0 & 0 & 0 & 0 & 0 & 0 & 0 & 0 & 0 & 0 & -1 \\
0 & 0 & 0 & 0 & 0 & 0 & 0 & 0 & 0 & 0 & 0 \\
0 & 0 & 0 & 0 & 0 & 0 & 0 & 0 & 0 & 0 & 0
\end{pmatrix}.$$

Let us define the morphism φ as the composite map:

$$\mathcal{A}(-2,s,t) \overset{\log}{\to} \mathfrak{a}(-2,s,t) \to \mathfrak{a}(-2,s,t)/\langle e_{11}\rangle \overset{\phi}{\to} \mathfrak{gl}(11,\mathbb{R}) \overset{\exp}{\to} Gl(11,\mathbb{R})$$

As the images of φ consist of unitriangular matrices, we can look at the Lie group $\mathcal{A}(-2, s, t)/Z(\mathcal{A}(-2, s, t))$ as being a simply transitive group of affine motions of $\mathbb{R}^{10}$.

In the next step we lift the map $\varphi : \Gamma \to Gl(11, \mathbb{R})$ to a faithful morphism $\psi : \Gamma \to P$. We present ψ by giving the images of the generators:

$$\psi(A_i) : \mathbb{R}^{11} \to \mathbb{R}^{11} : \begin{pmatrix} y_{11} \\ y_{10} \\ \vdots \\ y_2 \\ y_1 \end{pmatrix} \mapsto \begin{pmatrix} y_{11} + f_i(y_1, y_2, \ldots, y_{10}) \\ y_{10} \\ \varphi(A_i) \quad \vdots \\ y_2 \\ y_1 \end{pmatrix}.$$

Here, we see $\varphi(A_i)$ as an affine mapping of $\mathbb{R}^{10}$ and the f_i are polynomials of degree ≤ 2 given by:

$$f_1(y_1, \ldots, y_{10}) = \frac{99569925\, t + 17013024\, s^2 - 250}{90720000} y_2 - \frac{1}{40320} y_3$$

$$- \frac{1}{5040} y_4 - \frac{1}{720} y_5 - \frac{1}{120} y_6 - \frac{1}{24} y_7 - \frac{1}{6} y_8 - \frac{1}{2} y_9 - y_{10}$$

$$+ \frac{-5458475 + 7392820\, s + 1937376\, s^2 + 10623200\, t}{1680000} y_2 y_3$$

$$+ \frac{-142755 - 6860\, s + 3136\, s^2 - 38675\, t}{84000} y_2 y_4$$

$$+ \frac{1617 - 320\, s - 960\, t}{1920} y_2 y_5 - \frac{800 + 40\, s + 448\, s^2 - 1525\, t}{2000} y_2 y_6$$

$$+ \frac{-195 + 46\, s}{50} y_2 y_7 + \frac{321}{80} y_2 y_8$$

$$+ \frac{38025 - 2560\, s + 1792\, s^2 - 14100\, t}{16000} y_3 y_4 +$$

$$\frac{15075 - 1920\, s + 896\, s^2 - 7050\, t}{4000} y_3 y_5 + \frac{3\,(45 - 16\, s)}{50} y_3 y_6$$

$$- \frac{189}{16} y_3 y_7 + \frac{1377}{160} y_4 y_5 + \frac{1377}{80} y_4 y_6 -$$

$$\frac{6486150 - 22287100\, s - 5727456\, s^2 - 1128960\, s^3 - 32661825\, t}{10080000} y_2^2$$

$$+ \frac{11881800\, s t}{10080000} y_2^2 + \frac{-627615 - 13440\, s + 12544\, s^2 - 98700\, t}{336000} y_3^2 +$$

$$\frac{459}{160} y_4^2$$

$f_2(y_1, \ldots, y_{10})\ =$

$$\frac{-69500 - 3263904\,s^2 - 376320\,s^3 + 6044325\,t - 3960600\,s\,t}{3360000}\,y_3 +$$

$$\frac{21465\,s - 6272\,s^3 - 66010\,s\,t}{28000}\,y_4 +$$

$$\frac{-8843500 - 2685984\,s^2 + 7797825\,t}{1680000}\,y_5 + \frac{198\,s}{875}\,y_6 +$$

$$\frac{15775 - 6272\,s^2 - 34650\,t}{28000}\,y_7 + \frac{-28\,s}{25}\,y_8 + \frac{-19}{16}\,y_9 +$$

$$\frac{-767800 + 210464\,s^2 - 120825\,t}{160000}\,y_2 y_3 +$$

$$\frac{-433649\,s + 6272\,s^3 + 66010\,s\,t}{28000}\,y_2 y_4 +$$

$$\frac{519000 + 83104\,s^2 - 218825\,t}{40000}\,y_2 y_5 + \frac{517\,s}{14000}\,y_2 y_6 - \frac{4089}{560}\,y_2 y_7 +$$

$$\frac{-11391300 + 309792\,s^2 + 2217775\,t}{560000}\,y_3 y_4 + \frac{1671\,s}{350}\,y_3 y_5 +$$

$$\frac{21087}{1400}\,y_3 y_6 - \frac{5913}{560}\,y_4 y_5 +$$

$$\frac{-16554375 - 339872\,s + 150528\,s^3 + 1584240\,s\,t}{1344000}\,y_2^2 +$$

$$\frac{1671\,s}{700}\,y_3^2 + \frac{-33039\,s}{2800}\,y_4^2$$

$f_3(y_1, \ldots, y_{10})\ =$

$$\frac{3\,(2187 - 3584\,s)}{2800}\,y_4 - \frac{62901}{2800}\,y_5 + \frac{448\,s^2 - 1525\,t}{2000}\,y_6 + \frac{-23\,s}{25}\,y_7$$

$$-\frac{321}{80}\,y_8 + \frac{-41175 - 6272\,s^3 - 66010\,s\,t}{28000}\,y_2 y_3 +$$

$$\frac{11\,(-9184\,s^2 + 29075\,t)}{80000}\,y_2 y_4 + \frac{-19407\,s}{14000}\,y_2 y_5 - \frac{549}{560}\,y_2 y_6 +$$

$$\frac{3573\,s}{400}\,y_3 y_4 + \frac{76383}{1400}\,y_3 y_5 + \frac{-26336\,s^2 - 457825\,t}{80000}\,y_3^2 - \frac{2673}{112}\,y_4^2$$

$f_4(y_1, \ldots, y_{10})\ =$

$$\frac{-448\,s^2 + 3525\,t}{4000}\,y_5 + \frac{12\,s}{25}\,y_6 + \frac{189}{32}\,y_7 + \frac{26336\,s^2 + 457825\,t}{160000}\,y_2 y_3$$

$$-\frac{123183\,s}{28000}\,y_2 y_4 - \frac{35397}{1120}\,y_2 y_5 + \frac{6561}{1400}\,y_3 y_4 + \frac{448\,s^3 + 4715\,s\,t}{4000}\,y_2^2$$

$$f_5(y_1, \ldots, y_{10}) =$$

$$\frac{3157\,s^2}{15000}y_2^2 - \frac{12793\,t}{19200}y_2^2 - \frac{39\,s}{1750}y_2y_3 - \frac{2187}{1400}y_3^2 + \frac{891}{224}y_2y_4 - \frac{459}{160}y_6$$

$$f_6(y_1, \ldots, y_{10}) = \frac{6469\,s}{112000}y_2^2 + \frac{8073}{22400}y_2y_3$$

$$f_7(y_1, \ldots, y_{10}) = \frac{183}{22400}y_2^2$$

$$f_8(y_1, \ldots, y_{10}) = 0$$

$$f_9(y_1, \ldots, y_{10}) = 0$$

$$f_{10}(y_1, \ldots, y_{10}) = 0$$

$$f_{11}(y_1, \ldots, y_{10}) = \frac{1}{90720000}.$$

Theorem 4.5.17 *Let $\mathcal{A}(-2, s, t)$ be the Lie group which stands in one-to-one correspondence with the Lie algebra $\mathfrak{a}(-2, s, t)$. If $s, z \in \mathbb{Z}$, then $\mathcal{A}(-2, s, t)$ acts simply transitively and via polynomial diffeomorphisms of degree ≤ 2 on $\mathbb{R}^{11}$. So, if $s \neq 0$, then any lattice of $\mathcal{A}(-2, s, t)$ is a group of affine defect one.*

<u>Proof:</u> By the computations above we know that $\mathcal{A}(-2, s, t)$ contains a lattice Γ which embeds nicely into a Lie group P of polynomial diffeomorphisms of degree ≤ 2. Observe that this group P, introduced above, is nilpotent. This can be seen as follows. Let P_1 be the normal subgroup of P consisting of those polynomial diffeomorphisms for which $p_i \equiv 0$ ($1 \leq i \leq 10$). So P_1 consists of those polynomials of $\mathbb{R}^{11}$ which only affect 1 coordinate. Denote by P_2 the subgroup of P determined by those polynomials for which $p_{11} \equiv 0$. So P_2 is isomorphic to the group of unitriangular 11×11–matrices. There is a split short exact sequence

$$1 \to P_1 \to P \to P_2 \to 1.$$

P_2 is a nilpotent group, and so $\gamma_n(P) \subseteq P_1$ if n is big enough. It is now an exercise to see that further commutator subgroups will gradually get smaller and smaller until, at the end, they vanish.

Using [33, Proposition 2.5], the embedding $\Gamma \to P$ extends uniquely to an embedding of the Lie group $\mathcal{A}(-2, s, t) \to P$, making it acting properly discontinuously on $\mathbb{R}^{11}$, which was to be shown. ∎

Let us finally point out that the use of MATHEMATICA® [61] was quite crucial in setting up and performing the computations above. A set of procedures enabling the reader to verify these results is available on request from the author.

Chapter 5

The Cohomology of virtually nilpotent groups

5.1 The need of cohomology computations

For the classification of the infra–nilmanifolds (or AB–groups) we will make intensive use of the reduction lemma 2.4.2 (see section 2.4 "The first proof revisited"). This lemma tells us that an AC–group E having a Fitting subgroup N of nilpotency class c can be obtained as an extension

$$1 \to Z \to E \to E/Z \to 1.$$

Z is the free abelian group of finite rank defined as $Z = \sqrt[N]{\gamma_c(N)} = \gamma_c(G) \cap N$, where G denotes the Mal'cev completion of N. E/Z is also an AC–group, with Fitting subgroup N/Z of nilpotency class $c - 1$. The action of E/Z on Z, induced by the extension, is trivial when restricted to N/Z. This motivates the importance of having a way to compute $H^2(E/Z, Z)$ for actions of E/Z on Z which factor through E/N.

5.2 The cohomology for some virtually nilpotent groups

First, we develop the theory for crystallographic groups (i.e. the AC–groups with abelian Fitting subgroup) and at the end of the section, we make some considerations about how to generalize this theory to almost–crystallographic groups.

We already explained that an abstract crystallographic group Q is characterized algebraically by the fact that it contains a free abelian normal subgroup of finite rank, say n, which is maximal abelian and of finite index in Q, where n is the dimension of Q. So Q fits in a short exact sequence $0 \longrightarrow \mathbb{Z}^n \longrightarrow Q \longrightarrow F \longrightarrow 1$, where F is finite. Also the F-module structure of $\mathbb{Z}^n$ is faithful (i.e. F embeds in $\mathrm{Gl}(n, \mathbb{Z})$). Every group Q has a faithful affine representation $\rho : Q \to \mathrm{Aff}(\mathbb{R}^n)$ mapping $\mathbb{Z}^n$ into the group of pure translations. From now on, we fix such a representation ρ.

F is generated by a finite number of elements, which we denote by $\alpha_1, \alpha_2, \ldots, \alpha_k$. Elements of F will be written as a word in these generators; for each element α of F we fix a unique word $u(\alpha) = \alpha_{i_1}^{\pm 1} \alpha_{i_2}^{\pm 1} \ldots \alpha_{i_r}^{\pm 1}$ which represents it (take $u(1) = 1$). So, F can be looked at as

$$F : \; < \alpha_1, \alpha_2, \ldots, \alpha_k \; \| \; w_i(\alpha_1, \alpha_2, \ldots, \alpha_k) = 1 \; (1 \leq i \leq l) > .$$

$\mathbb{Z}^n$, the translation subgroup of Q, will be considered as determined by

$$\mathbb{Z}^n : \; < a_1, a_2, \ldots, a_n \; \| \; [a_i, a_j] = 1 \; (1 \leq j < i \leq n) > .$$

As a consequence, Q can be presented as

$$Q : \; < a_1, \ldots, a_n, \alpha_1, \ldots, \alpha_k \; \| \; [a_i, a_j] = 1 \; (1 \leq j < i \leq n)$$
$$w_i(\alpha_1, \alpha_2, \ldots, \alpha_k) = a_1^{l_{1,i}} a_2^{l_{2,i}} \ldots a_n^{l_{n,i}} \; (1 \leq i \leq l) \quad >$$
$$\alpha_i a_j \alpha_i^{-1} = w_{i,j}(a_1, a_2, \ldots, a_n) \; (1 \leq i \leq k, \; 1 \leq j \leq n)$$

and elements q of Q can be written uniquely as words

$$q = a_1^{q_1} a_2^{q_2} \ldots a_n^{q_n} u(\alpha) \; (q_j \in \mathbb{Z}, \; 1 \leq j \leq n, \; \alpha \in F).$$

We are interested in computing $H_\varphi^2(Q, \mathbb{Z}^m)$, for an action $\varphi : Q \to \mathrm{Aut}(\mathbb{Z}^m)$ which factors through the finite group F. From a conceptual point of view, let us observe the following result:

Proposition 5.2.1 *Assume Q is a crystallographic group acting on $\mathbb{Z}^m$ via $\varphi : Q \to \mathrm{Aut}(\mathbb{Z}^m)$ such that $\varphi_{|\mathbb{Z}^n} = 1$. Write* res $: H_\varphi^2(Q, \mathbb{Z}^m) \to H_{tr}^2(\mathbb{Z}^n, \mathbb{Z}^m)$ *for the restriction morphism. Then ker(res) is the torsion part of $H_\varphi^2(Q, \mathbb{Z}^m)$ and we can write*

$$H_\varphi^2(Q, \mathbb{Z}^m) \cong Im(res) \oplus ker(res),$$

where the torsion free part has rank $\leq mn(n-1)/2$.

Proof: The proof is immediate once we realize that $H_{tr}^2(\mathbb{Z}^n, \mathbb{Z}^m) \cong \mathbb{Z}^{mn(n-1)/2}$ is torsion free and that $\mathbb{Z}^n$ is of finite index in Q. ∎

In view of computing $H_\varphi^2(Q, \mathbb{Z}^m)$ in practice, let us consider an extension

$$0 \longrightarrow \mathbb{Z}^m \longrightarrow E \longrightarrow Q \longrightarrow 1$$

compatible with φ. E has a presentation

$$
\begin{aligned}
E: \ &< \alpha_1 \ldots, \alpha_k, a_1, \ldots, a_n, b_1, \ldots, b_m \ \| \qquad\qquad\qquad > \\
&[b_i, b_j] = 1 \ (1 \le j < i \le m) \\
&[a_i, a_j] = b_1^{k_{1,i,j}} b_2^{k_{2,i,j}} \ldots b_m^{k_{m,i,j}} \ (1 \le j < i \le n) \\
&[b_i, a_j] = 1 \ (1 \le i \le m, \ 1 \le j \le n) \\
&\alpha_i b_j \alpha_i^{-1} = \varphi(\alpha_i)(b_j) \ (1 \le i \le k, \ 1 \le j \le m) \\
&w_i(\alpha_1, \alpha_2, \ldots, \alpha_k) = a_1^{l_{1,i}} \ldots a_n^{l_{n,i}} b_1^{k_{1,i}} \ldots b_m^{k_{m,i}} \ (1 \le i \le l) \\
&\alpha_i a_j \alpha_i^{-1} = w_{i,j}(a_1, a_2, \ldots, a_n) b_1^{k'_{1,i,j}} \ldots b_m^{k'_{m,i,j}} \\
&(1 \le i \le k, 1 \le j \le n)
\end{aligned}
$$

$$(5.1)$$

which is completely determined by the integers $k_{r,i,j}$, $k_{r,i}$ and $k'_{r,i,j}$. As we already explained in the example of section 2.4 these integers cannot be chosen completely free, since there should be **computational consistency**. Let us look at another example to make this more clear.

Example 5.2.2

Take $Q \cong \mathbb{Z}^2 \rtimes \mathbb{Z}_2$ given as

$$Q: \ < a_1, a_2, \alpha \ \| \ [a_2, a_1] = 1, \ \alpha a_1 \alpha^{-1} = a_1, \ \alpha a_2 \alpha^{-1} = a_2^{-1}, \ \alpha^2 = 1 >$$

Here $F = \mathbb{Z}_2$. Take $\varphi: Q \to \mathrm{Aut}\,\mathbb{Z}$ such that it induces the non-trivial action of F on $\mathbb{Z}$. Then, an extension E has a presentation:

$$
\begin{aligned}
E: \ &< a_1, a_2, b, \alpha \ \| \ [a_2, a_1] = b^{k_1}, \ [b, a_1] = [b, a_2] = 1, \ \alpha a_1 \alpha^{-1} = a_1 b^{k_2} > \\
&\alpha a_2 \alpha^{-1} = a_2^{-1} b^{k_3}, \ \alpha b \alpha^{-1} = b^{-1}, \ \alpha^2 = b^{k_4}
\end{aligned}
$$

Now, note (in E) that $\alpha(\alpha^2) = (\alpha^2)\alpha \Rightarrow \alpha b^{k_4} = b^{k_4}\alpha \Rightarrow b^{-k_4}\alpha = b^{k_4}\alpha$ from which it follows that $k_4 = 0$ showing that not all choices of (k_1, k_2, k_3, k_4) can be accepted. A 4-tuple (k_1, k_2, k_3, k_4), which really determines an extension of Q by $\mathbb{Z}$ is said to be a computational consistent 4-tuple.

We return to the general set up. Fix an ordering for the parameters $k_{r,i,j}$, $k_{r,i}$ and $k'_{r,i,j}$ in (5.1). Assume there are p parameters. This means that E is determined completely by a p-tuple of integers which we denote from now on by K. We refer to $E(K)$ as the group E determined by K. As in the example, we refer to a p-tuple K for which there exists a group $E(K)$ (as an extension of $\mathbb{Z}^m$ by Q) as a **computational consistent** p-tuple.

The elements of $E(K)$ can be written uniquely as words

$$b_1^{p_1} \ldots b_m^{p_m} a_1^{q_1} \ldots a_n^{q_n} u(\alpha), \quad (p_i, q_j \in \mathbb{Z}).$$

Take a section $s : Q \longrightarrow E(K) : q = a_1^{q_1} a_2^{q_2} \ldots a_n^{q_n} u(\alpha) \mapsto b_1^0 \ldots b_m^0 q$, which we will refer to as **the standard section**.

The cocycle $f_K : Q \times Q \to \mathbb{Z}^m$, determined by the standard section is called a **standard cocycle**. The set of standard cocycles $\{f_K \| K$ computational consistent $\}$ will be denoted by $SZ_\varphi^2(Q, \mathbb{Z}^m)$.

Since every extension of $\mathbb{Z}^m$ by Q is equivalent to some $E(K)$, every cocycle in $Z_\varphi^2(Q, \mathbb{Z}^m)$ is cohomologous to a standard cocycle. So it will be sufficient to work with $SZ_\varphi^2(Q, \mathbb{Z}^m)$ to obtain the cohomology group. The following proposition shows that computing $H_\varphi^2(Q, \mathbb{Z}^m)$ based on these standard cocycles is possible without too much difficulties and, in fact, reduces to a linear problem.

Proposition 5.2.3 $H_\varphi^2(Q, \mathbb{Z}^m)$ *can be computed algorithmically, since*

1. $SZ_\varphi^2(Q, \mathbb{Z}^m)$ *is a subgroup of* $Z_\varphi^2(Q, \mathbb{Z}^m)$. *Moreover, if* K_1 *and* K_2 *are computational consistent, then* $K_1 + K_2$ *is computational consistent and* $f_{K_1 + K_2} = f_{K_1} + f_{K_2}$.

2. *A standard cocycle* f_K *is a coboundary if and only if*

    (a) *the parameters* $k_{r,i,j}$ *(in 5.1)* $(1 \le j < i \le n, 1 \le r \le m)$ *are zero*

    (b) K *is an integral solution of a well determined finite set of linear equations.*

Proof of 1: By remark 4.4.25 we know that every $E(K)$ has a faithful representation into $\mathrm{Aff}(\mathbb{R}^{m+n})$. Although not unique, this representation is of canonical type. We can now exploit this information to understand the proposition. Elements of $\mathrm{Aff}(\mathbb{R}^{m+n})$ will as usual be represented by a $(m+n+1) \times (m+n+1)$ matrix with bottom row $(0, 0, \ldots, 0, 1)$.

Now, assume $\lambda : E(K) \to \text{Aff}(\mathbb{R}^{m+n})$ is a canonical type representation for $E(K)$. Then,

$$\lambda(b_j) = \begin{pmatrix} I_m & \tilde{B}_j \\ 0 & I_{n+1} \end{pmatrix} = \begin{pmatrix} I_m & 0 & u_j \\ 0 & I_n & 0 \\ 0 & 0 & 1 \end{pmatrix} \overset{\text{not}}{=} B_j$$

where $u_j = (0, 0, \ldots, 0, 1, 0, \ldots, 0)^{\text{tr}}$ having 1 on the j-th spot. Furthermore,

$$\lambda(a_j) = \begin{pmatrix} \varphi(a_j) = I_m & \tilde{A}_j \\ 0 & \rho(a_j) \end{pmatrix} \overset{\text{not}}{=} A_j \quad \text{and}$$

$$\lambda(\alpha_j) = \begin{pmatrix} \varphi(\alpha_j) & \tilde{C}_j \\ 0 & \rho(\alpha_j) \end{pmatrix} \overset{\text{not}}{=} C_j.$$

In order to know λ, we compute the entries of $\tilde{A}_j$ and $\tilde{C}_j$ by requiring that the matrices A_j, B_j and C_j satisfy all the relations of (5.1) when substituted for a_j, b_j and α_j respectively.

Remark that the relations in (5.1), not involving components of K, are satisfied automatically (because of the type of λ). The other relations can be written in a form

$$w(a_1, a_2, \ldots, a_n, \alpha_1, \ldots, \alpha_k) = b_1^{k_1} \ldots b_m^{k_m},$$

where $k_1, \ldots, k_m$ are some of the components of K.
  Since

$$A_j^{-1} = \begin{pmatrix} I_m & -\tilde{A}_j \rho(a_j^{-1}) \\ 0 & \rho(a_j^{-1}) \end{pmatrix} \quad \text{and}$$

$$C_j^{-1} = \begin{pmatrix} \varphi(\alpha_j^{-1}) & -\varphi(\alpha_j^{-1}) \tilde{C}_j \rho(\alpha_j^{-1}) \\ 0 & \rho(\alpha_j^{-1}) \end{pmatrix},$$

we see that

$$w(A_1, A_2, \ldots, C_k) = \begin{pmatrix} I_m & P(\tilde{A}_1, \tilde{A}_2, \ldots, \tilde{C}_k) \\ 0 & I_{n+1} \end{pmatrix}$$

where P is a matrix whose entries are linear combinations of the entries of $\tilde{A}_i$ $(1 \le i \le n)$ and $\tilde{C}_j$ $(1 \le j \le k)$ with coefficients determined by $\varphi(\alpha_j)$, $\rho(\alpha_j)$ and $\rho(a_i)$. We have to claim that

$$P(\tilde{A}_1, \tilde{A}_2, \ldots, \tilde{C}_k) = \begin{pmatrix} 0 & 0 & \ldots & 0 & k_1 \\ 0 & 0 & \ldots & 0 & k_2 \\ \vdots & \vdots & & \vdots & \vdots \\ 0 & 0 & \ldots & 0 & k_m \end{pmatrix}.$$

Hence every relation of (5.1) gives rise to a set of linear equations, the variables being the entries of $\tilde{A}_j$ and $\tilde{C}_j$. Each of these equations is of the form

$$\sum_{j,l,m} f_{j,l,m}(\tilde{A}_j)_{l,m} + \sum_{j,l,m} g_{j,l,m}(\tilde{C}_j)_{l,m} = h(K) \tag{5.2}$$

with $h(K) = 0$ or $h(K) = k_i$ $(1 \leq i \leq m)$. (The coefficients $f_{j,l,m}$ and $g_{j,l,m}$, independent of $\tilde{A}_j$ and $\tilde{C}_j$, are well determined by φ and ρ.)

It is not hard to see that this system of equations has a solution for a particular choice of K if and only if this K is computational consistent. In fact, this is a practical way to determine all computational consistent K's.

Assume that $\tilde{A}_{j,1}$ (resp. $\tilde{A}_{j,2}$) and $\tilde{C}_{j,1}$ (resp. $\tilde{C}_{j,2}$) are solutions to (5.2) with $K = K_1$ (resp. $K = K_2$). Then it is clear that $\tilde{A}_{j,1} + \tilde{A}_{j,2}$ and $\tilde{C}_{j,1} + \tilde{C}_{j,2}$ form a solution to (5.2) for $K = K_1 + K_2$. As a consequence the set of computational consistent K's forms a subgroup of $\mathbb{Z}^p$ and hence is a free abelian group of rank $s \leq p$. Choose s generators $K_1, K_2, \ldots, K_s$ for this group and assume $\tilde{A}_j(K_i)$ and $\tilde{C}_j(K_i)$ are solutions to (5.2) for $K = K_i$ $(1 \leq i \leq s)$.

One can now write any computational consistent p-tuple K as

$$K = l_1 K_1 + l_2 K_2 + \cdots + l_s K_s \quad \text{for some (unique) } l_1, l_2, \cdots, l_s \in \mathbb{Z}$$

according to which $\tilde{A}_j(K) = \sum_{i=1}^{s} l_i \tilde{A}_j(K_i)$ and $\tilde{C}_j(K) = \sum_{i=1}^{s} l_i \tilde{C}_j(K_i)$ are solutions to (5.2).

Having obtained a faithful representation λ for each $E(K)$, we use λ to compute the corresponding special 2-cocycle $f_K(x,y)$, defined by $s(x).s(y).s(xy)^{-1} = f_K(x,y)$, $\forall x, y \in Q$. For $x, y \in Q$, the left side of this equality is a word in $a_1, a_2, \ldots, a_n, \alpha_1, \ldots, \alpha_k$ and so

$$\lambda(s(x).s(y).s(xy)^{-1}) = \begin{pmatrix} I_m & R(\tilde{A}_1, \tilde{A}_2, \ldots, \tilde{C}_k) \\ 0 & I_{n+1} \end{pmatrix}$$

where $R(\tilde{A}_1, \tilde{A}_2, \ldots, \tilde{C}_k)$ is a matrix having entries which are linear in $l_1, l_2, \ldots, l_s$. However since $R(\tilde{A}_1, \tilde{A}_2, \ldots, \tilde{C}_k) = (0, 0, \ldots, 0, f_K(x,y))$ this implies that $f_K(x,y)$ is linear in $l_1, \ldots, l_s$. And so we may conclude that $SZ^2_\varphi(Q, \mathbb{Z}^m)$ is a subgroup of $Z^2_\varphi(Q, \mathbb{Z}^m)$ generated by s elements.

<u>Proof of 2:</u> To understand part a), note that if $< f_K > = 0 \in H^2_\varphi(Q, \mathbb{Z}^m)$, then also $\text{res}(< f_K >) = 0 \in H^2_{tr}(\mathbb{Z}^n, \mathbb{Z}^m)$. It is well known that the $m \cdot$

$\dfrac{n(n-1)}{2}$ parameters $k_{r,i,j}$ uniquely identify the elements of $H^2_{tr}(\mathbb{Z}^n, \mathbb{Z}^m)$. This proves the claim.

Let us now try to understand b). Assume $f_K(x,y) = \delta\gamma(x,y) =^x \gamma(y) - \gamma(xy) + \gamma(x)$ for some 1-cochain $\gamma : Q \to \mathbb{Z}^m$. It is clear that, for each $\alpha \in Q$,

$$0 = f_K(a_1^{q_1} a_2^{q_2} \ldots a_n^{q_n}, u(\alpha)) = \gamma(a_1^{q_1} \ldots a_n^{q_n}) + \gamma(u(\alpha)) - \gamma(a_1^{q_1} \ldots a_n^{q_n} u(\alpha)),$$

which implies that

$$\gamma(a_1^{q_1} \ldots a_n^{q_n} u(\alpha)) = \gamma(a_1^{q_1} \ldots a_n^{q_n}) + \gamma(u(\alpha)).$$

In a similar way, one verifies easily that

$$\gamma(a_1^{q_1} \ldots a_n^{q_n}) = q_1 \gamma(a_1) + \cdots + q_n \gamma(a_n).$$

So, we have

$$\gamma(a_1^{q_1} \ldots a_n^{q_n} u(\alpha)) = q_1 \gamma(a_1) + \cdots + q_n \gamma(a_n) + \gamma(u(\alpha)) \tag{5.3}$$

showing that γ is completely determined by a finite subset of $\mathbb{Z}^m$ i.e.

$$\gamma(a_1), \gamma(a_2), \ldots, \gamma(a_n) \quad \text{and} \quad \gamma(u(\alpha)), \ \forall \alpha \in F.$$

Moreover, the affine representation λ for $E(K)$, allows us to compute the expression $f_K(a_1^{x_1} \ldots a_n^{x_n} u(\alpha), a_1^{y_1} \ldots a_n^{y_n} u(\beta))$ for fixed α and β and for arbitrary $x_1, \ldots, x_n, y_1, \ldots y_n$. Here it will be necessary to use the identity

$$A_i^x = \sum_{l=0}^{m+n} \binom{x}{l} (A_i - I_{m+n+1})^l$$

(I_{m+n+1} is the $(m+n+1)$–identity–matrix).

Now, the problem of finding all K such that $f_K \sim 0$ is transformed to finding all K for which the finite set of linear equations (in the variables $\gamma(a_1), \ldots, \gamma(a_n)$ and $\gamma(u(\alpha)) \ \forall \alpha \in F$)

$$f_K(a_1^{x_1} \ldots a_n^{x_n} u(\alpha), a_1^{y_1} \ldots a_n^{y_n} u(\beta)) = \delta\gamma(a_1^{x_1} \ldots a_n^{x_n} u(\alpha), a_1^{y_1} \ldots a_n^{y_n} u(\beta)).$$
$$\tag{5.4}$$

has an integral solution. We have one equation for each pair $(u(\alpha), u(\beta))$. Notice also that both sides of (5.4) are polynomials in the variables $x_1, \ldots, y_n$.

Although this problem seems quite difficult to solve in general, it is often possible to simplify a lot. Also the key information of the above observations is that the problem has been linearized. Let us continue this section with two examples:

Example 5.2.4 Take $Q \cong \mathbb{Z}^2 \rtimes \mathbb{Z}_2$ as

$$Q : \; < a, b, \alpha \parallel [b, a] = 1, \; \alpha a = a^{-1}\alpha, \; \alpha b = b^{-1}\alpha, \; \alpha^2 = 1 > .$$

$F = \mathbb{Z}_2$ and generated by α. Take $u(\alpha) = \alpha$. Consider the trivial action of Q on $\mathbb{Z}$. Then an extension E has a presentation of the form

$$E : \; < a, b, c, \alpha \parallel \begin{array}{ll} [b, a] = c^{k_1} & \alpha c = c\alpha \\ [c, a] = 1 & [c, b] = 1 \\ \alpha a = a^{-1}\alpha c^{k_2} & \alpha b = b^{-1}\alpha c^{k_3} \\ \alpha^2 = c^{k_4} \end{array} > .$$

Every group E is determined by four parameters $(k_1, k_2, k_3, k_4) = K$. In this case there are no consistency conditions on K since we know a faithful representation $\lambda : E(K) \to \mathrm{Aff}(\mathbb{R}^3)$:

$$\lambda(a) = \begin{pmatrix} 1 & 0 & \frac{-k_1}{2} & 0 \\ 0 & 1 & 0 & 1 \\ 0 & 0 & 1 & 0 \\ 0 & 0 & 0 & 1 \end{pmatrix} \qquad \lambda(b) = \begin{pmatrix} 1 & \frac{k_1}{2} & 0 & 0 \\ 0 & 1 & 0 & 0 \\ 0 & 0 & 1 & 1 \\ 0 & 0 & 0 & 1 \end{pmatrix}$$

$$\lambda(c) = \begin{pmatrix} 1 & 0 & 0 & 1 \\ 0 & 1 & 0 & 0 \\ 0 & 0 & 1 & 0 \\ 0 & 0 & 0 & 1 \end{pmatrix} \qquad \lambda(\alpha) = \begin{pmatrix} 1 & k_2 & k_3 & \frac{k_4}{2} \\ 0 & -1 & 0 & 0 \\ 0 & 0 & -1 & 0 \\ 0 & 0 & 0 & 1 \end{pmatrix} .$$

We could proceed by computing special cocycles via this representation, but in fact we can as well calculate $f_K(x, y)$ in a direct way. Let $x = a^{x_1} b^{x_2} \alpha^{x_3}$ (resp. $y = a^{y_1} b^{y_2} \alpha^{y_3}$), where x_3 (resp. y_3) is calculated modulo 2. Verifying

$$a^{x_1} b^{x_2} \alpha^{x_3} a^{y_1} b^{y_2} \alpha^{y_3} = c^{(-1)^{x_3} x_2 y_1 k_1 + x_3 y_1 k_2} a^{x_1 + (-1)^{x_3} y_1} b^{x_2} \alpha^{x_3} b^{y_2} \alpha^{y_3}$$

$$= c^{(-1)^{x_3} x_2 y_1 k_1 + x_3 y_1 k_2 + x_3 y_2 k_3}$$

$$a^{x_1 + (-1)^{x_3} y_1} b^{x_2 + (-1)^{x_3} y_2} \alpha^{x_3 + y_3}$$

$$= c^{(-1)^{x_3} x_2 y_1 k_1 + x_3 y_1 k_2 + x_3 y_2 k_3 + x_3 y_3 k_4}$$

$$a^{x_1 + (-1)^{x_3} y_1} b^{x_2 + (-1)^{x_3} y_2} \alpha^{(x_3 + y_3) \bmod 2},$$

we find $f(x, y) = (-1)^{x_3} x_2 y_1 k_1 + x_3 y_1 k_2 + x_3 y_2 k_3 + x_3 y_3 k_4$.
Now we investigate when $f(x, y) = \delta\gamma(x, y)$, for $\gamma(a^{x_1} b^{x_2} \alpha^{x_3}) = x_1 \gamma(a) + x_2 \gamma(b) + x_3 \gamma(\alpha)$ (see (5.3), x_3 is either 0 or 1). First we calculate

$$\delta\gamma(x, y) = (x_1 + y_1)\gamma(a) + (x_2 + y_2)\gamma(b) + (x_3 + y_3)\gamma(\alpha)$$

$$-(x_1 + (-1)^{x_3} y_1)\gamma(a) - (x_2 + (-1)^{x_3} y_2)\gamma(b)$$
$$-((x_3 + y_3) \bmod 2)\gamma(\alpha)$$
$$= (y_1 - (-1)^{x_3} y_1)\gamma(a) + (y_2 - (-1)^{x_3} y_2)\gamma(b)$$
$$+(x_3 + y_3 - (x_3 + y_3) \bmod 2)\gamma(\alpha).$$

In this case we get four equations of the type described in (5.4), namely

1. $x_3 = 0$, $y_3 = 0 \Rightarrow k_1 = 0$.

2. $x_3 = 1$, $y_3 = 0 \Rightarrow \gamma(a) = \frac{k_2}{2}$ and $\gamma(b) = \frac{k_3}{2} \Rightarrow k_2, k_3 \in 2\mathbb{Z}$.

3. $x_3 = 1$, $y_3 = 1 \Rightarrow \gamma(\alpha) = \frac{k_4}{2} \Rightarrow k_4 \in 2\mathbb{Z}$.

4. $x_3 = 0$, $y_3 = 1$. This equation does not imply new conditions.

This allows us to conclude with:

$$H^2(Q, \mathbb{Z}) = \mathbb{Z} \oplus (\mathbb{Z}_2)^3.$$

Example 5.2.5 Take $Q \cong \mathbb{Z}^2 \rtimes \mathbb{Z}_2$ as in the previous example.
Consider $\varphi : Q \to \mathrm{Aut}\mathbb{Z}^2$, given by $\varphi(\alpha) = \begin{pmatrix} 0 & 1 \\ 1 & 0 \end{pmatrix}$. Extensions $E(K)$
are of the form

$$E(K) : <a, b, c, d, \alpha \parallel \begin{array}{ll} [d,a] = 1 & [d,b] = 1 \\ [d,c] = 1 & [c,a] = 1 \\ [c,b] = 1 & [b,a] = c^{k_1} d^{l_1} \\ \alpha a = a^{-1} \alpha c^{k_2} d^{l_2} & \alpha b = b^{-1} \alpha c^{k_3} d^{l_3} \\ \alpha^2 = c^{k_4} d^{l_4} & \alpha c = d\alpha \\ \alpha d = c\alpha \end{array} \quad > .$$

However, not every $K = (k_1, k_2, k_3, k_4, l_1, l_2, l_3, l_4)$ is computational consistent. It is not difficult to get the following consistency conditions:

1. From the above presentation we deduce that

$$ba = abc^{k_1} d^{l_1} \Rightarrow \alpha ba \alpha^{-1} = \alpha abc^{k_1} d^{l_1} \alpha^{-1}$$
$$\Rightarrow b^{-1} a^{-1} c^{l_2 + l_3} d^{k_2 + k_3} = a^{-1} b^{-1} c^{l_1 + l_2 + l_3} d^{k_1 + k_2 + k_3}$$

and so it follows that $k_1 = l_1$ since $b^{-1} a^{-1} = a^{-1} b^{-1} c^{k_1} d^{l_1}$.

2. Notice that $\alpha a \alpha^{-1} = a^{-1} c^{l_2} d^{k_2}$ and thus $\alpha a^{-1} \alpha^{-1} = ac^{-l_2} d^{-k_2}$.
   Use this in

$$\alpha a = a^{-1} \alpha c^{k_2} d^{l_2} \Rightarrow \alpha a \alpha \alpha^{-1} = \alpha a^{-1} \alpha^{-1} \alpha^2 c^{k_2} d^{l_2} \alpha^{-1}$$
$$\Rightarrow ac^{k_4} d^{l_4} = ac^{-l_2 + k_4 + k_2} d^{-k_2 + l_4 + l_2} \text{ which implies } k_2 = l_2.$$

3. Similarly one finds that $k_3 = l_3$.

4. Also $\alpha.\alpha^2 = \alpha^2.\alpha \Rightarrow \alpha c^{k_4} d^{l_4} = c^{k_4} d^{l_4} \alpha \Rightarrow k_4 = l_4$.

These four conditions are sufficient for K to be computational consistent, since for each K we have the following faithful representation λ : $E \rightarrow \mathrm{Aff}(\mathbb{R}^4)$:

$$
a \mapsto \begin{pmatrix} 1 & 0 & 0 & \frac{-k_1}{2} & 0 \\ 0 & 1 & 0 & \frac{-k_1}{2} & 0 \\ 0 & 0 & 1 & 0 & 1 \\ 0 & 0 & 0 & 1 & 0 \\ 0 & 0 & 0 & 0 & 1 \end{pmatrix}
\qquad
b \mapsto \begin{pmatrix} 1 & 0 & \frac{k_1}{2} & 0 & 0 \\ 0 & 1 & \frac{k_1}{2} & 0 & 0 \\ 0 & 0 & 1 & 0 & 0 \\ 0 & 0 & 0 & 1 & 1 \\ 0 & 0 & 0 & 0 & 1 \end{pmatrix}
$$

$$
c \mapsto \begin{pmatrix} 1 & 0 & 0 & 0 & 0 \\ 0 & 1 & 0 & 0 & 1 \\ 0 & 0 & 1 & 0 & 0 \\ 0 & 0 & 0 & 1 & 0 \\ 0 & 0 & 0 & 0 & 1 \end{pmatrix}
\qquad
d \mapsto \begin{pmatrix} 1 & 0 & 0 & 0 & 1 \\ 0 & 1 & 0 & 0 & 0 \\ 0 & 0 & 1 & 0 & 0 \\ 0 & 0 & 0 & 1 & 0 \\ 0 & 0 & 0 & 0 & 1 \end{pmatrix}
$$

$$
\alpha \mapsto \begin{pmatrix} 0 & 1 & k_2 & k_3 & \frac{k_4}{2} \\ 1 & 0 & k_2 & k_3 & \frac{k_4}{2} \\ 0 & 0 & -1 & 0 & 0 \\ 0 & 0 & 0 & -1 & 0 \\ 0 & 0 & 0 & 0 & 1 \end{pmatrix}.
$$

Let $x = a^{x_1} b^{x_2} \alpha^{x_3}$ (resp. $y = a^{y_1} b^{y_2} \alpha^{y_3}$), where x_3 (resp. y_3) is equal to 0 or 1. $f_K(x,y)$ can be calculated from the matrix representation above. This results in

$$
f : Q \times Q \rightarrow \mathbb{Z}^2 : (x,y) \mapsto \begin{pmatrix} (-1)^{x_3} x_2 y_1 k_1 + x_3 y_1 k_2 + x_3 y_2 k_3 + x_3 y_3 k_4 \\ (-1)^{x_3} x_2 y_1 k_1 + x_3 y_1 k_2 + x_3 y_2 k_3 + x_3 y_3 k_4 \end{pmatrix}.
$$

Searching which $f = \delta \gamma$ for $\gamma : Q \rightarrow \mathbb{Z}^2$ easily implies $k_1 = 0$ (see also part 2.(a) in proposition 5.2.3). Moreover, if $k_1 = 0$ then $f_K = \delta \gamma$ for

$$
\gamma : Q \rightarrow \mathbb{Z}^2 : a^{x_1} b^{x_2} \alpha^{x_3} \mapsto x_1 \begin{pmatrix} k_2 \\ 0 \end{pmatrix} + x_2 \begin{pmatrix} k_3 \\ 0 \end{pmatrix} + x_3 \begin{pmatrix} k_4 \\ 0 \end{pmatrix}.
$$

We conclude:

$$
H_\varphi^2(Q, \mathbb{Z}^2) = \mathbb{Z}.
$$

Now, we do not longer restrict ourselves to the virtually abelian case, but we suppose that Q is an almost–crystallographic group, with a group N of rank n as its normal, maximal nilpotent subgroup. The way to compute $H^2(Q, \mathbb{Z})$ will be completely the same as the way we followed in the proof of proposition 5.2.3.

We suppose that a canonical type representation $\rho : Q \to \text{Aff}(\mathbb{R}^n)$ is provided for Q. (This is certainly possible if N is of class ≤ 3.) Again, we assume that an action $\varphi : Q \to \text{Aut}(\mathbb{Z}^m)$ is given, with $\varphi(N) = 1$. Moreover, we suppose that the following, rather restrictive, property is fulfilled, namely we assume that the homomorphism

$$H^2(Q, \mathbb{Z}^m) \longrightarrow H^2(Q, \text{Aff}(\mathbb{R}^n, \mathbb{R}^m))$$

is trivial. This assumption is needed, in order to be able to lift the canonical representation ρ of Q to any extension of Q by $\mathbb{Z}^m$, compatible with the given action. This condition is very complicated to investigate and involves the computational consistency conditions. We remark here however, that there is one fact which simplifies things quite a bit. We write this down in the following lemma.

Lemma 5.2.6 *Let G be a group, containing a normal subgroup H of finite index. Suppose there is an exact sequence of G–modules*

$$0 \to Z \xrightarrow{i} A \to A/Z \to 0$$

where A is also a vector space (over a field of characteristic 0). Then,

$$i_* : H^2(H, Z) \to H^2(H, A) \text{ is trivial}$$
$$\Downarrow$$
$$i_* : H^2(G, Z) \to H^2(G, A) \text{ is trivial.}$$

<u>Proof:</u> The proof is based on the following commutative diagram:

$$
\begin{array}{ccc}
H^2(G, Z) & \xrightarrow{i_*} & H^2(G, A) \\
\downarrow \text{res} & & \downarrow \text{res} \\
H^2(H, Z) & \xrightarrow{i_*} & H^2(H, A) \\
\downarrow \text{cor} & & \downarrow \text{cor} \\
H^2(G, Z) & \xrightarrow{i_*} & H^2(G, A)
\end{array}
$$

Since cor $\circ$ res=multiplication by the index $[G : H]$, we see that $[G : H]i_* = 0$. Now we use the fact that A is a vector space to conclude that i_* itself is 0. ∎

So in order to check the condition needed, we may restrict ourselves to the nilpotent group N. Once we have a group Q, satisfying the conditions above we can now repeat the procedure of proposition 5.2.3 to compute the group $H^2(Q, \mathbb{Z}^m)$.

As we know that any virtually nilpotent group admits a stable affine representation and by theorem 4.4.24, we may conclude that

Theorem 5.2.7 *Let Q be an almost–crystallographic group with a 2–step nilpotent Fitting subgroup. If $\varphi : Q \to \mathrm{Aut}\,(\mathbb{Z}^m)$ is any representation factoring through the holonomy group of Q, then there exists an algorithm to compute*

$$H^2_\varphi(Q, \mathbb{Z}^m).$$

In stead of writing out very carefully the proof of this theorem, it is more helpful to the reader to present an instructive example.

Example 5.2.8

Consider

$$Q \; : \; < a, b, c, \alpha \; \| \quad [b, a] = c^2 \quad [c, a] = [c, b] = 1 \; > .$$
$$\alpha a = a^{-1}\alpha \quad \alpha b = b^{-1}\alpha$$
$$\alpha^2 = c \qquad \alpha c = c\alpha$$

One can check that Q is an almost–crystallographic group and we propose the following stable representation ρ for Q:

$$\rho(a) = \begin{pmatrix} 1 & 0 & -1 & 0 \\ 0 & 1 & 0 & 1 \\ 0 & 0 & 1 & 0 \\ 0 & 0 & 0 & 1 \end{pmatrix} \quad \rho(b) = \begin{pmatrix} 1 & 1 & 0 & 0 \\ 0 & 1 & 0 & 0 \\ 0 & 0 & 1 & 1 \\ 0 & 0 & 0 & 1 \end{pmatrix}$$

$$\rho(c) = \begin{pmatrix} 1 & 0 & 0 & 1 \\ 0 & 1 & 0 & 0 \\ 0 & 0 & 1 & 0 \\ 0 & 0 & 0 & 1 \end{pmatrix} \quad \rho(\alpha) = \begin{pmatrix} 1 & 0 & 0 & 1/2 \\ 0 & -1 & 0 & 0 \\ 0 & 0 & -1 & 0 \\ 0 & 0 & 0 & 1 \end{pmatrix}.$$

We choose an action of Q on $\mathbb{Z}$ by letting a, b, c act trivially on $\mathbb{Z}$ and α non trivially.

As ρ is a stable affine representation, we know (theorem 4.4.24) that the map $H^2(Q, \mathbb{Z}) \to H^2(Q, \mathrm{Aff}(\mathbb{R}^3, \mathbb{R}^1))$ is trivial.

An extension of Q by $\mathbb{Z}$ inducing the action proposed above, has a presentation of the form:

$$E : \; < a, b, c, d, \alpha \; || \; \begin{array}{ll} [b,a] = c^2 d^{k_1} & [d,a] = 1 \\ [c,a] = d^{k_2} & [d,b] = 1 \\ [c,b] = d^{k_3} & [d,c] = 1 \\ \alpha a = a^{-1} \alpha d^{k_4} & \alpha^2 = c d^{k_7} \\ \alpha b = b^{-1} \alpha d^{k_5} & \alpha d = d^{-1} \alpha \\ \alpha c = c \alpha d^{k_6} & \end{array} \; > .$$

The first thing to do is to search the consistency conditions. This is done by checking for which k_i's we can construct a canonical type affine representation for E, based on ρ. The fact that we know that the morphism $H^2(Q, \mathbb{Z}) \to H^2(Q, \mathrm{Aff}(\mathbb{R}^3, \mathbb{R}^1))$ is zero, implies that we are able to construct a canonical type affine representation for a group E if and only if the parameters $(k_1, k_2, \ldots, k_7)$ are computational consistent. Using this procedure, we find the following equations, which should be satisfied:

$$\begin{cases} k_6 = -2k_7 \\ k_2 = 2k_4 \\ k_3 = 2k_5 \\ k_1 = 4(k_4 + k_5 + k_7). \end{cases}$$

We introduce the new symbols $l_1 = k_4, l_2 = k_5$ and $l_3 = k_7$. Now, an extension is determined by the three parameters l_1, l_2, l_3 and has a presentation

$$E : \; < a, b, c, d, \alpha \; || \begin{array}{lll} [b,a] = c^2 d^{2(l_1+l_2+l_3)} & [c,b] = d^{2l_2} & [c,a] = d^{2l_1} \\ [d,b] = 1 & [d,a] = 1 & [d,c] = 1 \\ \alpha a = a^{-1} \alpha d^{l_1} & \alpha b = b^{-1} \alpha d^{l_2} & \alpha d = d^{-1} \alpha \\ \alpha^2 = c d^{l_3} & \alpha c = c \alpha d^{-2l_3} & \end{array} \; > .$$

A representation for such a group E looks like

$$\lambda(a) = \begin{pmatrix} 1 & \frac{-4l_1}{3} & 0 & \frac{2l_1}{3} - l_3 & 0 \\ 0 & 1 & 0 & -1 & 0 \\ 0 & 0 & 1 & 0 & 1 \\ 0 & 0 & 0 & 1 & 0 \\ 0 & 0 & 0 & 0 & 1 \end{pmatrix} \quad \lambda(b) = \begin{pmatrix} 1 & \frac{-4l_2}{3} & \frac{-2l_2}{3} + l_3 & 0 & 0 \\ 0 & 1 & 1 & 0 & 0 \\ 0 & 0 & 1 & 0 & 0 \\ 0 & 0 & 0 & 1 & 1 \\ 0 & 0 & 0 & 0 & 1 \end{pmatrix}$$

$$\lambda(c) = \begin{pmatrix} 1 & 0 & \frac{2l_1}{3} & \frac{2l_2}{3} & 0 \\ 0 & 1 & 0 & 0 & 1 \\ 0 & 0 & 1 & 0 & 0 \\ 0 & 0 & 0 & 1 & 0 \\ 0 & 0 & 0 & 0 & 1 \end{pmatrix} \quad \lambda(d) = \begin{pmatrix} 1 & 0 & 0 & 0 & 1 \\ 0 & 1 & 0 & 0 & 0 \\ 0 & 0 & 1 & 0 & 0 \\ 0 & 0 & 0 & 1 & 0 \\ 0 & 0 & 0 & 0 & 1 \end{pmatrix}$$

$$\lambda(\alpha) = \begin{pmatrix} -1 & 2l_3 & \frac{-l_1}{3} & \frac{-l_2}{3} & 0 \\ 0 & 1 & 0 & 0 & \frac{1}{2} \\ 0 & 0 & -1 & 0 & 0 \\ 0 & 0 & 0 & -1 & 0 \\ 0 & 0 & 0 & 0 & 1 \end{pmatrix}.$$

Now we are ready to compute the cohomology group $H^2(Q,\mathbb{Z})$. Of course, we work with the section

$$s : Q \to E : a^x b^y c^z \alpha^\epsilon \mapsto a^x b^y c^z d^0 \alpha^\epsilon, \; x, y, z \in \mathbb{Z}, \epsilon \in \{0,1\}.$$

We compute the cocycles $f(m,n)$, $m, n \in Q$ with respect to this section and compare them with $\delta\gamma(m,n)$, where

$$\gamma : Q \to \mathbb{Z} : a^x b^y c^z \alpha^\epsilon \mapsto x\gamma(a) + y\gamma(b) + z\gamma(c) + \epsilon\gamma(\alpha).$$

This computation is done with the matrix representation and the result is summarized in the table below:

m	n	$f(m,n)$ ⸺⸺⸺⸺⸺⸺⸺⸺ $f(m,n) - \delta\gamma(m,n)$
$a^{x_1} b^{x_2} c^{x_3}$	$a^{y_1} b^{y_2} c^{y_3}$	$2l_3 x_2 y_1 + 2l_2 x_2^2 y_1 + 2l_1 x_3 y_1 + 2l_1 x_2 y_1^2 + 2l_2 x_3 y_2 + 4l_2 x_2 y_1 y_2$ <hr> $2\gamma(c)x_2 y_1 + 2l_3 x_2 y_1 + 2l_2 x_2^2 y_1 + 2l_1 x_3 y_1 + 2l_1 x_2 y_1^2 + 2l_2 x_3 y_2 + 4l_2 x_2 y_1 y_2$
$a^{x_1} b^{x_2} c^{x_3}$	$a^{y_1} b^{y_2} c^{y_3} \alpha$	$2l_3 x_2 y_1 + 2l_2 x_2^2 y_1 + 2l_1 x_3 y_1 + 2l_1 x_2 y_1^2 + 2l_2 x_3 y_2 + 4l_2 x_2 y_1 y_2$ <hr> $2\gamma(c)x_2 y_1 + 2l_3 x_2 y_1 + 2l_2 x_2^2 y_1 + 2l_1 x_3 y_1 + 2l_1 x_2 y_1^2 + 2l_2 x_3 y_2 + 4l_2 x_2 y_1 y_2$
$a^{x_1} b^{x_2} c^{x_3} \alpha$	$a^{y_1} b^{y_2} c^{y_3}$	$-(l_1 y_1) - 2l_3 x_2 y_1 - 2l_2 x_2^2 y_1 - 2l_1 x_3 y_1 + 2l_1 x_2 y_1^2 - l_2 y_2 - 2l_2 x_3 y_2 + 4l_2 x_2 y_1 y_2 + 2l_3 y_3$ <hr> $-(l_1 y_1) - 2\gamma(c)x_2 y_1 - 2l_3 x_2 y_1 - 2l_2 x_2^2 y_1 - 2l_1 x_3 y_1 + 2l_1 x_2 y_1^2 - l_2 y_2 - 2l_2 x_3 y_2 + 4l_2 x_2 y_1 y_2 + 2\gamma(c)y_3 + 2l_3 y_3$
$a^{x_1} b^{x_2} c^{x_3} \alpha$	$a^{y_1} b^{y_2} c^{y_3} \alpha$	$l_3 - l_1 y_1 - 2l_3 x_2 y_1 - 2l_2 x_2^2 y_1 - 2l_1 x_3 y_1 + 2l_1 x_2 y_1^2 - l_2 y_2 - 2l_2 x_3 y_2 + 4l_2 x_2 y_1 y_2 + 2l_3 y_3$ <hr> $\gamma(c) + l_3 - l_1 y_1 - 2\gamma(c)x_2 y_1 - 2l_3 x_2 y_1 - 2l_2 x_2^2 y_1 - 2l_1 x_3 y_1 + 2l_1 x_2 y_1^2 - l_2 y_2 - 2l_2 x_3 y_2 + 4l_2 x_2 y_1 y_2 + 2\gamma(c)y_3 + 2l_3 y_3$

The group of standard cocycles $SZ^2(Q,\mathbb{Z})$ is the free abelian group on three generators l_1, l_2 and l_3. The above table shows that a standard cocycle f of an extension determined by (l_1, l_2, l_3) is cohomologous to zero iff $l_1 = 0$ and $l_2 = 0$. (We have to take $\gamma(c) = -l_3$ to find a coborder $\delta\gamma$ which equals f). Therefore, we conclude that

$$H^2(Q, \mathbb{Z}) = \mathbb{Z}^3 / \text{grp}\,\{l_3\} \cong \mathbb{Z}^2$$

5.3 More about the cohomology of virtually a-belian groups

Again, let Q be any virtually finitely generated torsion free abelian group. (e.g. Q is a crystallographic group). Q fits in a short exact sequence

$$0 \longrightarrow \mathbb{Z}^n \longrightarrow Q \longrightarrow F \longrightarrow 1$$

such that F is a finite group. We suppose that F is generated by the elements $\alpha_1, \alpha_2, \ldots, \alpha_k$. We denote the action of F on $\mathbb{Z}^n$ via conjugation in Q, by $\rho : F \to \text{Aut}\,\mathbb{Z}^n$.

Theorem 5.3.1
For any action $\varphi : Q \to \text{Aut}\,\mathbb{Z}^m$ which factors through F (i.e. $\varphi(\mathbb{Z}^n) = 1$), the rank of $H_\varphi^2(Q, \mathbb{Z}^m)$ is completely determined by ρ and φ. In other words, for any group $< Q_1 >\in H_\rho^2(F, \mathbb{Z}^n)$ we have that

$$\text{rank}(H_\varphi^2(Q, \mathbb{Z}^m)) = \text{rank}(H_\varphi^2(Q_1, \mathbb{Z}^m)).$$

<u>Proof:</u>
Whenever we use "α", we will mean one of the α_l's of F. Any extension E of Q by $\mathbb{Z}^m$, compatible with φ has a presentation of the form

$$E : < \; a_1, a_2, \ldots, a_n, b_1, b_2, \ldots, b_m, \alpha_1, \alpha_2, \ldots, \alpha_k \; \| \qquad \qquad >$$
$$[a_i, a_j] = b_1^{k_{1,i,j}} b_2^{k_{2,i,j}} \ldots b_m^{k_{m,i,j}} \;\; (1 \le j < i \le n)$$
$$[b_i, b_j] = 1 \;\; (1 \le j < i \le m)$$
$$[b_i, a_j] = 1 \;\; (1 \le i \le m, 1 \le j \le n)$$
$$\alpha a_i \alpha^{-1} = a_1^{\beta_{1,i}} a_2^{\beta_{2,i}} \ldots a_n^{\beta_{n,i}} b_1^{l_{1,i}} \ldots b_m^{l_{m,i}} \;\; (1 \le i \le n)$$
$$\alpha b_i \alpha^{-1} = b_1^{\gamma_{1,i}} \ldots b_m^{\gamma_{m,i}} \;\; (1 \le i \le m)$$
$$\ldots$$

where $\rho(\alpha) = (\beta_{i,j})_{1 \le i,j \le n}$ and $\varphi(\alpha) = (\gamma_{i,j})_{1 \le i,j \le m}$.

There are some consistency conditions to be fulfilled by the parameters $k_{l,i,j}$, because

$$\alpha[a_i, a_j]\alpha^{-1} = \alpha b_1^{k_{1,i,j}} \ldots b_m^{k_{m,i,j}} \alpha^{-1}$$
$$= b_1^{\sum_{t=1}^{m} \gamma_{1,t} k_{t,i,j}} \; b_2^{\sum_{t=1}^{m} \gamma_{2,t} k_{t,i,j}} \; \ldots \; b_m^{\sum_{t=1}^{m} \gamma_{m,t} k_{t,i,j}}$$

and

$$\alpha[a_i, a_j]\alpha^{-1} = [\alpha a_i \alpha^{-1}, \alpha a_j \alpha^{-1}]$$
$$= [a_1^{\beta_{1,i}} \ldots a_n^{\beta_{n,i}}, a_1^{\beta_{1,j}} \ldots a_n^{\beta_{n,j}}]$$
$$= \prod_{n \geq p > q \geq 1} [a_p, a_q]^{(\beta_{p,i}\beta_{q,j} - \beta_{q,i}\beta_{p,j})}$$
$$= \prod_{l=1}^{n} \prod_{n \geq p > q \geq 1} b_l^{\det \begin{pmatrix} \beta_{q,j} & \beta_{q,i} \\ \beta_{p,j} & \beta_{p,i} \end{pmatrix} k_{l,p,q}} .$$

By comparison we find that the $k_{l,i,j}$ must satisfy a system of linear equations, completely determined by ρ and φ, namely

$$\forall i, j \ (1 \leq j < i \leq n), \ \forall l = 1, \ldots, m, \ \forall \alpha :$$

$$\sum_{n \geq p > q \geq 1} \det \begin{pmatrix} \beta_{q,j} & \beta_{q,i} \\ \beta_{p,j} & \beta_{p,i} \end{pmatrix} k_{l,p,q} = \sum_{t=1}^{m} \gamma_{l,t} k_{t,i,j} \qquad (5.5)$$

This system contains $km\dfrac{n(n-1)}{2}$ equations in $m\dfrac{n(n-1)}{2}$ unknowns $k_{l,i,j}$.

Although we did not describe all consistency conditions involved, we will be able to show that the rank of the subgroup of $\mathbb{Z}^{m\frac{n(n-1)}{2}}$ which forms the set of solutions of (5.5) is equal to the rank of $H^2_\varphi(Q, \mathbb{Z}^m)$.

Therefore it is enough to show that for every solution $(k_{t,i,j})$ of (5.5), there exists an extension E determined by a cocycle g, $<g> \in H^2_\varphi(Q, \mathbb{Z}^m)$, such that res $<g> \in H^2(\mathbb{Z}^n, \mathbb{Z}^m)$ determines the group

$$< a_1, a_2, \ldots, a_n, b_1, b_2, \ldots, b_m \ \| \qquad\qquad > . \qquad (5.6)$$
$$[a_i, a_j] = b_1^{lk_{1,i,j}} b_2^{lk_{2,i,j}} \ldots b_m^{lk_{m,i,j}} \ (1 \leq j < i \leq n)$$
$$[b_i, b_j] = 1 \ (1 \leq j < i \leq m)$$
$$[b_i, a_j] = 1 \ (1 \leq i \leq m, 1 \leq j \leq n)$$

for some $l \in \mathbb{N}_0$. We call N the group with a presentation (5.6) in which $l = 1$. If we choose the section s to be defined as

$$s : \mathbb{Z}^n \to N : a_1^{x_1} \ldots a_n^{x_n} \mapsto a_1^{x_1} \ldots a_n^{x_n} b_1^0 \ldots b_m^0$$

then the corresponding 2–cocycle f satisfies

$$f(a_1^{x_1} \ldots a_n^{x_n}, a_1^{y_1} \ldots a_n^{y_n}) = \left(\sum_{n \geq i > j \geq 1} k_{1,i,j} x_i y_j, \ldots, \sum_{n \geq i > j \geq 1} k_{m,i,j} x_i y_j \right).$$

With means of the group N we define the following function g:

$$g : \mathbb{Z}^n \times \mathbb{Z}^n \to \mathbb{Z}^m : (x, y) \mapsto [s(x), s(y)] \in \mathbb{Z}^m \subset N.$$

This functions g has the following properties:

1. g is a normalized 2–cocycle from $\mathbb{Z}^n$ to $\mathbb{Z}^m$.

2. g is cohomologous to $2f$, because $g = 2f + \delta\gamma$ for

$$\gamma : \mathbb{Z}^n \to \mathbb{Z}^m : a_1^{x_1} \ldots a_n^{x_n} \mapsto (\sum_{n \geq i > j \geq 1} k_{1,i,j} x_i x_j, \ldots, \sum_{n \geq i > j \geq 1} k_{m,i,j} x_i x_j).$$

3. $\forall \beta \in F : g(^\beta x, ^\beta y) =^\beta g(x, y)$. This is seen because the exactness of (5.5) guarantees this property for $\beta = \alpha$.

4. $\text{res} \circ \text{cor}\, g \sim |F|g \sim 2|F|f$
   We may explain this as follows. By [11, Prop. 9.5, p. 83] we know that $\text{res} \circ \text{cor}\, g \sim \sum_{\beta \in F} {}^\beta g$. Since for $x, y \in \mathbb{Z}^m$ we have that
   $(^\beta g)(x, y) =^\beta g(^{\beta^{-1}} x, ^{\beta^{-1}} y) = g(x, y)$, we may indeed conclude that $\text{res} \circ \text{cor}\, g \sim |F|g$.

Conclusion: The cocycle $\text{cor}\, g$ is the cocycle we were looking for.

5.4 Application to the construction of almost–crystallographic groups

• Suppose Q is a 2-dimensional crystallographic group and we try to compute $H^2_\varphi(Q, \mathbb{Z})$ for an action which factors through the holonomy group of Q. We are interested in those cases where rank $H^2(Q, \mathbb{Z}) > 0$. For if the rank $H^2(Q, \mathbb{Z}) = 0$, then all extensions will be virtually abelian and we really whish to construct virtually 2–step nilpotent groups (see section 5.1). So the only possibility is a rank $= 1$. Let α be any element of the holonomy group F of Q. The system of equations (5.5) now consists of only one equation. We write

$$\rho(\alpha) = \begin{pmatrix} l_1 & l_2 \\ l_3 & l_4 \end{pmatrix} \text{ and } \varphi(\alpha) = \varepsilon = \pm 1.$$

With these notations (5.5) becomes

$$\det \begin{pmatrix} l_1 & l_2 \\ l_3 & l_4 \end{pmatrix} k_{2,1} = \varepsilon k_{2,1}.$$

Or we may conclude that

$$\operatorname{rank}(H^2(Q,\mathbb{Z})) = 1 \Leftrightarrow \varepsilon = \det \begin{pmatrix} l_1 & l_2 \\ l_3 & l_4 \end{pmatrix}. \qquad (5.7)$$

• Now suppose that Q is a three–dimensional crystallographic group with holonomy group F and write

$$\rho(\alpha) = \begin{pmatrix} k_1 & l_1 & m_1 \\ k_2 & l_2 & m_2 \\ k_3 & l_3 & m_3 \end{pmatrix} \text{ and } \varphi(\alpha) = \varepsilon.$$

Now the system of equations (5.5) reads as

$$\det \begin{pmatrix} k_1 & l_1 \\ k_2 & l_2 \end{pmatrix} k_{2,1} + \det \begin{pmatrix} k_1 & l_1 \\ k_3 & l_3 \end{pmatrix} k_{3,1} + \det \begin{pmatrix} k_2 & l_2 \\ k_3 & l_3 \end{pmatrix} k_{3,2} = \varepsilon k_{2,1}$$

$$\det \begin{pmatrix} k_1 & m_1 \\ k_2 & m_2 \end{pmatrix} k_{2,1} + \det \begin{pmatrix} k_1 & m_1 \\ k_3 & m_3 \end{pmatrix} k_{3,1} + \det \begin{pmatrix} k_2 & m_2 \\ k_3 & m_3 \end{pmatrix} k_{3,2} = \varepsilon k_{3,1}$$

$$\det \begin{pmatrix} l_1 & m_1 \\ l_2 & m_2 \end{pmatrix} k_{2,1} + \det \begin{pmatrix} l_1 & m_1 \\ l_3 & m_3 \end{pmatrix} k_{3,1} + \det \begin{pmatrix} l_2 & m_2 \\ l_3 & m_3 \end{pmatrix} k_{3,2} = \varepsilon k_{3,2}.$$

And so the dimension of the vectorspace of solutions of this system of equations equals the rank of $H^2(Q,\mathbb{Z})$. Moreover, one can check that for crystallographic groups with holonomy $F = \mathbb{Z}_4$ or $F = \mathbb{Z}_6$ one has to take $\varepsilon = 1$ in order to have a non zero rank.

Therefore, we may conclude this section with the following theorem:

Theorem 5.4.1 *Let Q be a 3–dimensional crystallographic group. Suppose Q acts on $\mathbb{Z}$ via a morphism φ which factors through the holonomy group F of Q. If there exists an element $\alpha \in F$ of order $\neq 2$, acting non trivially on $\mathbb{Z}$, then*

$$\operatorname{rank}(H_\varphi^2(Q,\mathbb{Z})) = 0.$$

Chapter 6

Infra–nilmanifolds and their topological invariants

6.1 3-dimensional Almost–Bieberbach groups

As indicated already, a 3-dimensional AC-group E (which is not a crystallographic group) fits into a short exact sequence $1 \to N \to E \to F \to 1$, where N is the unique maximal nilpotent normal subgroup of E, and N is a lattice of the Heisenberg group H (see (2.1)). It is known that such N is isomorphic to one of the following groups, for some k:

$$N_k : < a, b, c \, \| \, [b, a] = c^k, \; [c, a] = [c, b] = 1 >, \;\; k \neq 0.$$

This group is realized as a uniform lattice of H if one takes

$$a = \begin{pmatrix} 1 & 0 & 0 \\ 0 & 1 & 1 \\ 0 & 0 & 1 \end{pmatrix}, \; b = \begin{pmatrix} 1 & 1 & 0 \\ 0 & 1 & 0 \\ 0 & 0 & 1 \end{pmatrix}, \; c = \begin{pmatrix} 1 & 0 & 1/k \\ 0 & 1 & 0 \\ 0 & 0 & 1 \end{pmatrix}.$$

Remark that N_k is isomorphic to N_{-k} and is nilpotent of class 2. It is easy to see that for each $k \neq 0$, $Z = \sqrt[N]{[N, N]} = grp\{c\} \cong \mathbb{Z}$. This implies that $N_k/Z \cong \mathbb{Z}^2$.

Suppose we have a 3-dimensional AC-group E containing N_k as its Fitting subgroup. Applying proposition 2.4.2, we know that E/Z is a 2-dimensional crystallographic group (i.e. wallpaper group) containing $N_k/Z \cong \mathbb{Z}^2$ as a maximal abelian normal subgroup (i.e. translation subgroup).

Conversely, fix N_k and assume we are given any 2-dimensional crystallographic group $E/Z \overset{\text{not.}}{=} Q$, i.e. $1 \to N_k/Z \cong \mathbb{Z}^2 \to Q \to F \to 1$.

Assume the F-module structure of $\mathbb{Z}^2$ is given by $\phi : F \to \mathrm{Gl}(2, \mathbb{Z})$. We can try to build up all possible AC-groups E, giving rise to a commutative diagram (2.6), with $N = N_k$.

Here we make the following observations which will be of great practical impact.

Observation 1. Since Z lies in the center of N_k, the action of $Q = E/Z$ on Z (induced by the sequence $1 \to Z \to E \to Q = E/Z \to 1$) must factor through the finite group F and so is completely determined by an F-action on Z.

Observation 2. An extension E of $Q = E/Z$ by Z, which is compatible with such an action, will be an AC-group containing N_k as maximal normal nilpotent group if and only if the restricted extension of $N_k/Z \cong \mathbb{Z}^2$ by Z is a group isomorphic to N_k.

Observation 3. On principle, the F-action on Z can be chosen. Since F is finite and $\mathrm{Aut}\, Z \cong \mathbb{Z}_2$, the choice is (very) limited. However, by (5.7) there is no choice at all.

In our search for the torsion free AC-groups (= the AB-groups), we need a criterion to detect torsion. Here we use the following well known lemma:

Lemma 6.1.1 *Let Q be any group. Assume that a Q-module structure on $\mathbb{Z}^m$ is given by $\varphi : Q \to \mathrm{Aut}(\mathbb{Z}^m)$. Write $n(q)$ for the order of a torsion element q in Q. Then, an element $< f > \in H^2_\varphi(Q, \mathbb{Z}^m)$ determines a group with torsion if and only if there exists a torsion element $q \in Q$ and an element $z \in \mathbb{Z}^m$, such that*

$$(1 + \varphi(q) + \cdots + \varphi(q)^{n(q)-1})z = f(q,q) + f(q,q^2) + \cdots + f(q, q^{n(q)-1}).$$

Moreover, the order of a torsion element in the extension determined by $< f >$ equals the order of its image in Q. ∎

As a consequence, we notice that the order of an element in a crystallographic group (and in an AC-group) is always a divisor of the order of its holonomy group F.

In our context of AC-groups, we apply the lemma with $m = 1$.

Corollary 6.1.2 *Consider an extension $1 \to \mathbb{Z} \to E \xrightarrow{j} Q \to 1$ where Q is a wallpaper group or an AC–group of dimension 3.*

1. *If there exists a torsion element in Q acting non-trivially on $\mathbb{Z}$ then E has torsion.*

2. *If every torsion element in Q acts trivially on $\mathbb{Z}$, then checking the presence of torsion in E can be done in a finite number of steps.*

<u>Proof:</u> Take $\tilde{q} \in E$ such that $q = j(\tilde{q}) \in Q$ is of order $n(q)$ and acts non-trivially on $\mathbb{Z}$. Let $\mathbb{Z}$ be generated by z. Then, $\tilde{q}z = z^{-1}\tilde{q}$. Since $\tilde{q}^{n(q)} = z^l \in \mathbb{Z}$, we can verify that

$$\tilde{q}\tilde{q}^{n(q)} = \tilde{q}^{n(q)}\tilde{q} \Rightarrow \tilde{q}z^l = z^l\tilde{q} \Rightarrow z^{-l}\tilde{q} = z^l\tilde{q} \Rightarrow l = 0.$$

This implies that $\tilde{q}$ has order equal to $n(q)$.

Now, assume that every torsion element in Q acts trivially on $\mathbb{Z}$ and that the extension E is determined by a cocycle $f : Q \times Q \to \mathbb{Z}$. Checking for torsion in E is equivalent to looking for an element q of finite order in Q such that

$$f(q,q) + f(q,q^2) + \cdots + f(q, q^{(n(q)-1)}) \equiv 0 \bmod n(q).$$

As noticed before, in some cases (e.g. when Q is crystallographic or AC of dimension 3) f can be chosen to be a standard cocycle and consequently it is a linear expression in a finite number of parameters $K = (k_1, k_2, \ldots, k_s)$. Now, the condition above shows that E will be torsion free or not, depending only on the values of the $k_i \bmod n$ ($1 \leq i \leq s$), where n is the order of the holonomy group F associated with Q. ∎

There will be more information on how to detect torsion in section 6.6 and in the appendix.

At this point, we are ready to start the computations to obtain all AC-groups of dimension 3. We refer to section 6.3 for an outline of the steps to follow in such a computation. For each of the 17 wallpaper groups Q, and each of the N_k, we determined the corresponding AC-groups. The results are summarized in chapter 7.

6.2 Classification of rank 4 nilpotent groups

Let N be a finitely generated, rank 4, torsion free nilpotent group. Then N is of class 1, 2 or 3. We will show that this nilpotency class completely determines the Mal'cev completion of N. In fact we give a standard commutator presentation for each isomorphism type. Also we show for more general commutator presentations how to obtain this standard commutator presentation. During this section we will use the convention that in a commutator presentation all commutators which are not explicitely written down are trivial. E.g. the presentation for the group N_k of the previous section is given by

$$N_k : < a, b, c \,\|\, [b, a] = c^k > .$$

6.2.1 N is abelian (class 1)

$$N \cong \mathbb{Z}^4, \quad N :< a, b, c, d \,\|\, \quad > .$$

N is a uniform lattice of $\mathbb{R}^4$.

6.2.2 N is of class 2

In order to show that any group N of rank 4 and of nilpotency class 2 has a center of rank 2, we prove the following proposition:

Proposition 6.2.1 *Suppose N is a nilpotent group with a presentation*

$$N :< a, b, c, d \,\|\, [b, a] = d^\alpha, \; [c, a] = d^\beta, \; [c, b] = d^\gamma > \qquad (6.1)$$

then N can be given a presentation

$$N :< a, b, c, d \,\|\, [b, a] = d^{(\alpha,\beta,\gamma)} >,$$

where (α, β, γ) denotes the (positive) greatest common divisor of α, β and γ.

Proof: Suppose N is given by a presentation (6.1). By eventually renaming the generators we may assume that $\alpha > 0$.

(a) Suppose $\beta \neq 0$ (If $\beta = 0$ go to part (b)).

   (a1) We choose a new set of generators for N, namely

$$a' = a, \; b' = b, \; c' = b^x c, \; d' = d \text{ for some } x \in \mathbb{Z}.$$

   We determine the new presentation of N by computing all the commutators of the generators (use [56, lemma 4.1 p.93]):
   $[d', a'] = [d', b'] = [d', c'] = 1$
   $[b', a'] = [b, a] = d^\alpha = d'^\alpha$

$$[c', a'] = [b^x c, a] = [b, a]^x [c, a] = d'^{\beta + \alpha x}$$
$$[c', b'] = [b^x c, b] = [b, b]^x [c, b] = d'^{\gamma}.$$

By choosing an appropriate x we can obtain $\beta + x\alpha = (\beta \bmod \alpha)$. Therefore N can also be presented as:

$$N :< a, b, c, d \parallel [b, a] = d^{\alpha}, \ [c, a] = d^{\beta \bmod \alpha}, \ [c, b] = d^{\gamma} >,$$

where we deleted the accents.

(a2) If $\beta \bmod \alpha \neq 0$ we again choose a new set of generators:

$$a' = a, \ b' = c^x b, \ c' = c, \ d' = d \text{ for some } x \in \mathbb{Z}.$$

Computing the commutators gives:
$$[d', a'] = [d', b'] = [d', c'] = 1$$
$$[b', a'] = [c^x b, a] = [c, a]^x [b, a] = d^{\alpha + x(\beta \bmod \alpha)}$$
$$[c', a'] = [c, a] = d'^{\beta \bmod \alpha}$$
$$[c', b'] = [c, c^x b] = [c, c]^x [c, b] = d'^{\gamma}.$$

This shows that we can reduce α to $\alpha \bmod (\beta \bmod \alpha)$.

By repeating steps (a1) and (a2) we finally get a presentation for N of the form:

Situation 1:

$$N :< a, b, c, d \parallel [b, a] = d^{(\alpha, \beta)}, \ [c, b] = d^{\gamma} >$$

or

Situation 2:

$$N :< a, b, c, d \parallel [c, a] = d^{(\alpha, \beta)}, \ [c, b] = d^{\gamma} > .$$

In *Situation 2*, now consider the new set of generators:

$$a' = a, \ b' = c, \ c' = b^{-1}, \ d' = d,$$

and a new presentation of N is obtained:

$$N :< a, b, c, d \parallel [b, a] = d^{(\alpha, \beta)}, \ [c, b] = d^{\gamma} >$$

which is *Situation 1*.

(b) The starting point for step (b) is a presentation for N as in *Situation 1* of step (a). By repeating simular steps as (a1) and (a2) (now adjusting generators a and c in stead of b and c) one finally finds:

$$N :< a, b, c, d \, \| \, [b, a] = d^{(\alpha,\beta,\gamma)} > .$$

∎

Corollary 6.2.2 *Every 2-step nilpotent group of rank 4, has a center of rank 2.*

Proposition 6.2.3 *Let N be a group with a presentation*

$$N :< a, b, c, d \, \| \, [b, a] = c^\alpha d^\beta >$$

then N can also be presented as

$$N :< a, b, c, d \, \| \, [b, a] = d^{(\alpha,\beta)} > .$$

Proof: Let $(\alpha, \beta) = k\alpha + l\beta$ for $k, l \in \mathbb{Z}$. We consider the following set of new generators for N:

$$a' = a, \; b' = b, \; c' = c^l d^{-k}, \; d' = c^{\frac{\alpha}{(\alpha,\beta)}} d^{\frac{\beta}{(\alpha,\beta)}}.$$

It is now easy to see that

$$N :< a', b', c', d' \, \| \, [b', a'] = d'^{(\alpha,\beta)} > .$$

∎

Corollary 6.2.4 *Every finitely generated 2-step nilpotent group of rank 4 has a presentation of the form*

$$N :< a, b, c, d \, \| \, [b, a] = d^k >,$$

for some $k \in \mathbb{Z}$, $k > 0$. Moreover this presentation is uniquely determined by N and therefore we call it the **standard presentation** *of N.*

Proof: The existence of such a presentation is guaranteed by the preceeding propositions. The uniqueness is also clear since such a presentation shows that $N \cong N_k \times \mathbb{Z}$, where $N_k \; :< a, b, d \, \| \, [b, a] = d^k >$. By e.g. looking at the quotient $Z(N)/[N, N]$ of characteristic subgroups of N, we see that $N_k \times \mathbb{Z} \cong N_l \times \mathbb{Z}$ iff $k = \pm l$.

∎

Corollary 6.2.5 *Every 2-step nilpotent group of rank 4 can be seen as a uniform lattice of $H \times \mathbb{R}$, where H denotes the 3–dimensional Heisenberg group.*

6.2.3 N is of class 3

Given a group N of nilpotency class 3 and rank 4, we may assume that N has a presentation of the form:

$$N :< a, b, c, d \parallel [b, a] = c^\alpha d^\beta \quad > \qquad (6.2)$$
$$[c, a] = d^\gamma$$
$$[c, b] = d^\delta$$

with $\alpha > 0$ and $\gamma \neq 0$ or $\delta \neq 0$.

Proposition 6.2.6 *A group N with a presentation (6.2) can also be given any of the two presentations:*

$$N :< a, b, c, d \parallel [b, a] = c^\alpha d^{\pm\beta \bmod(\alpha,\gamma,\delta)} \quad > .$$
$$[c, a] = d^{(\gamma,\delta)}$$
$$[c, b] = 1$$

<u>Proof:</u> Suppose N is given by a presentation (6.2).

(a) Suppose $\gamma \neq 0$ and $\delta \neq 0$.

 (a1) We choose a new set of generators:

$$a' = b^x a, \ b' = b, \ c' = c, \ d' = d,$$

for which
$[d', a'] = [d', b'] = [d', c'] = 1$
$[c', a'] = [c, b^x a] = [c, b]^x [c, a] = d'^{\gamma + x\delta}$
$[c', b'] = [c, b] = d'^\delta$
$[b', a'] = [b, b^x a] = b^{-1} a^{-1} b^{-x} b b^x a = [b, a] = c'^\alpha d'^\beta$.
By an appropriate choice of x we may reduce γ to $\gamma \bmod \delta$.

 (a2) Suppose $\gamma \bmod \delta \neq 0$.
Take as new generators:

$$a' = a, \ b' = a^x b, \ c' = c, \ d' = d$$

so that
$[d', a'] = [d', b'] = [d', c'] = 1$
$[c', a'] = d'^{\gamma \bmod \delta}$
$[c', b'] = [c, a]^x [c, b] = d'^{\delta + x(\gamma \bmod \delta)}$
$[b', a'] = [a^x b, a] = b^{-1} a^{-x} a^{-1} a^x b a = [b, a] = c'^\alpha d'^\beta$,
showing we can reduce δ modulo $\gamma \bmod \delta$.

By repeating steps (a1) and (a2) a finite number of times we find

Situation 1:

$$N :< a, b, c, d \parallel \begin{array}{l} [b, a] = c^\alpha d^\beta \\ [c, a] = d^{(\gamma, \delta)} \\ [c, b] = 1 \end{array} >$$

or

Situation 2:

$$N :< a, b, c, d \parallel \begin{array}{l} [b, a] = c^\alpha d^\beta \\ [c, a] = 1 \\ [c, b] = d^{(\gamma, \delta)} \end{array} > .$$

In case of *Situation 2*, consider the generators

$$a' = b, \ b' = a, \ c' = c^{-1}, \ d' = d^{-1},$$

which transform *Situation 2* into *Situation 1*.

(b) In *Situation 1* consider the generators

$$a' = a^{-1}, \ b' = b, \ c' = c^{-1}, \ d' = d$$

$[b', a'] = [b, a^{-1}]$.

Since $ba = abc^\alpha d^\beta$, $a^{-1}b = c^\alpha d^\beta ba^{-1}$, and so $ba^{-1} = c^{-\alpha} a^{-1} b d^{-\beta}$, implying:

$[b', a'] = c'^\alpha d'^{-\beta + \alpha(\gamma, \delta)}$, while

$[c', a'] = d^{(\gamma, \delta)}$, $[c', b'] = [d', a'] = [d', b'] = [d', c'] = 1$.

This leads to *Situation 1'*:

$$N :< a, b, c, d \parallel \begin{array}{l} [b, a] = c^\alpha d^{-\beta + \alpha(\gamma, \delta)} \\ [c, a] = d^{(\gamma, \delta)} \\ [c, b] = 1 \end{array} > .$$

(c) (c1) The starting point for this step is either *Situation 1* (call $\beta' = \beta$) or *Situation 1'* (call $\beta' = -\beta + \alpha(\gamma, \delta)$).
The new generators:

$$a' = a, \ b' = c^x b, \ c' = c, d' = d$$

give:

$$[d', a'] = [d', b'] = [d', c'] = [c', b'] = 1, \quad [c', a'] = d'^{(\gamma, \delta)}$$
$$[b', a'] = [c^x b, a] = b^{-1} c^{-x} a^{-1} c^x a a^{-1} b a \ ,$$
$$= [c, a]^x [b, a]$$
$$= c'^\alpha d'^{\beta' + x(\gamma, \delta)}$$

showing that we can reduce β' modulo (γ, δ).

(c2) The starting point is the result of (c1). We consider

$$a' = a, \ b' = b, \ c' = cd^x, \ d' = d$$

to obtain

$$[d', a'] = [d', b'] = [d', c'] = [c', b'] = 1$$
$$[c', a'] = [c, a] = d'^{(\gamma, \delta)}$$
$$[b', a'] = [b, a] = c'^\alpha d'^{\beta' - x\alpha}.$$

So it is possible to reduce β' modulo α.

By a good reduction in (c1) and (c2) one may reduce β' modulo (α, γ, δ). This shows

$$N :< a, b, c, d \ \| \ \begin{array}{l} [b, a] = c^\alpha d^{\pm \beta \bmod(\alpha, \gamma, \delta)} \\ [c, a] = d^{(\gamma, \delta)} \\ [c, b] = 1 \end{array} > .$$

Conversely,

Proposition 6.2.7 *Suppose*

$$N :< a, b, c, d \ \| \ \begin{array}{l} [b, a] = c^\alpha d^\beta \\ [c, a] = d^\gamma \\ [c, b] = 1 \end{array} > \cong M :< a, b, c, d \ \| \ \begin{array}{l} [b, a] = c^{\alpha'} d^{\beta'} \\ [c, a] = d^{\gamma'} \\ [c, b] = 1 \end{array} >$$

$$(6.3)$$

with $\alpha, \gamma, \alpha', \gamma' > 0$, *then*

$$\alpha' = \alpha, \ \gamma' = \gamma, \ and \ \beta' = \pm \beta \bmod (\alpha, \gamma).$$

<u>Proof:</u> Suppose $\varphi : N \to M$ is an isomorphism of groups, then φ is given by:

$$\varphi(d) = d^{\pm 1}$$
$$\varphi(c) = c^{\pm 1} d^k$$
$$\varphi(a) = a^{m_1} b^{m_2} w(c, d)$$
$$\varphi(b) = a^{l_1} b^{l_2} w(c, d)$$

where $w(c, d)$ denotes a not further specified word in c and d and

$$\begin{pmatrix} m_1 & m_2 \\ l_1 & l_2 \end{pmatrix} \in \mathrm{Aut}\,(\mathbb{Z}^2).$$

$$\begin{aligned}
\varphi[c, b] &= 1, \\
[\varphi(c), \varphi(b)] &= [c^{\pm 1} d^k, a^{l_1} b^{l_2} w(c, d)] \\
&= [c, a]^{\pm l_1} \\
&= d^{\pm \gamma' l_1}.
\end{aligned}$$

Therefore $l_1 = 0 \Rightarrow m_1 = \pm 1, \; l_2 = \pm 1.$

$$\begin{aligned}
\varphi[c, a] &= \varphi(d^\gamma) = d^{\pm \gamma}, \\
[\varphi(c), \varphi(a)] &= [c^{\pm 1}, a^{m_1} b^{m_2}] \\
&= d^{\pm m_1 \gamma'} = d^{\pm \gamma'}.
\end{aligned}$$

$\Rightarrow \gamma = \gamma'.$

$$\begin{aligned}
\varphi[b, a] &= \varphi(c^\alpha d^\beta) = c^{\pm \alpha} d^{k\alpha \pm \beta}, \\
[b^{l_2} w(c, d), a^{m_1} b^{m_2} w(c, d)] &= [b^{l_2} c^p, a^{m_1} b^{m_2} c^q] \\
&= c^{-p} b^{\mp 1} c^{-q} b^{-m_2} a^{\mp 1} b^{\pm 1} c^p a^{\pm 1} b^{m_2} c^q \\
&= c^{-q} b^{-m_2} [b^{\pm 1}, a^{\pm 1}] b^{m_2} c^q d^{\pm p\gamma} \\
&= c^{\pm \alpha'} d^{\pm \beta' + r\gamma}.
\end{aligned}$$

$\Rightarrow \alpha' = \alpha.$
$\Rightarrow \pm \beta = \beta' + s(\gamma, \alpha) \Rightarrow \beta = \pm \beta' \mod (\gamma, \alpha).$ ∎

We call a presentation for a rank 4, 3–step nilpotent group N of the form (6.3), with $\beta \leq (\alpha, \gamma)/2$ the **Standard presentation** for N.

By playing the same game as in proposition 6.2.6 on the continuous level, one can see that every torsion free 3-step nilpotent group of rank 4 is a uniform lattice of $\mathbb{R}^3 \rtimes \mathbb{R}$ where the action of 1 is given by:

$$\begin{pmatrix} 1 & 1 & 0 \\ 0 & 1 & 1 \\ 0 & 0 & 1 \end{pmatrix}$$

6.3 4–dimensional Almost–Bieberbach groups

•First we look at the 4–dimensional Almost–Bieberbach groups E whose Fitting subgroup N is a group of nilpotency class 2. In view of corollary 6.2.4, N has a presentation of the form

$$N :< a, b, c, d \parallel [b, a] = d^k, c \text{ and } d \text{ central} >$$

for some integer $k \neq 0$. Since $\sqrt[N]{[N, N]} = grp\{d\}$, such a group E fits in a short exact sequence $1 \to \mathbb{Z} \to E \to Q \to 1$, with Q 3–dimensional crystallographic. The possible actions of Q on $\mathbb{Z}$ are described in section 5.4. The construction of all possible E's is analogous to the 3–dimensional case. For more specific information, we refer to the appendix, where all computations involved in dimension 4 are carried out for a concrete case.

At this point we limit ourselves to an indication of the steps involved in the construction of AB–groups:

1. **Choose an AC–group Q.**
   The basic idea behind our theory was the reduction lemma 2.4.2. This shows that AC–groups E having a normal maximal nilpotent subgroup of nilpotency class $c > 1$, are built up from an AC–group Q with a normal maximal nilpotent subgroup N of class $c - 1$. Fix such a Q. We have to look at all extensions of Q by some torsion free abelian group Z compatible with an action of $Q/N = F$ on Z. The choice of the action can be limited by the results of section 5.4, corollary 6.1.2 and by the theory developed in the rest of this section.

2. **Determination of the computational consistent presentations.**
   Once the building blocks for the application of the reduction lemma are provided we have to determine the computational consistent presentations of the extensions of Q by Z, compatible with the chosen action. The way to do this is explained and illustrated in section 5.2. These computations also realize these extensions as matrix groups.

3. **Computation of $H^2(Q, Z)$.**
   Still following section 5.2, we now compute $H^2(Q, Z)$, which is interesting as being a first (rough) indication of the isomorphism types of the AC–groups obtained.

4. **Investigation of torsion.**
   The next step is to determine which of the extensions E are torsion

free. Of great help for this is corollary 6.1.2, section 6.6 and the discussion made in the appendix.

5. Determination of the isomorphism types.

In many cases we will find several torsion free extensions E of Q by Z. In order to really classify AB–groups, we will have to search the isomorphism classes of the groups obtained. This can be done fairly easy using the matrix representations obtained for the groups E. We give some more information in chapter 7.

• Now we look at the other possibility. The standard presentation for a 3-step nilpotent group N of rank 4, shows us that $\sqrt{[N, [N, N]]} = grp\{d\} = Z$. Therefore, if a group E is an AB-group of dimension 4, having such a N as maximal nilpotent subgroup, then E/Z is an AC-group of dimension 3 (and not a crystallographic group).

Hence, we are looking for extensions of 3-dimensional AC–groups Q by $\mathbb{Z}$ inducing a (restricted) extension $1 \to Z \cong \mathbb{Z} \to N \to N/Z \to 1$.

Again, we may derive some information concerning the actions we have to consider. Let Q be such an AC-group of dim. 3, then,

$$Q = < a, b, c, \alpha\,(,\beta) \;\|\; \begin{aligned} &[b, a] = c^k \\ &[c, a] = 1 \\ &[c, b] = 1 \\ &\alpha a \alpha^{-1} = a^{l_1} b^{l_2} c^{l_3} \\ &\alpha b \alpha^{-1} = a^{m_1} b^{m_2} c^{m_3} \\ &\alpha c \alpha^{-1} = c^{l_1 m_2 - l_2 m_1} \end{aligned} > \qquad (6.4)$$

$$\cdots$$

with $k > 0$.

Consider an action of Q on $\mathbb{Z} = grp\{d\}$, such that a, b and c act trivially, say $^\alpha d = d^e$ with $e = \pm 1$. An extension compatible with this action can be presented as

$$E = < a, b, c, d, \alpha\,(,\beta) \;\|\; \begin{aligned} &[b, a] = c^k d^{r_1}, \quad [d, a] = 1 \\ &[c, a] = d^{r_2} \quad\;\; [d, b] = 1 \\ &[c, b] = d^{r_3} \quad\;\; [d, c] = 1 \\ &\alpha a \alpha^{-1} = a^{l_1} b^{l_2} c^{l_3} d^{r_4} \\ &\alpha b \alpha^{-1} = a^{m_1} b^{m_2} c^{m_3} d^{r_5} \\ &\alpha c \alpha^{-1} = c^{l_1 m_2 - l_2 m_1} d^{r_6} \\ &\alpha d = d^e \alpha \end{aligned} >$$

$$\cdots$$

We are only interested in extensions for which $(r_2, r_3) \neq (0, 0)$. Let us first of all notice that:

$$
\begin{aligned}
\alpha[c, a]\alpha^{-1} &= [\alpha c \alpha^{-1}, \alpha a \alpha^{-1}] \\
&= [c^{l_1 m_2 - l_2 m_1}, a^{l_1} b^{l_2} c^{l_3}] \\
&= d^{(r_2 l_1 + r_3 l_2)(l_1 m_2 - l_2 m_1)}
\end{aligned}
$$

while on the other hand

$$
\alpha d^{r_2} \alpha^{-1} = d^{e r_2}.
$$

This leads to

$$
(r_2 l_1 + r_3 l_2) \det \begin{pmatrix} l_1 & l_2 \\ m_1 & m_2 \end{pmatrix} = e r_2. \tag{6.5}
$$

From $\alpha[c, b]\alpha^{-1} = \alpha d^{r_3} \alpha^{-1}$ we deduce

$$
(r_2 m_1 + r_3 m_2) \det \begin{pmatrix} l_1 & l_2 \\ m_1 & m_2 \end{pmatrix} = e r_3. \tag{6.6}
$$

We distinguish two cases

Case 1. $e = \det \begin{pmatrix} l_1 & l_2 \\ m_1 & m_2 \end{pmatrix}$.

Equation (6.5) and equation (6.6) then look like

$$
\begin{cases} r_2(l_1 - 1) + r_3 l_2 = 0 \\ r_2 m_1 + r_3(m_2 - 1) = 0 \end{cases}
$$

having a non-zero solution (r_2, r_3) iff $\det \begin{pmatrix} l_1 - 1 & l_2 \\ m_1 & m_2 - 1 \end{pmatrix} = 0$.

Remark: the solution can be seen as an eigenvector with eigenvalue 1.

Case 2. $e = -\det \begin{pmatrix} l_1 & l_2 \\ m_1 & m_2 \end{pmatrix}$.

Equation (6.5) and equation (6.6) then look like

$$
\begin{cases} r_2(l_1 + 1) + r_3 l_2 = 0 \\ r_2 m_1 + r_3(m_2 + 1) = 0 \end{cases}
$$

having a non-zero solution (r_2, r_3) iff $\det \begin{pmatrix} l_1 + 1 & l_2 \\ m_1 & m_2 + 1 \end{pmatrix} = 0$.

Remark: the solution can be seen as an eigenvector with eigenvalue -1.

Proposition 6.3.1 *An AC-group with a presentation (6.4) can be used to build up an AC-group of dimension 4, containing a 3-step nilpotent group as maximal nilpotent subgroup, only if*

$$\begin{pmatrix} l_1 & l_2 \\ m_1 & m_2 \end{pmatrix}$$

has an eigenvector with eigenvalue=± 1.

Corollary 6.3.2
The AC-groups listed in section 7.1 with number 10,11,...,17 cannot be used as building blocks for 4-dimensional AC-groups, containing a 3-step nilpotent group as maximal nilpotent subgroup.

6.4   On the Betti numbers of Infra–nilmanifolds

In this section, we involve ourselves with the computation of all Betti numbers of an infra–nilmanifold of dimension ≤ 4. Since we are dealing with aspherical manifolds, any computation can be done on the level of the fundamental group and so we will actually determine Betti numbers of AB–groups.

The Euler characteristic of an (infra–nilmanifold with fundamental group the) AB–group E may be computed as

$$\chi(E) = \sum_{i=0}^{\dim E} (-1)^i \mathrm{rank}\,(H_i(E, \mathbb{Z}))$$

where $\mathbb{Z}$ is to be considered as a trivial E–module.

Remark 6.4.1 *In this section, we will frequently speak about the rank (Hirsh number) of a group. This number will always be defined since we only deal with polycyclic-by-finite groups.*

The following theorem seems to be well known.

Theorem 6.4.2 *Let E be any AB–group, then*

$$\chi(E) = 0.$$

Proof:
We first proof this theorem for any finitely generated, torsion free nilpotent group N. Note that $\chi(\mathbb{Z}) = 0$. Now let N be of rank $k \geq 2$, then N fits in a short exact sequence

$$1 \to \mathbb{Z} \to N \to N' \to 1,$$

with N' a torsion free nilpotent group of rank $k - 1$. It is known that $\chi(N) = \chi(\mathbb{Z})\chi(N')$ (e.g. see [11]) from which it follows that $\chi(N) = 0$. Now, we consider a general AB–group E, with maximal normal nilpotent group N, and use the property that $\chi(E) = \chi(N)/[E : N]$ to see that indeed $\chi(E) = 0$.

∎

The i-th Betti number β_i of an AB–group E is defined as

$$\beta_i(E) = \text{rank } H_i(E, \mathbb{Z})$$

and so the Euler Characteristic is in fact the alternating sum of the Betti numbers. For the rest of this section we will be concerned with some of the Betti numbers of an AB–group. However, before we continue this investigation let us recall some facts needed further on. For more details, the reader is referred to [11].

Let E be any group for which there is a compact $K(E, 1)$–manifold Y. We call X the universal covering space of Y. X is an orientable space, and we denote by D, the orientation module of X. Thus D is an infinite cyclic group in which 1 and -1 correspond to the two possible orientations of X. The action of E on X induces an action of E on D in the following way: an element $e \in E$ acts as $+1$ (resp. -1) if the action of e on X is orientation–preserving (resp. orientation-reversing). E acts trivially on D iff Y is an orientable space. In case Y is orientable we will also say that E is orientable. D is generally considered as a right E–module.

Theorem 6.4.3 *[11, p. 220] For all integers i, all E–modules A and using the conventions introduced above we have that*

$$H^i(E, A) \cong H_{n-i}(E, D \otimes A).$$

We remark that $D \otimes A$ is meant to be $D \otimes_{\mathbb{Z}} A$ equiped with the diagonal E-action $^e(d \otimes a) = d^{e^{-1}} \otimes {}^e a$.
We may as well consider D as a left E–module, and in this case we find the following:

Lemma 6.4.4 *For all integers i and all E–modules A we have*

$$H^i(E, D) \cong H_{n-i}(E, \mathbb{Z})$$

where $\mathbb{Z}$ is to be considered as a trivial E–module.

Proof:
We use the previous theorem with $A = D$ considered as a left E–module.
This implies that $H^i(E, D) \cong H_{n-i}(E, D \otimes D)$. Now we claim that as
E–modules $D \otimes D \cong \mathbb{Z}$. Indeed, consider the map

$$\lambda : D \otimes D \to \mathbb{Z} : d_1 \otimes d_2 \mapsto d_1 d_2.$$

This map is an E–morphism since,

$$\lambda(^e(d_1 \otimes d_2)) = \lambda(d_1^{e^{-1}} \otimes {}^e d_2) = \lambda(d_1 d_2 \otimes 1) = d_1 d_2 =^e \lambda(d_1 \otimes d_2)$$

which was to be shown. ∎

We now concentrate on the case where $X = G$, a connected and
simply connected nilpotent Lie group of class c. For any diffeomorphism
$\lambda : G \to G$ we define

$$\text{or}\,(\lambda) = \begin{cases} +1 & \lambda \text{ is an orientation-preserving map,} \\ -1 & \lambda \text{ is an orientation-reversing map.} \end{cases}$$

We are especially interested in the action of $G \rtimes \text{Aut}\, G$ on G. Remember
that an element $(g, \varphi) \in G \rtimes \text{Aut}\, G$ acts on G, in the following way: for
any $h \in G :\ {}^{(g,\varphi)}h = g\varphi(h)$. Since $(g, \varphi) = (g, 1)(1, \varphi)$ and or $(g, 1) = +1$,
it is seen that or $(g, \varphi) = $ or $(1, \varphi) = $ or (φ) (Aut G acts in a natural
way on G). Therefore we fix a $\varphi \in \text{Aut}\, G$ and we try to provide a way
to conclude whether φ is orientation-preserving or not. Note that for
each integer i $(1 \leq i \leq c)$, φ induces a morphism on $\gamma_i(G)$ and so on
$\gamma_i(G)/\gamma_{i+1}(G) \cong \mathbb{R}^{k_i}$ for some k_i. So, the induced morphism can, after a
choice of basis for $\mathbb{R}^{k_i}$, be represented by means of a $k_i \times k_i$–matrix $A_i(\varphi)$.
This means that we can attach to each automorphism φ of G, a series of
matrices, $A_1(\varphi), A_2(\varphi), \ldots, A_c(\varphi)$. Each matrix $A_i(\varphi)$ is well determined
up to conjugation in Aut $\mathbb{R}^{k_i}$. Now, if we define for $a \in \mathbb{R}_0$, sgn $(a) = \pm 1$,
according to the sign of a, we have the following theorem:

Theorem 6.4.5 *Using the notations above:*

$$\text{or}\,(\varphi) = \text{sgn} \prod_{i=1}^{c} (\det A_i(\varphi)) \ \ \forall \varphi \in \text{Aut}\, G.$$

Proof:
We will prove this by induction on the nilpotency class c of G. It is
obvious that for $c = 1$, $G \cong \mathbb{R}^{k_1}$ and the theorem holds.

Now suppose $c > 1$, so we get a commutative diagram with induced morphisms φ_c and $\bar{\varphi}$:

$$
\begin{array}{ccccccccc}
1 & \to & \gamma_c(G) & \to & G & \to & G/\gamma_c(G) & \to & 1 \\
  &   & \downarrow \varphi_c & & \downarrow \varphi & & \downarrow \bar{\varphi} & & \\
1 & \to & \gamma_c(G) & \to & G & \to & G/\gamma_c(G) & \to & 1
\end{array}
$$

It is obvious that or $(\varphi) = $ or (φ_c).or $(\bar{\varphi})$. Together with the fact that $A_i(\bar{\varphi}) = A_i(\varphi)$ $(1 \le i \le c - 1)$ this allows us to conclude that

$$\text{or}\,(\varphi) = \text{sgn}(\det A_1(\varphi)).\text{sgn}(\det A_2(\varphi)).....\text{sgn}(\det A_c(\varphi)).$$

■

Now we return to infra–nilmanifolds. Let E be any AB–group, with maximal normal nilpotent group N, s.t. E fits in a short exact sequence

$$1 \to N \to E \to F \to 1.$$

We define a sequence of free abelian groups Z_i $(1 \le i \le c)$ by

$$Z_i = \frac{\sqrt[N]{\gamma_i(N)}}{\sqrt[N]{\gamma_{i+1}(N)}}.$$

All these abelian groups may be considered as F–modules, where the action is induced by conjugation in E. Since $Z_i \cong \mathbb{Z}^{k_i}$, for some k_i, we may choose a set of k_i generators for Z_i, and then represent the action of F on Z_i via a morphism

$$B_i : F \to \text{Aut}\, Z_i : \bar{\alpha} \mapsto B_i(\bar{\alpha}).$$

Thus, according to each $\bar{\alpha} \in F$ there is a set of matrices $B_1(\bar{\alpha}), B_2(\bar{\alpha}), \ldots,$ $B_c(\bar{\alpha})$ such that each matrix $B_i(\bar{\alpha})$ is well determined up to conjugation in $\text{Aut}\, \mathbb{Z}^{k_i}$.

Theorem 6.4.6 *Using the notations introduced above, E acts on the orientation module D in the following way:*

$$\forall \alpha \in E : \quad \text{or}\,(\alpha) = \prod_{i=1}^{c} \det B_i(\bar{\alpha})$$

where $\bar{\alpha}$ denotes the projection of α in F.

Proof:
Since E is an AB–group we can realize it as a subgroup of $G \rtimes \text{Aut}\, G$, where G is the Mal'cev completion of N, more precisely, there exists a commutative diagram, with a monomorphism ψ

$$
\begin{array}{ccccccccc}
1 & \to & N & \to & E & \to & F & \to & 1 \\
 & & \downarrow 1_N & & \downarrow \psi & & \downarrow \bar{\psi} & & \\
1 & \to & G & \to & G \rtimes \text{Aut}\, G & \to & \text{Aut}\, G & \to & 1
\end{array}
$$

Recall that $\psi(E) \cap G = N$. For any $\alpha \in E$ (projecting onto $\bar{\alpha} \in F$), we can write $\psi(\alpha) = (g_\alpha, \varphi_\alpha)$. As before, we may associate to φ_α a set of matrices $A_i(\varphi_\alpha)$ $(1 \le i \le c)$, such that or $(\alpha) = \text{sgn} \prod_{i=1}^{c} \det A_i(\varphi_\alpha)$. It is enough to show that we can take $A_i(\varphi_\alpha) = B_i(\bar{\alpha})$. By an induction argument it is sufficient to deal with the case $i = c$. Since Z_c is a uniform lattice of $\gamma_c(G)$ we can take as a basis for $\gamma_c(G)$ the same set of k_c generators of Z_c as we used to establish the matrix $B_c(\bar{\alpha})$. Let $z \in Z_c$ be any of these generators then we have

$$\psi(\alpha z \alpha^{-1}) = \psi(B_c(\bar{\alpha})z) = (B_c(\bar{\alpha})z, 1) \tag{6.7}$$

and

$$
\begin{aligned}
\psi(\alpha)z\psi(\alpha^{-1}) &= (g_\alpha, \varphi_\alpha)(z, 1)(\varphi_\alpha^{-1}(g_\alpha^{-1}), \varphi_\alpha^{-1}) \\
&= (g_\alpha \varphi_\alpha(z) g_\alpha^{-1}, 1) \\
&= (\varphi_\alpha(z), 1). \tag{6.8}
\end{aligned}
$$

Comparing (6.7) and (6.8) for any generator of Z_c allows us to conclude that for this basis of $\gamma_c(G)$, $A_c(\varphi_\alpha) = B_c(\bar{\alpha})$, which finishes the proof.

∎

Remark 6.4.7 *This theorem also shows that if there exists a canonical representation* $\lambda : E \to \text{Aff}(\mathbb{R}^n)$*, then the action of* $\alpha \in E$ *on* D*, corresponds to the determinant of the linear part of* $\lambda(\alpha)$*.*

Now, we really start computing Betti numbers for an infra–nilmanifold.

Theorem 6.4.8 *Let E be any AB–group, then $\beta_0(E) = 1$.*

Proof:
$H_0(E, \mathbb{Z}) = \mathbb{Z}$, for any group E and a trivial E–module $\mathbb{Z}$.

∎

Theorem 6.4.9 *Let E be any AB–group of dimension n, then*

$$\beta_n(E) = \begin{cases} 1 & \Leftrightarrow E \text{ is orientable,} \\ 0 & \Leftrightarrow E \text{ is non-orientable.} \end{cases}$$

Proof:
We are searching the rank of $H_n(E, \mathbb{Z})$. By lemma 6.4.4 $H_n(E, \mathbb{Z}) = H^0(E, D) = D^E$. And so, if D is a trivial E–module (E orientable) then $\beta_n(E) = 1$ else $\beta_n(E) = 0$. ∎

Remark 6.4.10 *The preceeding 2 theorems are true for any group E for which there exists a $K(E, 1)$–manifold.*

In order to find $\beta_1(E)$ we have to calculate

$$\text{rank}(H_1(E, \mathbb{Z})) = \text{rank}(E/[E, E]).$$

Theorem 6.4.11 *Let E be any AB–group, with a maximal normal nilpotent subgroup N, then*

$$\text{rank } H_1(E/\sqrt[N]{\gamma_2(N)}, \mathbb{Z}) = \text{rank } H_1(E, \mathbb{Z}).$$

Proof:
Consider the commutative diagram

$$\begin{array}{ccccccccc}
1 & \to & N & \to & E & \to & F & \to & 1 \\
& & \downarrow p & & \downarrow p & & \downarrow \bar{p} & & \\
1 & \to & N/\sqrt[N]{\gamma_2(N)} & \to & E/\sqrt[N]{\gamma_2(N)} & \to & F & \to & 1
\end{array}$$

If we denote $E' = E/\sqrt[N]{\gamma_2(N)}$, then p induces an epimorphism

$$q : \frac{E}{[E, E]} \twoheadrightarrow \frac{E'}{[E', E']}.$$

Of course, torsion elements are mapped onto torsion elements, and conversely, suppose $\bar{e} \in E/[E, E]$ ($e \in E$) s.t. $q(\bar{e})$ is a torsion element. So there exists a $k \in \mathbb{N}_0$ such that $q(\bar{e})^k = 1$. This is equivalent to $p(e)^k \in [E', E']$. It follows that $e^k \in [E, E]\sqrt[N]{\gamma_2(N)}$. We write $e^k = e_1 n_1$ with $e_1 \in [E, E]$ and $n_1 \in N$ such that there exists an $l \in \mathbb{N}_0$ for which $n_1^l \in [N, N]$. We compute $e^{kl} = e_1 n_1 e_1 n_1 \ldots e_1 n_1 = e_1^l n_1^l e_2$, for some $e_2 \in [E, E]$. Therefore we may conclude that $e^{kl} \in [E, E]$ and so $\bar{e}$ is a torsion element in $E/[E, E]$.

Because q is an epimorphism, and q maps only torsion elements onto torsion elements, we are allowed to conclude that the rank of $E/[E, E]$ equals the rank of $E'/[E', E']$.

∎

We remark here that $\dfrac{E}{\sqrt[N]{\gamma_2(N)}}$ is a crystallographic group in which the maximal abelian group is exactly $\dfrac{N}{\sqrt[N]{\gamma_2(N)}}$. So we reduced the problem of finding $\beta_1(E)$ to computing the rank$(H_1(Q, \mathbb{Z}))$ for a crystallographic group Q. This is strongly simplified by the following theorem:

Theorem 6.4.12 *Let $1 \to \mathbb{Z}^k \to Q \to F \to 1$ be any extension of groups in which F is finite. Then*

$$\operatorname{rank}\frac{Q}{[Q, Q]} = \operatorname{rank}\frac{\mathbb{Z}^k}{[Q, \mathbb{Z}^k]}.$$

Proof:
The short exact sequence $1 \to \mathbb{Z}^k \to Q \to F \to 1$ leads to the exact sequence

$$H_2(F, \mathbb{Z}) \to \frac{\mathbb{Z}^k}{[Q, \mathbb{Z}^k]} \to \frac{Q}{[Q, Q]} \to \frac{F}{[F, F]} \to 1.$$

Since $H_2(F, \mathbb{Z})$ and $F/[F, F]$ are finite, we must have the same ranks for $\mathbb{Z}^k/[Q, \mathbb{Z}^k]$ and $Q/[Q, Q]$.

∎

Theorem 6.4.13 *Let $1 \to \mathbb{Z}^k \to Q \to F \to 1$ be any extension of groups in which F is finite. Suppose this extension induces an action $\varphi : F \to \operatorname{Aut} \mathbb{Z}^k$. Let $\bar{\alpha}_1, \bar{\alpha}_2, \ldots, \bar{\alpha}_s$ be a set of generators for F. Then*

$$\operatorname{rank}\frac{\mathbb{Z}^k}{[Q, \mathbb{Z}^k]} = k - \operatorname{rank}\left(\varphi(\bar{\alpha}_1) - I, \varphi(\bar{\alpha}_2) - I, \ldots, \varphi(\bar{\alpha}_s) - I\right)$$

in which I is the $k \times k$–identity matrix.

We remark here that we used the term "rank" to indicate two different concepts: at the left–hand side of the equality we mean the rank of a group, while at the right–hand side we want to indicate the rank of a matrix, obtained by the juxtaposition of s square matrices.

Proof:

$\operatorname{rank} \dfrac{\mathbb{Z}^k}{[Q, \mathbb{Z}^k]} = k - \operatorname{rank}([Q, \mathbb{Z}^k])$. We denote by $a_1, a_2, \ldots, a_k$ the k generators for $\mathbb{Z}^k$ and we choose s elements $\alpha_1, \alpha_2, \ldots, \alpha_s \in Q$ which project respectively to $\bar{\alpha}_1, \ldots, \bar{\alpha}_s \in F$. $[Q, \mathbb{Z}^k]$ is the subgroup of $\mathbb{Z}^k$ generated by all elements $[q, z]$ ($q \in Q$, $z \in \mathbb{Z}^k$). We remark first of all that

$$\forall q \in Q, \ \forall z_1, z_2 \in \mathbb{Z}^k : \ [q, z_1 z_2] = [q, z_1][q, z_2]. \tag{6.9}$$

This means that $[Q, \mathbb{Z}^k]$ is generated by all elements of the form $[q, a_i^{-1}]$ ($q \in Q$, $1 \le i \le k$). We also see that

$$\forall z, z_1 \in \mathbb{Z}^k, \ \forall q \in Q : \ [zq, z_1] = q^{-1} z^{-1} z_1^{-1} z q z_1 = [q, z_1] \tag{6.10}$$

and

$$\forall \alpha, \beta \in Q, \ \forall z \in \mathbb{Z}^k : \ [\alpha\beta, z] = \beta^{-1}[\alpha, z]\beta[\beta, z] = [\alpha, z][\beta, [\alpha, z]]^{-1}[\beta, z]. \tag{6.11}$$

(6.9),(6.10) and (6.11) show that $[Q, \mathbb{Z}^k]$ is generated by all elements $[\alpha_i^{-1}, a_j^{-1}]$ ($1 \le i \le s$, $1 \le j \le k$). When we use the notation $a_j = (0, \ldots, 0, 1, 0, \ldots, 0)^{\mathrm{Tr}}$ (1 on the j-th spot) then

$$[\alpha_i^{-1}, a_j^{-1}] = \alpha_i a_j \alpha_i^{-1} a_j^{-1} = \varphi(\bar{\alpha}_i) \begin{pmatrix} 0 \\ \vdots \\ 1 \\ \vdots \\ 0 \end{pmatrix} - \begin{pmatrix} 0 \\ \vdots \\ 1 \\ \vdots \\ 0 \end{pmatrix}.$$

So for a fixed i, the rank of the subgroup generated by all $[\alpha_i^{-1}, a_j^{-1}]$ is equal to the maximal number of linear independent columns of $\varphi(\bar{\alpha}_i) - I$. Now letting i vary from 1 to s, leads to the desired result. ∎

Remark 6.4.14 *We notice that the rank of $H_1(Q, \mathbb{Z})$ depends only on the action $\varphi : F \to \operatorname{Aut} \mathbb{Z}^k$ and not on the group Q itself.*

Theorem 6.4.15 *Let E be an n-dimensional AB-group with Fitting subgroup N and let $F = E/N$ be generated by $\bar{\alpha}_1, \ldots \bar{\alpha}_s$. We denote by φ the action induced by the extension $1 \to N \to E \to F \to 1$ on $N/\sqrt[N]{\gamma_2(N)} \cong \mathbb{Z}^k$ (for some k). Then*

$$\beta_{n-1}(E) = k - \operatorname{rank}\left(\operatorname{or}(\alpha_1)\varphi(\bar{\alpha}_1) - I, \ldots, \operatorname{or}(\alpha_s)\varphi(\bar{\alpha}_s) - I\right)$$

where I is the $k \times k$-identity matrix and α_i is a lift of $\bar{\alpha}_i$ to E.

Proof:

$\beta_{n-1}(E) = \operatorname{rank} H_{n-1}(E, \mathbb{Z}) = \operatorname{rank} H^1(E, D)$ (lemma 6.4.4). Now we use the restriction–inflation 5-term exact sequence

$$0 \to H^1(F, D^N) \to H^1(E, D) \to H^1(N, D)^F \to H^2(F, D^N) \to H^2(E, D).$$

Notice that $H^1(F, D^N)$ and $H^2(F, D^N)$ are finite, and so the rank of $H^1(E, D)$ equals the rank of $H^1(N, D)^F$. We remark also that $D \cong \mathbb{Z}$ is a trivial N–module, and so $H^1(N, D)$ is the set of morphisms of N to $\mathbb{Z}$. We take a closer look at this set. Let $\lambda : N \to \mathbb{Z}$ be a morphism of groups. Then, for $n_1, n_2 \in N$ we have that $\lambda([n_1, n_2]) = 0$. Therefore we see that each morphism λ leads to a unique morphism

$$\bar{\lambda} : \frac{N}{\sqrt[N]{\gamma_2(N)}} \to \mathbb{Z}.$$

Conversely, each $\bar{\lambda}$ has a lift $\lambda : N \to \mathbb{Z}$. Therefore we may identify $H^1(N, D)$ with the set of morphisms from $N/\sqrt[N]{\gamma_2(N)}$ to $\mathbb{Z}$. Such a morphism $\bar{\lambda}$ can be seen as a k-tuple of integers $(\lambda_1, \lambda_2, \ldots, \lambda_k)$, in a way that for the natural set of generators $a_1, a_2, \ldots, a_k$ of $N/\sqrt[N]{\gamma_2(N)} \cong \mathbb{Z}^k$

$$\bar{\lambda}(a_i) = (\lambda_1, \lambda_2, \ldots, \lambda_n) \begin{pmatrix} 0 \\ \vdots \\ 1 \\ \vdots \\ 0 \end{pmatrix} = \lambda_i.$$

Since for a 1–cocycle $\lambda : N \to \mathbb{Z}$ the action of $\bar{\alpha}_i^{-1}$ on λ is defined as $(^{\bar{\alpha}_i^{-1}}\lambda)(n) = \operatorname{or}(\alpha_i^{-1})\lambda(\alpha_i n \alpha_i^{-1})$, the corresponding action of F on $\bar{\lambda}$ is defined as

$$(^{\bar{\alpha}_i^{-1}}\bar{\lambda})(z) = \operatorname{or}(\alpha_i^{-1})\bar{\lambda}(\varphi(\alpha_i)z) = \operatorname{or}(\alpha_i)\bar{\lambda}(\varphi(\alpha_i)z)$$

Therefore the rank of the elements which are fixed under this action equals the dimension of the space of solutions $(\lambda_1, \ldots, \lambda_k)$ such that for every i

$$(\lambda_1, \ldots, \lambda_k) = \operatorname{or}(\alpha_i)(\lambda_1, \ldots, \lambda_k)\varphi(\bar{\alpha}_i).$$

Therefore the rank of this group is

$$k - \operatorname{rank}(\operatorname{or}(\alpha_1)\varphi(\bar{\alpha}_1) - I, \ldots, \operatorname{or}(\alpha_s)\varphi(\bar{\alpha}_s) - I).$$

■

Remark 6.4.16 *The morphism* $\varphi : F \to \operatorname{Aut} \mathbb{Z}^k$ *used in the theorem 6.4.13 and theorem 6.4.15 is in fact determined by the matrices* $B_1(\bar{\alpha})$ *introduced above. Therefore, we can summarize the results obtained in this section as follows:*

> Let E be any AB–group of dimension n. Suppose that the holonomy group F of E is generated by $\bar{\alpha}_1, \ldots, \bar{\alpha}_s$, then:

$$\beta_0(E) = 1.$$

$$\beta_n(E) = \begin{cases} 1 \Leftrightarrow \forall \alpha \in E : \text{ or } (\alpha) = 1, \\ 0 \Leftrightarrow \exists \alpha \in E : \text{ or } (\alpha) = -1. \end{cases}$$

$$\beta_1(E) = \operatorname{rank} Z_1 - \operatorname{rank}(B_1(\bar{\alpha}_1) - I, B_1(\bar{\alpha}_2) - I, \ldots, B_1(\bar{\alpha}_s) - I).$$

$$\beta_{n-1}(E) = \operatorname{rank} Z_1 - \operatorname{rank}(\operatorname{or}(\alpha_1)B_1(\bar{\alpha}_1) - I, \ldots, \operatorname{or}(\alpha_s)B_1(\bar{\alpha}_s) - I).$$

Remark 6.4.17 *Note that indeed, for an orientable AB–group E, we do have* $\beta_0(E) = \beta_n(E)$ *and* $\beta_1(E) = \beta_{n-1}(E)$. *Note also that we now have enough information to compute all Betti numbers of a 4–dimensional AB–group E, because* $\beta_2(E)$ *can be found from the identity* $\chi(E) = 0 = \beta_0(E) - \beta_1(E) + \beta_2(E) - \beta_3(E) + \beta_4(E)$.

Example 6.4.18

Consider the following AB–group E:

$$E : < a, b, c, d, \alpha \parallel \begin{array}{ll} [b, a] = d^2 & [d, a] = 1 \\ [c, a] = 1 & [d, b] = 1 \\ [c, b] = 1 & [d, c] = 1 \\ \alpha a = a^{-1}\alpha d & \alpha^2 = b \\ \alpha b = b\alpha & \alpha d = d^{-1}\alpha \\ \alpha c = c^{-1}\alpha \end{array} > .$$

$N = \operatorname{grp}\{a, b, c, d\}$ is nilpotent of class 2 and $F = \{1, \bar{\alpha}\} \cong \mathbb{Z}_2$.

$$Z_2 = \frac{\sqrt[N]{[N, N]}}{\sqrt[N]{[N, [N, N]]}} = \frac{\operatorname{grp}\{d\}}{1} \cong \mathbb{Z}.$$

The relation $\alpha d = d^{-1}\alpha$ implies that $B_2(\bar{\alpha}) = (-1)$.

$$Z_1 = \frac{\sqrt[N]{N}}{\sqrt[N]{[N, N]}} = \frac{N}{\operatorname{grp}\{d\}} \cong \mathbb{Z}^3.$$

From $\alpha a = a^{-1}\alpha d$, $\alpha b = b\alpha$ and $\alpha c = c^{-1}\alpha$ we deduce that

$$B_1(\tilde{\alpha}) = \begin{pmatrix} -1 & 0 & 0 \\ 0 & 1 & 0 \\ 0 & 0 & -1 \end{pmatrix}.$$

We compute $\text{or}(\alpha) = \det B_1(\tilde{\alpha}).\det B_2(\tilde{\alpha}) = -1$. This shows that E is non-orientable. So $\beta_0(E) = 1$ and $\beta_4(E) = 0$.

$$\beta_1(E) = 3 - \text{rank} \left(\begin{pmatrix} -1 & 0 & 0 \\ 0 & 1 & 0 \\ 0 & 0 & -1 \end{pmatrix} - \begin{pmatrix} 1 & 0 & 0 \\ 0 & 1 & 0 \\ 0 & 0 & 1 \end{pmatrix} \right) = 1$$

while

$$\beta_3(E) = 3 - \text{rank} \left(\begin{pmatrix} 1 & 0 & 0 \\ 0 & -1 & 0 \\ 0 & 0 & 1 \end{pmatrix} - \begin{pmatrix} 1 & 0 & 0 \\ 0 & 1 & 0 \\ 0 & 0 & 1 \end{pmatrix} \right) = 2.$$

So, $\beta_2(E)$ has to be equal to 2 in order to have $\chi(E) = 1 - 1 + \beta_2(E) - 2 + 0 = 0$.

6.5   Seifert invariants of 3-dimensional infra–nil-manifolds

In this section we will compute the Seifert invariants for all 3–dimensional infra–nilmanifolds. This result is not new, since L. Moser [53] obtained these invariants already by a topological construction. Nevertheless, we wanted to include this section, because we are able to recover the Seifert invariants in an algebraic way by manipulating the fundamental group. In the following sections, we are then able to extend this algebraic approach to more general classes of groups.

Let M be a 3–dimensional infra–nilmanifold, with $\Pi_1(M) = E$. We denote by N the maximal nilpotent subgroup of E and $Z = Z(N) = \sqrt[N]{[N,N]}$. We know that $Q = E/Z$ is a 2-dimensional crystallographic group. Following Conner and Raymond [14], we may look at M as being

$$M = \mathbb{R}^3/E = (T^1 \times \mathbb{R}^2)/Q$$

after first deviding out the $\mathbb{Z}$–part of the action. As a result of this we find an (injective) Seifert Fibering $M = (T^1 \times \mathbb{R}^2)/Q \longrightarrow \mathbb{R}^2/Q$ with base orbifold $\mathbb{R}^2/Q$ and typical fiber the circle T^1. It is known that the fiber above the orbit of $w \in \mathbb{R}^2$ (=a point of the orbit space $\mathbb{R}^2/Q$) is homeomorphic to $(T^1 \times \{w\})/Q_w$, where Q_w, the isotropy group of w, acts freely on T^1. The following lemma is interesting to note.

Lemma 6.5.1 *Let F be a finite group acting freely on a k-dimensional torus T^k, then T^k/F is a manifold with fundamental group isomorphic to a k-dimensional Bieberbach group.*

<u>Proof:</u> It is obvious that the orbit space is a manifold M. The group of covering transformations $A(T^k, p)$ corresponding to the covering $p : T^k \to T^k/F$ is isomorphic to F and moreover, it is known that $A(T^k, p) \cong \Pi_1(M)/p_*(\Pi_1(T^k))$ where p_* is a monomorphism. Since $\mathbb{R}^k$ is the universal covering space of M, M is a $K(\Pi_1(M), 1)$ manifold, we may conclude that $\Pi_1(M)$ is torsion free. As a conclusion, we see that $\Pi_1(M)$ is torsion free, finitely generated virtually free abelian, and therefore $\Pi_1(M)$ is a Bieberbach group. ∎

Corollary 6.5.2 *Let F be a finite group acting freely on T^1, then F is cyclic.*

<u>Proof:</u> $F \cong \Pi_1(M)/\mathbb{Z}$ where $\Pi_1(M)$ is a 1-dimensional Bieberbach group. So $\Pi_1(M) \cong \mathbb{Z}$ which implies F being cyclic. ∎

If we think of a 2-dimensional crystallographic group Q as being a subgroup of the rigid motions of $\mathbb{R}^2$, then we have the following possibilities for the finite cyclic subgroups F of Q:
- $F \cong \mathbb{Z}_m$ and acts on $\mathbb{R}^2$ as rotations over an angle of $2\Pi/m$.
- $F \cong \mathbb{Z}_2$ and acts on $\mathbb{R}^2$ as reflexion through a line.

Lemma 6.5.3 *Suppose Q is a 2-dimensional crystallographic group, and suppose there exists a point $x \in \mathbb{R}^2$ such that $Q_x \cong \mathbb{Z}_2$ acts on $\mathbb{R}^2$ as reflexion through an axis. There is no extension of the form $1 \to \mathbb{Z} \to E \to Q \to 1$ where E is an Almost–Bieberbach group (and no Bieberbach group).*

<u>Proof</u> Since $Q_x \cong \mathbb{Z}_2$ acts orientation reversing on $\mathbb{R}^2$, the only action of Q on $\mathbb{Z}$ to be considered is one for which Q_x acts non-trivially on $\mathbb{Z}$. But then there exists a torsion element of Q acting non-trivially, and so any E will have torsion. ∎

Using the terminology of [55] we may conclude that we are in the situation in which there are no fixed points (this should mean that some fiber reduces to a point) and there are no special exceptional orbits (because all singular points are isolated). In particular, this means that we deal with Seifert Fiber Spaces as in Seifert's original definition. This is a 3–manifold which satisfies:

1. The manifold decomposes into a collection of simple closed curves called fibers so that each point lies on a unique fiber.

2. Each fiber has a tubular neighbourhood V consisting entirely of fibers so that V is homeomorphic to a "fibered solid torus".

A trivial fibered solid torus is defined as being $S^1 \times D^2$, where the fibers are the circles $S^1 \times \{y\}$ for any $y \in D^2$. A general fibered solid torus is obtained from a trivial one as follows: choose a number $p \in \mathbb{N}_0$ and a number $0 < q < p$ relative prime to p. Cut a trivial fibered solid torus open along $\{x\} \times D^2$ for some $x \in S^1$, rotate one of the discs so obtained trough q/p of a full twist and glue both ends together. This fibered solid torus is said to have local Seifert invariants (α, β), where $\alpha = p$ and $0 < \beta < \alpha$, $\beta q \equiv 1 \bmod \alpha$.

We also know that all 3-dimensional infra–nilmanifolds are orientable, therefore we may attach to each one a series of Seifert invariants

$$\{b; (\epsilon, g); (\alpha_1, \beta_1), \ldots, (\alpha_n, \beta_n)\} \tag{6.12}$$

where

- $\epsilon = o_1$ if $\mathbb{R}^2/Q$ is orientable and $\epsilon = n_2$ otherwise.

- g denotes the genus of the surface $\mathbb{R}^2/Q$.

- n is the number of singular orbits.

- (α_i, β_i) are pairs of relatively prime integers, with $0 < \beta_i < \alpha_i$, called the local Seifert invariants and indicating to which fibered solid torus the environment of a singular fiber is homeomorphic to.

- $b \in \mathbb{Z}$.

These invariants depend on the chosen orientation of the infra–nilmanifold and a change of orientation transforms the above set of invariants (6.12) into

$$\{-b - n; (\epsilon, g); (\alpha_1, \alpha_1 - \beta_1), \ldots, (\alpha_n, \alpha_n - \beta_n)\}.$$

The structure of the fundamental group of a Seifert Fiber Space with invariants (6.12) is well known and a standard presentation for this fundamental group is given by:

1. Case $\epsilon = o_1$

$$\Pi :< a_1, b_1, \ldots, a_g, b_g, q_1, \ldots, q_n, h \parallel \qquad\qquad >.$$
$$[a_i, h] = [b_i, h] = 1 \ (1 \leq i \leq g)$$
$$[q_i, h] = 1 \ (1 \leq i \leq n)$$
$$q_i^{\alpha_i} h^{\beta_i} = 1 \ (1 \leq i \leq n)$$
$$q_1 q_2 \ldots q_n [a_1, b_1] \ldots [a_g, b_g] = h^b$$

2. Case $\epsilon = n_2$

$$\Pi :< v_1, \ldots, v_g, q_1, \ldots, q_n, h \parallel \qquad\qquad >.$$
$$v_i h v_i^{-1} = h^{-1} \ (1 \leq i \leq g)$$
$$[q_i, h] = 1 \ (1 \leq i \leq n)$$
$$q_i^{\alpha_i} h^{\beta_i} = 1 \ (1 \leq i \leq n)$$
$$q_1 q_2 \ldots q_n v_1^2 \ldots v_g^2 = h^b$$

A Seifert Fiber Space with "enough" singular fibers is called a large manifold and is completely determined by its invariants (up to a change of orientation). More precisely, an orientable Seifert Fiber Space is said to be small if it satisfies one of the conditions below:

1. $o_1, g = 0, n \leq 2$.

2. $o_1, g = 0, n = 3, \frac{1}{\alpha_1} + \frac{1}{\alpha_2} + \frac{1}{\alpha_3} > 1$.

3. The set of invariants equals $\{-2; (o_1, 0); (2, 1), (2, 1), (2, 1), (2, 1)\}$.

4. $o_1, g = 1, n = 0$.

5. $n_2, g = 1, n \leq 1$

If an orientable Seifert Fiber Space is not small, it is called large.

Theorem 6.5.4 *Let M and M' be large Seifert manifolds. The following statements are equivalent:*

1. *M and M' have the same set of invariants (possibly after a change of orientation).*

2. *M and M' are homeomorphic*

    3. M and M' have isomorphic fundamental groups.

This theorem provides another way to classify (non equivalent) Almost–Bieberbach groups in $H^2(Q, \mathbb{Z})$ up to isomorphism. We now start computing the Seifert invariants of the 3-dimensional infra–nilmanifolds by transforming the fundamental groups obtained in the table of section 7.1, into standard ones. We use the numbers and notations of our table.

1. $\mathbb{R}^2/Q = T^2$ with no singular points. So $\epsilon = o_1$, $g = 1$, $n = 0$.
For $k > 0$ we take the following new set of $< (k) >$:

$$a_1 = a, b_1 = b, h = c.$$

It is obvious that $< (k) >$ has the presentation

$$< (k) > : < a_1, a_2, h \parallel [a_1, b_1] = h^{-k} > .$$

Conclusion: the invariants of $< (k) >$ are

$$\boxed{\{-k; (o_1, 1)\}}$$

2. $\mathbb{R}^2/Q = S^2$ with 4 singular points. (We leave the verification of this to the reader or refer to [63]). So $\epsilon = o_1$, $g = 1$, $n = 0$.
For $k > 0$ we take the following new set of generators for $< (2k, 0, 0, 1) >$:

$$q_1 = \alpha c^{-1}, \ q_2 = a\alpha c^{-1}, \ q_3 = aba c^{k-1}, \ q_4 = bac^{-1}, \ h = c.$$

It is obvious that this set generates $< (2k, 0, 0, 1) >$ since

$$a = q_2 q_1 h, \ b = q_4 q_1 h, \ c = h, \ \alpha = q_1 h.$$

The following relations are fulfilled for the new generators:

$$[q_1, h] = [q_2, h] = [q_3, h] = [q_4, h] = 1,$$

$$q_1^2 h = q_2^2 h = q_3^2 h = q_4^2 h = 1, \ q_1 q_2 q_3 q_4 = h^{-2+k} \qquad (6.13)$$

This can be seen by calculating these by hand or by using the matrix representation of $< (2k, 0, 0, 1) >$. These relations determine the group structure of $< (2k, 0, 0, 1) >$ completely since all other relations can be computed for the relations (6.13), e.g.:

$$
\begin{aligned}
[b, a] &= h^{-1} q_1^{-1} q_4^{-1} h^{-1} q_1^{-1} q_2^{-1} q_4 q_1 h q_2 q_1 h \\
&= h^{-1} q_1 h q_4 h h^{-1} q_1 h q_2 h q_4 q_1 h q_2 q_1 h \quad (q_i^2 h = 1 \Rightarrow q_i^{-1} = q_i h) \\
&= q_1 q_4 q_1 q_2 q_4 q_1 q_2 q_1 h^4 \quad ([q_i, h] = 1) \\
&= q_1 q_4 q_1 q_2 q_4 q_1 q_2 q_3 q_4 q_4^{-1} q_3^{-1} q_1 h^4 \\
&= q_1 q_4 q_1 q_2 q_3 q_1 h^{k+3} \quad (q_1 q_2 q_3 q_4 = h^{k-2}) \\
&= q_1 q_4 q_4^{-1} q_1 h^{2k+1} \\
&= h^{2k} \\
&= c^{2k}.
\end{aligned}
$$

The other relations are proved in the same way. Therefore we may conclude that the invariants of $< (2k, 0, 0, 1) >$ equal

$$
\boxed{\{ k - 2; (o_1, 0); (2, 1), (2, 1), (2, 1), (2, 1) \}}
$$

For all other cases, we restrict ourselves to the results and omit all (almost trivial) calculations.

---

4. $\mathbb{R}^2 / Q = K$ (the Klein Bottle) without singular points.
As a new set of generators for $< (k, 0) > (k > 0)$, we take

$$
v_1 = \alpha, \quad v_2 = a^{-1} b \alpha, \quad h = c
$$

and find that

$$
< (k, 0) > : \; < v_1, v_2, h \parallel v_1 h v_1^{-1} = h^{-1}, \; v_2 h v_2^{-1} = h^{-1}, \; v_1^2 v_2^2 = h^{-k} > .
$$

Conclusion: The set of Sefeirt Invariants of $< (k, 0) >$ equals

$$
\boxed{\{ -k; (n_2, 2) \}}
$$

---

8. $\mathbb{R}^2 / Q = \mathbb{P}^2$ (the projective plane) with two singular points.
As a new set of generators for $< (2k, 0, 0, 1) > (k > 0)$, we take

$$
q_1 = \alpha c^{-1}, \quad q_2 = a \alpha c^{-1-k}, \quad v_1 = \beta, \quad h = c
$$

and find that

$$< (2k,0,0,1) >\ :\ < v_1, q_1, q_2, h \parallel \begin{array}{l} v_1 h v_1^{-1} = h^{-1} \\ [q_1, h] = [q_2, h] = 1 \\ q_1^2 h = q_2^2 h = 1 \\ q_1 q_2 v_1^2 = h^{k-1} \end{array} \quad >.$$

Conclusion: The invariants for $< (2k,0,0,1) >$ are

$$\boxed{\{k-1; (n_2, 1); (2,1), (2,1)\}}$$

---

10. $\mathbb{R}^2 / Q = S^2$ with 3 singular points.

• As a new set of generators for $< (4k,0,0,1) > (k > 0)$, we take

$$q_1 = \alpha c^{-1}, \ q_2 = a\alpha^2 c^{-1}, \ q_3 = a\alpha c^{k-1}, \ h = c$$

and find that

$$< (4k,0,0,1) >\ :\ < q_1, q_2, q_3, h \parallel \begin{array}{l} [q_1, h] = [q_2, h] = [q_3, h] = 1 \\ q_1^4 h^3 = q_2^2 h = q_3^4 h^3 = 1, \ q_1 q_2 q_3 = h^{k-2} \end{array} \quad >.$$

Conclusion: The invariants for $< (4k,0,0,1) >$ are

$$\boxed{\{k-2; (o_1, 0); (4,3), (4,3), (2,1)\}}$$

• For $< (4k,0,0,3) > (k > 0)$, we take

$$q_1 = \alpha c^{-1}, \ q_2 = a\alpha^2 c^{-2}, \ q_3 = a\alpha c^{k-1}, \ h = c$$

and find that

$$< (4k,0,0,3) >\ :\ < q_1, q_2, q_3, h \parallel \begin{array}{l} [q_1, h] = [q_2, h] = [q_3, h] = 1 \\ q_1^4 h = q_2^2 h = q_3^4 h = 1, \ q_1 q_2 q_3 = h^{k-1} \end{array} \quad >.$$

Conclusion: The invariants for $< (4k,0,0,3) >$ are

$$\boxed{\{k-1; (o_1, 0); (4,1), (4,1), (2,1)\}}$$

• For $< (4k+2,0,0,1) > (k \geq 0)$, we take

$$q_1 = \alpha c^{-1}, \ q_2 = a\alpha^2 c^{-1}, \ q_3 = a\alpha c^k, \ h = c$$

and find that

$$< (4k+2, 0, 0, 1) > \; : \; < q_1, q_2, q_3, h \; \| \; \begin{matrix} [q_1, h] = [q_2, h] = [q_3, h] = 1 \\ q_1^4 h^3 = q_2^2 h = q_3^4 h = 1, \; q_1 q_2 q_3 = h^{k-1} \end{matrix} >$$

and for $< (4k + 2, 0, 0, 3) > (k \geq 0)$, we take

$$q_1 = \alpha c^{-1}, \; q_2 = a\alpha^2 c^{-2}, \; q_3 = a\alpha c^{k-1}, \; h = c$$

and find that

$$< (4k+2, 0, 0, 3) > \; = \; < q_1, q_2, q_3, h \; \| \; \begin{matrix} [q_1, h] = [q_2, h] = [q_3, h] = 1 \\ q_1^4 h = q_2^2 h = q_3^4 h^3 = 1, \; q_1 q_2 q_3 = h^{k-1} \end{matrix} > .$$

Conclusion: The invariants for $< (4k + 2, 0, 0, 1) >$ and for $< (4k + 2, 0, 0, 3) >$ are

$$\boxed{\{k - 1; (o_1, 0); (4, 3), (4, 1), (2, 1)\}}$$

Remark that this implies indeed that $< (4k + 2, 0, 0, 1) > \; \cong \; < (4k + 2, 0, 0, 3) >$ as indicated in the table of section 7.1.

---

13. $\mathbb{R}^2/Q = S^2$ with 3 singular points.

- As a new set of generators for $< (3k, 0, 0, 1) > (k > 0)$, we take

$$q_1 = \alpha c^{-1}, \; q_2 = a\alpha c^{k-1}, \; q_3 = ab\alpha c^{3k-1}, \; h = c$$

$$< (3k, 0, 0, 1) > \; = \; < q_1, q_2, q_3, h \; \| \; \begin{matrix} [q_1, h] = [q_2, h] = [q_3, h] = 1 \\ q_1^3 h^2 = q_2^3 h^2 = q_3^3 h^2 = 1, \; q_1 q_2 q_3 = h^{k-2} \end{matrix} > .$$

Conclusion: The invariants for $< (3k, 0, 0, 1) >$ are

$$\boxed{\{k - 2; (o_1, 0); (3, 2), (3, 2), (3, 2)\}}$$

- For $< (3k, 0, 0, 2) > (k > 0)$, we take

$$q_1 = \alpha c^{-1}, \; q_2 = a\alpha c^{k-1}, \; q_3 = ab\alpha c^{3k-1}, \; h = c$$

$$< (3k, 0, 0, 2) > \; = \; < q_1, q_2, q_3, h \; \| \; \begin{matrix} [q_1, h] = [q_2, h] = [q_3, h] = 1 \\ q_1^3 h = q_2^3 h = q_3^3 h = 1, \; q_1 q_2 q_3 = h^{k-1} \end{matrix} > .$$

Conclusion: The invariants for $< (3k, 0, 0, 2) >$ are

$$\boxed{\{k - 1; (o_1, 0); (3, 1), (3, 1), (3, 1)\}}$$

- For $< (3k + 1, 1, 0, 1) > (k \geq 0)$, we take

$$q_1 = \alpha c^{-1}, \; q_2 = a\alpha^2 c^{k-1}, \; q_3 = ab\alpha c^{3k}, \; h = c$$

$$< (3k+1,1,0,1) >=< q_1, q_2, q_3, h \; \| \; \begin{array}{l} [q_1, h] = [q_2, h] = [q_3, h] = 1 \\ q_1^3 h^2 = q_2^3 h = q_3^3 h = 1, \; q_1 q_2 q_3 = h^{k-1} \end{array} \; > .$$

Conclusion: The invariants for $< (3k + 1, 1, 0, 1) >$ are

$$\boxed{\{k - 1; (o_1, 0); (3, 2), (3, 1), (3, 1)\}}$$

Remark: the same set of invariants is found for $< (3k+1, 0, 0, 2) >$ and for $< (3k + 1, 2, 0, 2) >$. • For $< (3k + 2, 1, 0, 1) > (k \geq 0)$, we take

$$q_1 = \alpha c^{-1}, \; q_2 = a\alpha^2 c^{k-1}, \; q_3 = ab\alpha c^{3k+1}, \; h = c$$

$$< (3k+2,1,0,1) >=< q_1, q_2, q_3, h \; \| \; \begin{array}{l} [q_1, h] = [q_2, h] = [q_3, h] = 1 \\ q_1^3 h^2 = q_2^3 h^2 = q_3^3 h = 1, \; q_1 q_2 q_3 = h^{k-1} \end{array} \; > .$$

Conclusion: The invariants for $< (3k + 2, 1, 0, 1) >$ are

$$\boxed{\{k - 1; (o_1, 0); (3, 2), (3, 2), (3, 1)\}}$$

Remark: the same set of invariants is found for $< (3k+2, 0, 0, 1) >$ and for $< (3k + 2, 2, 0, 2) >$.

---

16. $\mathbb{R}^2/Q = S^2$ with 3 singular points.
- As a new set of generators for $< (6k, 0, 0, 1) > (k > 0)$, we take

$$q_1 = \alpha c^{-1}, \; q_2 = a\alpha^3 c^{3k-1}, \; q_3 = a\alpha^2 c^{4k-1}, \; h = c$$

$$< (6k, 0, 0, 1) > : < q_1, q_2, q_3, h \; \| \; \begin{array}{l} [q_1, h] = [q_2, h] = [q_3, h] = 1 \\ q_1^6 h^5 = q_2^2 h = q_3^3 h^2 = 1, \; q_1 q_2 q_3 = h^{k-2} \end{array} \; > .$$

Conclusion: The invariants for $< (6k, 0, 0, 1) >$ are

$$\boxed{\{k - 2; (o_1, 0); (6, 5), (3, 2), (2, 1)\}}$$

- For $< (6k, 0, 0, 5) > (k > 0)$, we take

$$q_1 = \alpha c^{-1}, \; q_2 = a\alpha^3 c^{3k-3}, \; q_3 = a\alpha^2 c^{4k-2}, \; h = c$$

$$< (6k, 0, 0, 5) > : < q_1, q_2, q_3, h \; \| \; \begin{array}{l} [q_1, h] = [q_2, h] = [q_3, h] = 1 \\ q_1^6 h = q_2^2 h = q_3^3 h = 1, \; q_1 q_2 q_3 = h^{k-1} \end{array} \; > .$$

Conclusion: The invariants for $< (6k, 0, 0, 5) >$ are

$$\boxed{\{k - 1; (o_1, 0); (6, 1), (3, 1), (2, 1)\}}$$

• As a new set of generators for $< (6k + 2, 0, 0, 5) > (k \geq 0)$, we take

$$q_1 = \alpha c^{-1}, \ q_2 = a\alpha^3 c^{3k-2}, \ q_3 = a\alpha^2 c^{4k-1}, \ h = c$$

$< (6k+2, 0, 0, 5) > : \ < q_1, q_2, q_3, h \ \| \ \begin{array}{l} [q_1, h] = [q_2, h] = [q_3, h] = 1 \\ q_1^6 h = q_2^2 h = q_3^3 h^2 = 1, \ q_1 q_2 q_3 = h^{k-1} \end{array} \ >.$

Conclusion: The invariants for $< (6k + 2, 0, 0, 5) >$ are

$$\boxed{\{k - 1; (o_1, 0); (6, 1), (3, 2), (2, 1)\}}$$

• For $< (6k + 4, 0, 0, 1) > (k \geq 0)$, we take

$$q_1 = \alpha c^{-1}, \ q_2 = a\alpha^3 c^{3k+1}, \ q_3 = a\alpha^2 c^{4k+2}, \ h = c$$

$< (6k+4, 0, 0, 1) > : \ < q_1, q_2, q_3, h \ \| \ \begin{array}{l} [q_1, h] = [q_2, h] = [q_3, h] = 1 \\ q_1^6 h^5 = q_2^2 h = q_3^3 h = 1, \ q_1 q_2 q_3 = h^{k-1} \end{array} \ >.$

Conclusion: The invariants for $< (6k + 4, 0, 0, 1) >$ are

$$\boxed{\{k - 1; (o_1, 0); (6, 5), (3, 1), (2, 1)\}}$$

We summarize the computations of this section in the following theorem:

Theorem 6.5.5
Let M be any 3–dimensional infra–nilmanifold with fundamental group E, holonomy group F and with underlying crystallographic group Q, then M has a set of Seifert invariants according to the table below ($k > 0$):

	Set of Seifert Invariants	F	$\mathbb{R}^2/Q$	Fitt(E)
I.	$\{-k; (o_1, 1)\}$	$\{1\}$	T^2	N_k
II.	$\{k - 2; (o_1, 0); (2, 1), (2, 1), (2, 1), (2, 1)\}$	$\mathbb{Z}_2$	S^2	N_{2k}
III.	$\{-k; (n_2, 2)\}$	$\mathbb{Z}_2$	K	N_{2k}
IV.	$\{k - 1; (n_2, 1); (2, 1), (2, 1)\}$	$\mathbb{Z}_2 \times \mathbb{Z}_2$	$\mathbb{P}^2$	N_{4k}
V.	$\{k - 2; (o_1, 0); (4, 3), (4, 3), (2, 1)\}$			N_{4k}
	$\{k - 1; (o_1, 0); (4, 1), (4, 1), (2, 1)\}$	$\mathbb{Z}_4$	S^2	N_{4k}
	$\{k - 1; (o_1, 0); (4, 3), (4, 1), (2, 1)\}$			N_{4k+2}
VI.	$\{k - 2; (o_1, 0); (3, 2), (3, 2), (3, 2)\}$			N_{3k}
	$\{k - 1; (o_1, 0); (3, 1), (3, 1), (3, 1)\}$	$\mathbb{Z}_3$	S^2	N_{3k}
	$\{k - 1; (o_1, 0); (3, 2), (3, 1), (3, 1)\}$			N_{3k+1}
	$\{k - 1; (o_1, 0); (3, 2), (3, 2), (3, 1)\}$			N_{3k+2}
VII.	$\{k - 2; (o_1, 0); (6, 5), (3, 2), (2, 1)\}$			N_{6k}
	$\{k - 1; (o_1, 0); (6, 1), (3, 1), (2, 1)\}$	$\mathbb{Z}_6$	S^2	N_{6k}
	$\{k - 1; (o_1, 0); (6, 1), (3, 2), (2, 1)\}$			N_{6k+2}
	$\{k - 1; (o_1, 0); (6, 5), (3, 1), (2, 1)\}$			N_{6k+4}

where N_k : $< a, b, c \parallel [b, a] = c^k,\ [c, a] = [c, b] = 1 >$.

6.6 Investigation of torsion

In this section we will define a set of invariants for a large class of Almost–crystallographic (and other) groups, which can be seen as a kind of generalization of the local Seifert invariants computed in the previous section. These invariants no longer suffice to determine the isomorphism type of an almost–crystallographic group (even not in dimension 3) but are a very nice tool to detect whether there is torsion or not.

To compute the local Seifert invariants of a 3–dimensinonal AB–group E, we needed for each singular point of the quotient space $Q\backslash\mathbb{R}^2$ ($Q = E/Z$) a generator q_i of the group E. The projection of q_i in Q is a generator of the (cyclic) isotropy group $\mathbb{Z}_{\alpha_i}$ at the singular point. We were then especially interested in the result of $q_i^{\alpha_i}$ which was of the form h^β, where h was the generator of a (unique) normal, infinite cyclic subgroup of the fundamental group. We now try to generalize this situation as much as possible.

Let Q be any group acting properly discontinuously (on the left) on a topological space X. For $x \in X$, we will write Q_x for the (finite) isotropy group at x. The orbit type of Qx is the set of all isotropy groups which can be found along the orbit and is denoted by (Q_x). Of course $(Q_x) = \{Q_{qx} \parallel q \in Q\} = \{qQ_xq^{-1} \parallel q \in Q\}$, the conjugacy class of Q_x in Q. We will assume that there are only finitely many orbit types. This is e.g. always the case when Q is a polycyclic–by–finite group, since those groups only have a finite number of conjugacy classes of finite subgroups [59]. By a set of representatives for the orbit types, we will mean a collection of finite subgroups $Q_{x_1}, Q_{x_2}, \ldots, Q_{x_n}$ such that any orbit type can uniquely be written as a (Q_{x_i}) for a (unique) x_i. We want to investigate extensions E of the form

$$1 \longrightarrow \mathbb{Z} \longrightarrow E \longrightarrow Q \longrightarrow 1$$

compatible with a chosen action of Q on $\mathbb{Z} = grp\{h\}$. Fix such an extension E and let q be an element of E which projects onto $\bar{q} \in Q_{x_i}$ for some $x_i \in X$. By $\alpha_{\bar{q}}$ we will denote the order of $\bar{q}$. So $q^{\alpha_{\bar{q}}} = h^\beta$ for some integer β. We use the notation $\beta_{\bar{q}} = \beta \bmod \alpha_{\bar{q}}$, so $0 \leq \beta_{\bar{q}} < \alpha_{\bar{q}}$.

Lemma 6.6.1 *The ordered pair* $(\alpha_{\bar{q}}, \beta_{\bar{q}})$ *is independent of the chosen lift* $q \in E$, *used in computing it.*

<u>Proof:</u> Let $q' = qh^k$ be another lift of $\bar{q}$. We distinguish two cases. First let us suppose that the action of $\bar{q}$ on h is trivial, then $q'^{\alpha_{\bar{q}}} = (qh^k)^{\alpha_{\bar{q}}} = h^\beta h^{k\alpha_{\bar{q}}}$, showing that $\beta_{\bar{q}}$ is independent of the choice of lift of $\bar{q}$. If the action of $\bar{q}$ on $\mathbb{Z}$ is nontrivial, then $\alpha_{\bar{q}}$ is even and so $q'^{\alpha_{\bar{q}}} = (qh^k)^{2\frac{\alpha_{\bar{q}}}{2}} = (q^2)^{\frac{\alpha_{\bar{q}}}{2}} = q^{\alpha_{\bar{q}}} = h^\beta$, which proves the lemma. ∎

Remark 6.6.2 *In section 6.1, we even showed that in case the action of* $\bar{q}$ *on* $\mathbb{Z}$ *is nontrivial, then* $\beta = \beta_{\bar{q}} = 0$.

For each $Q_{x_i} = \{1, \bar{q}_1, \bar{q}_2, \ldots, \bar{q}_k\}$ we now define the local invariants of this orbit type to be

$$S_E(Q_{x_i}) = < (\alpha_{\bar{q}_1}, \beta_{\bar{q}_1}), (\alpha_{\bar{q}_2}, \beta_{\bar{q}_2}), \ldots, (\alpha_{\bar{q}_k}, \beta_{\bar{q}_k}) > .$$

This $S_E(Q_{x_i})$ is to be considered as a set of ordered 2–tuples, the order in which they appear not being important, but identical 2–tuples are listed more then ones. The following lemma shows that the local invariants of an orbit type do not depend nor on the choice of the point x_i of the orbit, nor on the generator chosen for $\mathbb{Z}$.

Lemma 6.6.3 *If $(Q_{x_i}) = (Q_{y_i})$ for some $y_i \in X$, then $S_E(Q_{x_i}) = S_E(Q_{y_i})$. Moreover, Q_{x_i} does not depend upon the generator chosen for $\mathbb{Z}$.*

Proof: Suppose $(Q_{x_i}) = (Q_{y_i})$ (This does not mean that x_i and y_i belong to the same orbit), then $Q_{y_i} = gQ_{x_i}g^{-1}$ for some $g \in Q$. So $\alpha_{\bar{q}} = \alpha_{g\bar{q}g^{-1}}$ for all $\bar{q} \in Q_{x_i}$. Take any lift $\tilde{g}$ of g in E, then $(\tilde{g}q\tilde{g}^{-1})^{\alpha_q} = \tilde{g}q^{\alpha_q}\tilde{g}^{-1} = \tilde{g}q^{\beta}\tilde{g}^{-1}$. So, if g acts trivial on $\mathbb{Z}$ we instantly see that we have a one to one correspondence between $S_E(Q_{x_i})$ and $S_E(Q_{y_i})$. But if g acts non trivial, then there is also a one to one correspondence, since in this case one sees that $(\alpha_{\bar{q}^{-1}}, \beta_{\bar{q}^{-1}}) = (\alpha_{g\bar{q}g^{-1}}, \beta_{g\bar{q}g^{-1}})$. Moreover, a change of generator $(h \longrightarrow h^{-1})$ for $\mathbb{Z}$ only changes $(\alpha_{\bar{q}}, \beta_{\bar{q}})$ into $(\alpha_{\bar{q}}, \alpha_{\bar{q}} - \beta_{\bar{q}})$ for each $\bar{q}$. But this does not effect the whole set $S_E(Q_{x_i})$ since $(\alpha_{\bar{q}}, \alpha_{\bar{q}} - \beta_{\bar{q}}) = (\alpha_{\bar{q}^{-1}}, \beta_{\bar{q}^{-1}})$. ∎

Definition 6.6.4 *Let Q be any group acting properly discontinuously on a topological space X with finitely many orbit types $(Q_{x_1}), (Q_{x_2}), \ldots, (Q_{x_n})$. Let E be any extension of Q by $\mathbb{Z}$, then we define a set of invariants $S(E)$ as*

$$S(E) = \{S_E(Q_{x_1}), S_E(Q_{x_2}), \ldots, S_E(Q_{x_n})\}.$$

Remark that this definition is independent of the chosen representants of the orbit types and of the chosen generator for $\mathbb{Z}$. The following theorem shows the importance of these invariants.

Theorem 6.6.5 *Let E be an extension of a group Q as described above where the action of Q is on a space $X \approx \mathbb{R}^n$ for some n, then E is torsion free if and only if for every 2–tuple (α, β) of $S(E)$, α and β are mutually prime.*

Proof: Suppose that there exists a pair (α, β) in $S(E)$, such that α and β are not mutually prime. We suppose that they came from $\bar{q} \in Q$. So there is a lift $q \in E$ of $\bar{q}$ for which $q^\alpha = h^\beta$, where h denotes the generator of $\mathbb{Z}$ as before. We remark that if $\bar{q}$ acts non trivially on $\mathbb{Z}$, $\beta = 0$ and so there is torsion. Therefore we may assume that the action of $\bar{q}$ on $\mathbb{Z}$ is trivial. Because of the fact that α and β are not mutually prime there exists a $\gamma, \delta \in \mathbb{Z}$, with $0 < \gamma < \alpha$ such that $\gamma\beta + \delta\alpha = 0$ and so $(q^\gamma h^\delta)^\alpha = h^{\beta\gamma + \delta\alpha} = 1$. This shows that E has torsion.

Conversely, suppose E has torsion. Then there is an element $q \in E$ of order α, with α a prime number. The projection $\bar{q}$ of q in Q is also of order α and therefore we may conclude that the action of $\bar{q}$ on $\mathbb{R}^n$ has some fixed point x Thus $\bar{q} \in Q_x$ and $(\alpha, 0)$ belongs to $S(E)$. ∎

Remark 6.6.6

In stead of using representatives of orbit types, one can also avoid the topological aspect by considering all (or all maximal) finite subgroups of Q. Analogous results can be formulated.

Chapter 7

Classification survey

7.1   3-dimensional AC–groups

The table we present here describes all possible 3-dimensional AC-groups E. One table entry contains several items, which we explain now. First of all, the entries are ordened according to the ordening of the wallpaper groups (Q) as found in ([10]) (and so as found in the International Tables for Crystallography).

Each table-entry contains a presentation for the family of 3-dimensional AC-groups corresponding to the indicated group Q. This presentation depends of at most 4 parameters k_1, k_2, k_3 and k_4. Always, the subgroup generated by the symbols a, b and c is the maximal nilpotent subgroup N_k contained in E.

In each entry we present a faithful affine representation $\lambda : E \rightarrow \mathrm{Aff}(\mathbb{R}^3)$. λ is given by its images of the generators. However, as we use stable affine representation we have for every AC-group E in the table that

$$\lambda(a) = \begin{pmatrix} 1 & 0 & -\frac{k}{2} & 0 \\ 0 & 1 & 0 & 1 \\ 0 & 0 & 1 & 0 \\ 0 & 0 & 0 & 1 \end{pmatrix} \quad \lambda(b) = \begin{pmatrix} 1 & \frac{k}{2} & 0 & 0 \\ 0 & 1 & 0 & 0 \\ 0 & 0 & 1 & 1 \\ 0 & 0 & 0 & 1 \end{pmatrix} \quad \lambda(c) = \begin{pmatrix} 1 & 0 & 0 & 1 \\ 0 & 1 & 0 & 0 \\ 0 & 0 & 1 & 0 \\ 0 & 0 & 0 & 1 \end{pmatrix}$$

where the values for k is determined by $[b, a] = c^k$.

Each table-entry also contains $H^2_\varphi(Q, \mathbb{Z})$ (φ, as indicated in observation 3 in section 6.1). This group has been computed using the method indicated in chapter 5 and depends on the number of parameters k_i in the presentation of E. An element (cohomology class) in this group is written as $< (k_1, k_2, k_3, k_4) >$. Furthermore, in each entry we show which groups are AB-groups and we indicate the isomorphism classes of these AB-groups.

The general set up of one table entry is as follows:

Number of Q Symbol of Q
Presentation for E depending on k_1, k_2, k_3, k_4
The images under λ for the generators other then a, b, c
$H^2(Q, \mathbb{Z})$ in terms of k_1, k_2, k_3, k_4
Eventually:
AB-groups:
The cohomology classes corresponding to AB-groups
and isomorphism type information for these.

1. $Q = p1$
$$E : \; < a, b, c \parallel [b, a] = c^{k_1} \; [c, a] = [c, b] = 1 >$$
$$H^2(Q, \mathbb{Z}) = \mathbb{Z}$$

AB-groups:
$k > 0, \; E = < (k) > \; (\cong N_k)$

---

2. $Q = p2$
$$E : \; < a, b, c, \alpha \parallel \begin{array}{ll} [b, a] = c^{k_1} & \alpha c = c\alpha \\ [c, a] = 1 & [c, b] = 1 \\ \alpha a = a^{-1} \alpha c^{k_2} & \alpha b = b^{-1} \alpha c^{k_3} \\ \alpha^2 = c^{k_4} \end{array} \; >$$

$$\lambda(\alpha) = \begin{pmatrix} 1 & k_2 & k_3 & \frac{k_4}{2} \\ 0 & -1 & 0 & 0 \\ 0 & 0 & -1 & 0 \\ 0 & 0 & 0 & 1 \end{pmatrix}$$

$$H^2(Q, \mathbb{Z}) = \mathbb{Z} \oplus (\mathbb{Z}_2)^3 = \mathbb{Z}^4 / A, \; A = \{(k_1, k_2, k_3, k_4) \parallel k_1 = 0, \; k_2, k_3, k_4 \in 2\mathbb{Z}\}$$

AB-groups:
$k > 0, \; k \equiv 0 \bmod 2, \; E = < (k, 0, 0, 1) >$

---

3. $Q = pm$
$$E : \; < a, b, c, \alpha \parallel \begin{array}{ll} [b, a] = c^{k_1} & \alpha c = c^{-1} \alpha \\ [c, a] = 1 & [c, b] = 1 \\ \alpha a = a\alpha c^{k_2} & \alpha b = b^{-1}\alpha \\ \alpha^2 = 1 \end{array} \; >$$

$$\lambda(\alpha) = \begin{pmatrix} -1 & -k_2 & 0 & 0 \\ 0 & 1 & 0 & 0 \\ 0 & 0 & -1 & 0 \\ 0 & 0 & 0 & 1 \end{pmatrix}$$

$$H^2(Q,\mathbb{Z}) = \mathbb{Z} \oplus \mathbb{Z}_2 = \mathbb{Z}^2/A, \ A = \{(k_1,k_2) \,\|\, k_1 = 0, \ k_2 \in 2\mathbb{Z}\}$$

---

4. $Q = pg$

$$E: \ < a,b,c,\alpha \,\| \ \begin{array}{ll} [b,a] = c^{2k_1} & \alpha c = c^{-1}\alpha \\ [c,a] = 1 & [c,b] = 1 \\ \alpha a = a\alpha c^{-2k_2} & \alpha b = b^{-1}\alpha c^{-k_1} \\ \alpha^2 = ac^{k_2} \end{array} \quad >$$

$$\lambda(\alpha) = \begin{pmatrix} -1 & 2k_2 & \frac{k_1}{2} & 0 \\ 0 & 1 & 0 & \frac{1}{2} \\ 0 & 0 & -1 & 0 \\ 0 & 0 & 0 & 1 \end{pmatrix}$$

$$H^2(Q,\mathbb{Z}) = \mathbb{Z} = \mathbb{Z}^2/A, \ A = \{(k_1,k_2) \,\|\, k_1 = 0, \ k_2 \in \mathbb{Z}\}$$

AB-groups:
$k > 0, \ k \equiv 0 \bmod 2, \ E = < (k/2, 0) >$

---

5. $Q = cm$

$$E: \ < a,b,c,\alpha \,\| \ \begin{array}{ll} [b,a] = c^{k_1} & \alpha c = c^{-1}\alpha \\ [c,a] = 1 & [c,b] = 1 \\ \alpha a = b\alpha c^{k_2} & \alpha b = a\alpha c^{k_2} \\ \alpha^2 = 1 \end{array} \quad >$$

$$\lambda(\alpha) = \begin{pmatrix} -1 & -k_2 & -k_2 & 0 \\ 0 & 0 & 1 & 0 \\ 0 & 1 & 0 & 0 \\ 0 & 0 & 0 & 1 \end{pmatrix}$$

$$H^2(Q,\mathbb{Z}) = \mathbb{Z} = \mathbb{Z}^2/A, \ A = \{(k_1,k_2) \,\|\, k_1 = 0, \ k_2 \in \mathbb{Z}\}$$

---

6. $Q = p2mm$

$$E: \ < a,b,c,\alpha,\beta \,\| \ \begin{array}{ll} [b,a] = c^{k_1} & \\ [c,a] = 1 & [c,b] = 1 \\ \alpha c = c\alpha & \beta c = c^{-1}\beta \\ \alpha a = a^{-1}\alpha c^{k_2} & \alpha b = b^{-1}\alpha c^{k_3} \\ \alpha^2 = c^{k_4} & \beta^2 = 1 \\ \beta a = a\beta c^{k_2} & \beta b = b^{-1}\beta \\ \alpha\beta = \beta\alpha c^{-k_4} \end{array} \quad >$$

$$\lambda(\alpha) = \begin{pmatrix} 1 & k_2 & k_3 & \frac{k_4}{2} \\ 0 & -1 & 0 & 0 \\ 0 & 0 & -1 & 0 \\ 0 & 0 & 0 & 1 \end{pmatrix} \quad \lambda(\beta) = \begin{pmatrix} -1 & -k_2 & 0 & 0 \\ 0 & 1 & 0 & 0 \\ 0 & 0 & -1 & 0 \\ 0 & 0 & 0 & 1 \end{pmatrix}$$

$$H^2(Q,\mathbb{Z}) = \mathbb{Z} \oplus (\mathbb{Z}_2)^3 = \mathbb{Z}^4/A, \ A = \{(k_1, k_2, k_3, k_4) \ \| \ k_1 = 0, \ k_2, k_3, k_4 \in 2\mathbb{Z}\}$$

---

7. $Q = p2mg$

$$E : \ < a, b, c, \alpha, \beta \ \| \ [b,a] = c^{2k_1} \qquad\qquad\qquad\qquad >$$

$$\begin{aligned} &[c,a] = 1 & [c,b] = 1 \\ &\alpha c = c\alpha & \beta c = c^{-1}\beta \\ &\alpha a = a^{-1}\alpha c^{k_2} & \alpha b = b^{-1}\alpha c^{-2(k_3+k_4)} \\ &\alpha^2 = c^{k_3} & \beta^2 = 1 \\ &\beta a = a\beta c^{k_2 - k_1} & \beta b = b^{-1}\beta \\ &\alpha\beta = b^{-1}\beta\alpha c^{k_4} \end{aligned}$$

$$\lambda(\alpha) = \begin{pmatrix} 1 & k_2 & -2(k_3+k_4) & \frac{k_3}{2} \\ 0 & -1 & 0 & 0 \\ 0 & 0 & -1 & 0 \\ 0 & 0 & 0 & 1 \end{pmatrix} \quad \lambda(\beta) = \begin{pmatrix} -1 & \frac{k_1}{2} - k_2 & 0 & 0 \\ 0 & 1 & 0 & 0 \\ 0 & 0 & -1 & \frac{1}{2} \\ 0 & 0 & 0 & 1 \end{pmatrix}$$

$$H^2(Q,\mathbb{Z}) = \mathbb{Z} \oplus (\mathbb{Z}_2)^2 = \mathbb{Z}^4/A,$$

$$A = \{(k_1, k_2, k_3, k_4) \ \| \ k_1 = 0, \ k_2, k_3 \in 2\mathbb{Z}, \ k_4 \in \mathbb{Z}\}$$

---

8. $Q = p2gg$

$$E : \ < a, b, c, \alpha, \beta \ \| \ [b,a] = c^{2k_1} \qquad\qquad\qquad\qquad >$$

$$\begin{aligned} &[c,a] = 1 & [c,b] = 1 \\ &\alpha c = c\alpha & \beta c = c^{-1}\beta \\ &\alpha a = a^{-1}\alpha c^{k_1 + 2k_3} & \alpha b = b^{-1}\alpha c^{-k_1 + 2k_2 - 2k_3} \\ &\alpha^2 = c^{k_4} & \beta^2 = ac^{-k_3} \\ &\beta a = a\beta c^{2k_3} & \beta b = b^{-1}\beta c^{-k_1} \\ & & \alpha\beta = a^{-1}b^{-1}\beta\alpha c^{-(k_4 + k_1 + k_2)} \end{aligned}$$

$$\lambda(\alpha) = \begin{pmatrix} 1 & k_1 + 2k_3 & 2k_2 - k_1 - 2k_3 & \frac{k_4}{2} \\ 0 & -1 & 0 & 0 \\ 0 & 0 & -1 & 0 \\ 0 & 0 & 0 & 1 \end{pmatrix}$$

$$\lambda(\beta) = \begin{pmatrix} -1 & -\frac{k_1}{2} - 2k_3 & \frac{k_1}{2} & 0 \\ 0 & 1 & 0 & \frac{1}{2} \\ 0 & 0 & -1 & \frac{1}{2} \\ 0 & 0 & 0 & 1 \end{pmatrix}$$

$$H^2(Q,\mathbb{Z}) = \mathbb{Z} \oplus \mathbb{Z}_2 = \mathbb{Z}^4/A,$$

$$A = \{(k_1, k_2, k_3, k_4) \,\|\, k_1 = 0, \ k_4 \in 2\mathbb{Z}, \ k_2, k_3 \in \mathbb{Z}\}$$

AB-groups:
$k > 0$, $k \equiv 0 \bmod 4$, $E = <(k/2, 0, 0, 1)>$.

---

9. $Q = c2mm$

$$E : \ <a, b, c, \alpha, \beta \,\| \ [b, a] = c^{k_1} \quad\quad\quad\quad\quad\quad >$$

$$[c, a] = 1 \quad\quad [c, b] = 1$$
$$\alpha c = c\alpha \quad\quad \beta c = c^{-1}\beta$$
$$\alpha a = a^{-1}\alpha c^{k_2} \quad \alpha b = b^{-1}\alpha c^{2k_3 - k_2}$$
$$\alpha^2 = c^{k_4} \quad\quad \beta^2 = 1$$
$$\beta a = b\beta c^{k_3} \quad \beta b = a\beta c^{k_3}$$
$$\alpha\beta = \beta\alpha c^{-k_4}$$

$$\lambda(\alpha) = \begin{pmatrix} 1 & k_2 & -k_2 + 2k_3 & \frac{k_4}{2} \\ 0 & -1 & 0 & 0 \\ 0 & 0 & -1 & 0 \\ 0 & 0 & 0 & 1 \end{pmatrix} \quad \lambda(\beta) = \begin{pmatrix} -1 & -k_3 & -k_3 & 0 \\ 0 & 0 & 1 & 0 \\ 0 & 1 & 0 & 0 \\ 0 & 0 & 0 & 1 \end{pmatrix}$$

$$H^2(Q, \mathbb{Z}) = \mathbb{Z} \oplus (\mathbb{Z}_2)^2 = \mathbb{Z}^4/A,$$

$$A = \{(k_1, k_2, k_3, k_4) \,\|\, k_1 = 0, \ k_4, k_2 \in 2\mathbb{Z}, \ k_3 \in \mathbb{Z}\}$$

---

10. $Q = p4$

$$E : \ <a, b, c, \alpha \,\| \ [b, a] = c^{k_1} \quad \alpha c = c\alpha \quad\quad\quad >$$

$$[c, a] = 1 \quad [c, b] = 1$$
$$\alpha a = b\alpha c^{k_2} \quad \alpha b = a^{-1}\alpha \, c^{k_3}$$
$$\alpha^4 = c^{k_4}$$

$$\lambda(\alpha) = \begin{pmatrix} 1 & k_2 & k_3 & \frac{k_4}{4} \\ 0 & 0 & -1 & 0 \\ 0 & 1 & 0 & 0 \\ 0 & 0 & 0 & 1 \end{pmatrix}$$

$$H^2(Q, \mathbb{Z}) = \mathbb{Z} \oplus \mathbb{Z}_2 \oplus \mathbb{Z}_4 = \mathbb{Z}^4/A,$$

$$A = \{(k_1, \ldots, k_4) \,\|\, k_1 = 0, \ (k_2 + k_3) \in 2\mathbb{Z}, \ k_4 \in 4\mathbb{Z}\}$$

AB-groups:
$k > 0$, $k \equiv 0 \bmod 2$, $E = <(k, 0, 0, 1)>$.
$k > 0$, $k \equiv 0 \bmod 4$, $E = <(k, 0, 0, 3)>$.
Remark: if $k = 4l + 2$, for some $l \in \mathbb{Z}$ then $<(k, 0, 0, 1)> \cong <(k, 0, 0, 3)>$.
$\quad\quad\quad k = 4l$, for some $l \in \mathbb{Z}$ then $<(k, 0, 0, 1)> \not\cong <(k, 0, 0, 3)>$.

11. $Q = p4mm$

$$E: \ <a, b, c, \alpha, \beta \ \| \ \ [b, a] = c^{k_1}$$
$$[c, a] = 1 \qquad\qquad [c, b] = 1$$
$$\alpha c = c\alpha \qquad\qquad \beta c = c^{-1}\beta$$
$$\alpha a = ba c^{k_2} \qquad\qquad \alpha b = a^{-1}\alpha c^{k_3}$$
$$\alpha^4 = c^{k_4} \qquad\qquad \beta^2 = 1$$
$$\beta a = a\beta c^{k_2 + k_3} \qquad \beta b = b^{-1}\beta$$
$$\alpha\beta = \beta\alpha^3 c^{-k_4} \ \ >$$

$$\lambda(\alpha) = \begin{pmatrix} 1 & k_2 & k_3 & \frac{k_4}{4} \\ 0 & 0 & -1 & 0 \\ 0 & 1 & 0 & 0 \\ 0 & 0 & 0 & 1 \end{pmatrix} \qquad \lambda(\beta) = \begin{pmatrix} -1 & -(k_2 + k_3) & 0 & 0 \\ 0 & 1 & 0 & 0 \\ 0 & 0 & -1 & 0 \\ 0 & 0 & 0 & 1 \end{pmatrix}$$

$$H^2(Q, \mathbb{Z}) = \mathbb{Z} \oplus \mathbb{Z}_2 \oplus \mathbb{Z}_4 = \mathbb{Z}^4/A,$$
$$A = \{(k_1, \ldots, k_4) \ \| \ k_1 = 0, (k_2 + k_3) \in 2\mathbb{Z}, k_4 \in 4\mathbb{Z}\}$$

---

12. $Q = p4gm$

$$E: \ <a, b, c, \alpha, \beta \ \| \ \ [b, a] = c^{2k_1}$$
$$[c, a] = 1 \qquad\qquad [c, b] = 1$$
$$\alpha c = c\alpha \qquad\qquad \beta c = c^{-1}\beta$$
$$\alpha a = ba c^{k_2} \qquad\qquad \alpha b = a^{-1}\alpha c^{k_1 - k_2 - 2k_3}$$
$$\alpha^4 = c^{k_4} \qquad\qquad \beta^2 = ac^{k_3}$$
$$\beta a = a\beta c^{-2k_3} \qquad \beta b = b^{-1}\beta c^{-k_1}$$
$$\alpha\beta = a^{-1}\beta\alpha^3 c^{k_3 - k_4} \ \ >$$

$$\lambda(\alpha) = \begin{pmatrix} 1 & k_2 & k_1 - k_2 - 2k_3 & \frac{k_4}{4} \\ 0 & 0 & -1 & 0 \\ 0 & 1 & 0 & 0 \\ 0 & 0 & 0 & 1 \end{pmatrix} \qquad \lambda(\beta) = \begin{pmatrix} -1 & -\frac{k_1}{2} + 2k_3 & \frac{k_1}{2} & 0 \\ 0 & 1 & 0 & \frac{1}{2} \\ 0 & 0 & -1 & \frac{1}{2} \\ 0 & 0 & 0 & 1 \end{pmatrix}$$

$$H^2(Q, \mathbb{Z}) = \mathbb{Z} \oplus \mathbb{Z}_4 = \mathbb{Z}^4/A,$$
$$A = \{(k_1, k_2, k_3, k_4) \ \| \ k_1 = 0, \ k_2, k_3 \in \mathbb{Z}, \ k_4 \in 4\mathbb{Z}\}$$

---

13. $Q = p3$

$$E: \ <a, b, c, \alpha \ \| \ \ [b, a] = c^{k_1} \qquad \alpha c = c\alpha$$
$$[c, a] = 1 \qquad [c, b] = 1$$
$$\alpha a = ba c^{k_2} \qquad \alpha b = a^{-1}b^{-1}\alpha \, c^{k_3}$$
$$\alpha^3 = c^{k_4} \ \ >$$

$$\lambda(\alpha) = \begin{pmatrix} 1 & k_2 & -\frac{k_1}{2} + k_3 & \frac{k_4}{3} \\ 0 & 0 & -1 & 0 \\ 0 & 1 & -1 & 0 \\ 0 & 0 & 0 & 1 \end{pmatrix}$$

$$H^2(Q, \mathbb{Z}) = \mathbb{Z} \oplus (\mathbb{Z}_3)^2 = \mathbb{Z}^4 / A,$$

$$A = \{(k_1, k_2, k_3, k_4) \parallel k_1 = 0, \ (k_2 - k_3) \in 3\mathbb{Z} \ k_4 \in 3\mathbb{Z}\}$$

AB-groups:

$k > 0, \ k \equiv 0 \bmod 3, \ E = <(k, 0, 0, 1)>$

$k > 0, \ k \equiv 0 \bmod 3, \ E = <(k, 0, 0, 2)>$

$k > 0, \ k \not\equiv 0 \bmod 3, \ E = <(k, 1, 0, 1)>$

Remark:

If $k \equiv 1 \bmod 3$, $E = <(k, 0, 0, 2)> \cong <(k, 1, 0, 1)> \cong <(k, 2, 0, 2)>$.

If $k \equiv 2 \bmod 3$, $E = <(k, 0, 0, 1)> \cong <(k, 1, 0, 1)> \cong <(k, 2, 0, 2)>$.

---

14. $Q = p3m1$

$$E : \ < a, b, c, \alpha, \beta \parallel \ [b, a] = c^{k_1} \qquad\qquad >$$
$$[c, a] = 1 \qquad\qquad [c, b] = 1$$
$$\alpha c = c\alpha \qquad\qquad \beta c = c^{-1}\beta$$
$$\alpha a = ba c^{k_2} \qquad\qquad \alpha b = a^{-1}b^{-1}\alpha c^{k_3}$$
$$\alpha^3 = c^{k_4} \qquad\qquad \beta^2 = 1$$
$$\beta a = b^{-1}\beta c^{k_2} \qquad\qquad \beta b = a^{-1}\beta c^{-k_2}$$
$$\alpha\beta = \beta\alpha^2 c^{-k_4}$$

$$\lambda(\alpha) = \begin{pmatrix} 1 & k_2 & -\frac{k_1}{2} + k_3 & \frac{k_4}{3} \\ 0 & 0 & -1 & 0 \\ 0 & 1 & -1 & 0 \\ 0 & 0 & 0 & 1 \end{pmatrix} \qquad \lambda(\beta) = \begin{pmatrix} -1 & -k_2 & k_2 & 0 \\ 0 & 0 & -1 & 0 \\ 0 & -1 & 0 & 0 \\ 0 & 0 & 0 & 1 \end{pmatrix}$$

$$H^2(Q, \mathbb{Z}) = \mathbb{Z} \oplus (\mathbb{Z}_3)^2 = \mathbb{Z}^4 / A,$$

$$A = \{(k_1, k_2, k_3, k_4) \parallel k_1 = 0, (k_2 - k_3) \in 3\mathbb{Z}, k_4 \in 3\mathbb{Z}\}$$

---

15. $Q = p31m$

$$E : \ < a, b, c, \alpha, \beta \parallel \ [b, a] = c^{k_1} \qquad\qquad >$$
$$[c, a] = 1 \qquad\qquad [c, b] = 1$$
$$\alpha c = c\alpha \qquad\qquad \beta c = c^{-1}\beta$$
$$\alpha a = ba c^{k_1 - 2k_3 + 3k_2} \qquad\qquad \alpha b = a^{-1}b^{-1}\alpha c^{k_3}$$
$$\alpha^3 = c^{k_4} \qquad\qquad \beta^2 = 1$$
$$\beta a = b\beta c^{k_2} \qquad\qquad \beta b = a\beta c^{k_2}$$
$$\alpha\beta = \beta\alpha^2 c^{-k_4}$$

$$\lambda(\alpha) = \begin{pmatrix} 1 & k_1 - 2k_3 + 3k_2 & -\frac{k_1}{2} + k_3 & \frac{k_4}{3} \\ 0 & 0 & -1 & 0 \\ 0 & 1 & -1 & 0 \\ 0 & 0 & 0 & 1 \end{pmatrix} \qquad \lambda(\beta) = \begin{pmatrix} -1 & -k_2 & -k_2 & 0 \\ 0 & 0 & 1 & 0 \\ 0 & 1 & 0 & 0 \\ 0 & 0 & 0 & 1 \end{pmatrix}$$

$$H^2(Q, \mathbb{Z}) = \mathbb{Z} \oplus \mathbb{Z}_3 = \mathbb{Z}^4 / A,$$

$$A = \{(k_1, k_2, k_3, k_4) \parallel k_1 = 0, \ k_2, k_3 \in \mathbb{Z}, \ k_4 \in 3\mathbb{Z}\}$$

16. $Q = p6$

$$E : \ <a, b, c, \alpha \ \| \ \begin{array}{ll} [b, a] = c^{k_1} & \alpha c = c\alpha \\ [c, a] = 1 & [c, b] = 1 \\ \alpha a = ab\alpha c^{k_2} & \alpha b = a^{-1}\alpha c^{k_3} \\ \alpha^6 = c^{k_4} \end{array} \ >$$

$$\lambda(\alpha) = \begin{pmatrix} 1 & -\frac{k_1}{2} + k_2 & k_3 & \frac{k_4}{6} \\ 0 & 1 & -1 & 0 \\ 0 & 1 & 0 & 0 \\ 0 & 0 & 0 & 1 \end{pmatrix}$$

$$H^2(Q, \mathbb{Z}) = \mathbb{Z} \oplus \mathbb{Z}_6 = \mathbb{Z}^4 / A,$$

$$A = \{(k_1, k_2, k_3, k_4) \ \| \ k_1 = 0, \ k_2, k_3 \in \mathbb{Z}, \ k_4 \in 6\mathbb{Z}\}$$

AB-groups:
$k > 0, \ k \equiv 0 \bmod 6, \ E = < (k, 0, 0, 1) > \in H^2(Q, \mathbb{Z})$.
$k > 0, \ k \equiv 4 \bmod 6, \ E = < (k, 0, 0, 1) > \in H^2(Q, \mathbb{Z})$.
$k > 0, \ k \equiv 0 \bmod 6, \ E = < (k, 0, 0, 5) > \in H^2(Q, \mathbb{Z})$.
$k > 0, \ k \equiv 2 \bmod 6, \ E = < (k, 0, 0, 5) > \in H^2(Q, \mathbb{Z})$.

---

17. $Q = p6mm$

$$E : \ < a, b, c, \alpha, \beta \ \| \ \begin{array}{ll} [b, a] = c^{k_1} \\ [c, a] = 1 & [c, b] = 1 \\ \alpha c = c\alpha & \beta c = c^{-1}\beta \\ \alpha a = ab\alpha c^{k_2} & \alpha b = a^{-1}\alpha c^{k_3} \\ \alpha^6 = c^{k_4} & \beta^2 = 1 \\ \beta a = b\beta c^{k_3} & \beta b = a\beta c^{k_3} \\ \alpha\beta = \beta\alpha^5 c^{-k_4} \end{array} \ >$$

$$\lambda(\alpha) = \begin{pmatrix} 1 & -\frac{k_1}{2} + k_2 & k_3 & \frac{k_4}{6} \\ 0 & 1 & -1 & 0 \\ 0 & 1 & 0 & 0 \\ 0 & 0 & 0 & 1 \end{pmatrix} \quad \lambda(\beta) = \begin{pmatrix} -1 & -k_3 & -k_3 & 0 \\ 0 & 0 & 1 & 0 \\ 0 & 1 & 0 & 0 \\ 0 & 0 & 0 & 1 \end{pmatrix}$$

$$H^2(Q, \mathbb{Z}) = \mathbb{Z} \oplus \mathbb{Z}_6 = \mathbb{Z}^4 / A,$$

$$A = \{(k_1, k_2, k_3, k_4) \ \| \ k_1 = 0, \ k_2, k_3 \in \mathbb{Z}, \ k_4 \in 6\mathbb{Z}\}$$

In the next table we summarize the above results by indicating how many infra–nilmanifolds are essentially covered by the nilmanifold with fundamental group

$$N_k \; : \; < a, b, c \, \| \, [b, a] = c^k > .$$

This number depends of the value of k mod 12.

Almost–Bieberbach groups of dim=3												
k mod 12	0	1	2	3	4	5	6	7	8	9	10	11
1.	1	1	1	1	1	1	1	1	1	1	1	1
2.	1	0	1	0	1	0	1	0	1	0	1	0
4.	1	0	1	0	1	0	1	0	1	0	1	0
8.	1	0	0	0	1	0	0	0	1	0	0	0
10.	2	0	1	0	2	0	1	0	2	0	1	0
13.	2	1	1	2	1	1	2	1	1	2	1	1
16.	2	0	1	0	1	0	2	0	1	0	1	0
Total	10	2	6	3	8	2	8	2	8	3	6	2

7.2 4-dimensional AB–groups, with 2–step nilpotent Fitting subgroup

The table we present here describes all possible 4-dimensional AB-groups E with a maximal normal nilpotent group of class 2. For many 3-dimensional crystallographic groups, we know in advance, e.g. by application of section 5.4 and of corollary 6.1.2, that they cannot give rise to an AB–group. So we did only include one entry for each "interesting" group Q. There are only few Q which cannot be excluded in advance and which are not the underlying crystallographic group of a 4–dimensional AB–group.

One table entry contains several items, which we explain now. First of all, the entries are listed according to the ordering of the crystallographic groups (Q) as found in ([10]). We also indicate the number of Q as in the International Tables for Crystallography (I.T.).

Each table-entry contains a presentation for the family of 4-dimensional AC-groups corresponding to the indicated group Q. This presentation depends of at most 7 parameters $k_1, k_2, \ldots k_7$. Always, the subgroup generated by the symbols a, b, c and d is the maximal nilpotent subgroup N contained in E.

In each entry we present a faithful affine representation $\lambda : E \to \text{Aff}(\mathbb{R}^4)$. λ is given by its images of the generators. However, as we use stable representations we have for every AC-group E in the table that

$$\lambda(a) = \begin{pmatrix} 1 & 0 & \frac{-l_1}{2} & \frac{-l_2}{2} & 0 \\ 0 & 1 & 0 & 0 & 1 \\ 0 & 0 & 1 & 0 & 0 \\ 0 & 0 & 0 & 1 & 0 \\ 0 & 0 & 0 & 0 & 1 \end{pmatrix} \quad \lambda(b) = \begin{pmatrix} 1 & \frac{l_1}{2} & 0 & \frac{-l_3}{2} & 0 \\ 0 & 1 & 0 & 0 & 0 \\ 0 & 0 & 1 & 0 & 1 \\ 0 & 0 & 0 & 1 & 0 \\ 0 & 0 & 0 & 0 & 1 \end{pmatrix}$$

$$\lambda(c) = \begin{pmatrix} 1 & \frac{l_2}{2} & \frac{l_3}{2} & 0 & 0 \\ 0 & 1 & 0 & 0 & 0 \\ 0 & 0 & 1 & 0 & 0 \\ 0 & 0 & 0 & 1 & 1 \\ 0 & 0 & 0 & 0 & 1 \end{pmatrix} \quad \text{and} \quad \lambda(d) = \begin{pmatrix} 1 & 0 & 0 & 0 & 1 \\ 0 & 1 & 0 & 0 & 0 \\ 0 & 0 & 1 & 0 & 0 \\ 0 & 0 & 0 & 1 & 0 \\ 0 & 0 & 0 & 0 & 1 \end{pmatrix},$$

where the values for l_1, l_2 and l_3 are determined by

$$[b, a] = d^{l_1}, \quad [c, a] = d^{l_2}, \quad [c, b] = d^{l_3}.$$

Each table-entry also contains $H^2_\varphi(Q, \mathbb{Z})$ (φ, can be read of the presentation of the extensions E). This group depends on the parameters k_i in the presentation of E. An element (cohomology class) in this group is written as $< (k_1, k_2, \ldots, k_7) >$. Furthermore, in each entry we show which groups are AB-groups and we indicate the isomorphism classes of these AB-groups.

The general set up of one table entry is as follows:

Number of Q as found in I.T. Symbol of Q
Presentation for E depending on $k_1, k_2, \ldots, k_7$
The images under λ for the generators other then a, b, c, d
$H^2(Q, \mathbb{Z})$ in terms of $k_1, k_2, \ldots, k_7$
AB-groups:
The cohomology classes corresponding to AB-groups
and isomorphism type information for these.

1. $Q = P1$

$$E : < a, b, c, d \ || \ \begin{array}{ll} [b, a] = d^{k_1} & [d, a] = 1 \\ [c, a] = d^{k_2} & [d, b] = 1 \\ [c, b] = d^{k_3} & [d, c] = 1 \end{array} >$$

$$H^2(Q, \mathbb{Z}) = \mathbb{Z}^3$$

AB-groups:
$\forall k > 0, \ < (k, 0, 0) >$
Remark: $\forall k_1, k_2, k_3 \in \mathbb{Z} : \ < (k_1, k_2, k_3) > \cong < ((k_1, k_2, k_3), 0, 0) >$.

---

2. $Q = P\bar{1}$

$$E : < a, b, c, d, \alpha \ || \ \begin{array}{ll} [b, a] = d^{k_1} & [d, a] = 1 \\ [c, a] = d^{k_2} & [d, b] = 1 \\ [c, b] = d^{k_3} & [d, c] = 1 \\ \alpha a = a^{-1} \alpha d^{k_4} & \alpha^2 = d^{k_7} \\ \alpha b = b^{-1} \alpha d^{k_5} & \alpha d = d \alpha \\ \alpha c = c^{-1} \alpha d^{k_6} & \end{array} >$$

$$\lambda(\alpha) = \begin{pmatrix} 1 & k_4 & k_5 & k_6 & \frac{k_7}{2} \\ 0 & -1 & 0 & 0 & 0 \\ 0 & 0 & -1 & 0 & 0 \\ 0 & 0 & 0 & -1 & 0 \\ 0 & 0 & 0 & 0 & 1 \end{pmatrix}$$

$$H^2(Q, \mathbb{Z}) = \mathbb{Z}^3 \oplus (\mathbb{Z}_2)^4 = \mathbb{Z}^7 / A,$$

$$A = \{(k_1, \ldots, k_7) \, || \, k_1 = k_2 = k_3 = 0, \ k_4, k_5, k_6, k_7 \in 2\mathbb{Z}\}$$

AB-groups:
$\forall k > 0, \ k \equiv 0 \bmod 2, \ < (k, 0, 0, 0, 0, 0, 1) >$
Remark: $< (2k, 2l, 2m, 0, 0, 0, 1) > \cong < (2(k, l, m), 0, 0, 0, 0, 0, 1) > \ \forall k, l, m \in \mathbb{Z}$

3. $Q = P2$

$$E : < a, b, c, d, \alpha \, \| \quad [b,a] = 1 \qquad\qquad [d,a] = 1 \quad >$$
$$[c,a] = d^{k_1} \qquad\qquad [d,b] = 1$$
$$[c,b] = 1 \qquad\qquad [d,c] = 1$$
$$\alpha a = a^{-1}\alpha d^{k_2} \qquad \alpha^2 = d^{k_4}$$
$$\alpha b = b\alpha \qquad\qquad \alpha d = d\alpha$$
$$\alpha c = c^{-1}\alpha d^{k_3}$$

$$\lambda(\alpha) = \begin{pmatrix} 1 & k_2 & 0 & k_3 & \frac{k_4}{2} \\ 0 & -1 & 0 & 0 & 0 \\ 0 & 0 & 1 & 0 & 0 \\ 0 & 0 & 0 & -1 & 0 \\ 0 & 0 & 0 & 0 & 1 \end{pmatrix}$$

$$H^2(Q, \mathbb{Z}) = \mathbb{Z} \oplus (\mathbb{Z}_2)^3 = \mathbb{Z}^4/A, \ A = \{(k_1, \ldots, k_4)\|k_1 = 0, \ k_2, k_3, k_4 \in 2\mathbb{Z}\}$$

AB-groups:
$\forall k > 0, \ k \equiv 0 \bmod 2, \ < (k, 0, 0, 1) >$

---

4. $Q = P2_1$
Trivial Action:

$$E : < a, b, c, d, \alpha \, \| \quad [b,a] = 1 \qquad\qquad [d,a] = 1 \quad >$$
$$[c,a] = d^{k_1} \qquad\qquad [d,b] = 1$$
$$[c,b] = 1 \qquad\qquad [d,c] = 1$$
$$\alpha a = a^{-1}\alpha d^{k_2} \qquad \alpha^2 = b d^{k_4}$$
$$\alpha b = b\alpha \qquad\qquad \alpha d = d\alpha$$
$$\alpha c = c^{-1}\alpha d^{k_3}$$

$$\lambda(\alpha) = \begin{pmatrix} 1 & k_2 & 0 & k_3 & \frac{k_4}{2} \\ 0 & -1 & 0 & 0 & 0 \\ 0 & 0 & 1 & 0 & \frac{1}{2} \\ 0 & 0 & 0 & -1 & 0 \\ 0 & 0 & 0 & 0 & 1 \end{pmatrix}$$

$$H^2(Q, \mathbb{Z}) = \mathbb{Z} \oplus (\mathbb{Z}_2)^2 = \mathbb{Z}^4/A,$$

$$A = \{(k_1, \ldots, k_4)\|k_1 = 0, \ k_2, k_3 \in 2\mathbb{Z}, \ k_4 \in \mathbb{Z}\}$$

AB-groups:
$\forall k > 0, \ < (k, 0, 0, 0) >$
$\forall k > 0, \ k \equiv 0 \bmod 2, \ < (k, 1, 0, 0) >$
Remark:
If $k \equiv 0 \bmod 2, < (k, 0, 0, 0) > \not\cong < (k, 1, 0, 0) > \cong < (k, 0, 1, 0) > \cong < (k, 1, 1, 0) >$
If $k \not\equiv 0 \bmod 2, < (k, 0, 0, 0) > \cong < (k, 1, 0, 0) > \cong < (k, 0, 1, 0) > \cong < (k, 1, 1, 0) >$

Action: $^{\alpha}d = d^{-1}$

$$E : < a, b, c, d, \alpha \ \| \quad \begin{matrix} [b, a] = d^{2k_1} & [d, a] = 1 \\ [c, a] = 1 & [d, b] = 1 \\ [c, b] = d^{2k_2} & [d, c] = 1 \\ \alpha a = a^{-1}\alpha d^{k_1} & \alpha^2 = bd^{k_3} \\ \alpha b = b\alpha d^{-2k_3} & \alpha d = d^{-1}\alpha \\ \alpha c = c^{-1}\alpha d^{-k_2} & \end{matrix} \quad >$$

$$\lambda(\alpha) = \begin{pmatrix} -1 & -\frac{k_1}{2} & 2k_3 & \frac{k_2}{2} & 0 \\ 0 & -1 & 0 & 0 & 0 \\ 0 & 0 & 1 & 0 & \frac{1}{2} \\ 0 & 0 & 0 & -1 & 0 \\ 0 & 0 & 0 & 0 & 1 \end{pmatrix}$$

$H^2(Q, \mathbb{Z}) = \mathbb{Z}^2 = \mathbb{Z}^3/A, \ A = \{(k_1, k_2, k_3)\|k_1 = k_2 = 0, \ k_3 \in \mathbb{Z}\}$

AB-groups:

$\forall k > 0, \ < (k, 0, 0) >$

Remark: $< (k, l, 0) > \cong < ((k, l), 0, 0) >$

---

5. $Q = C2$

$$E : < a, b, c, d, \alpha \ \| \quad \begin{matrix} [b, a] = 1 & [d, a] = 1 \\ [c, a] = d^{k_1} & [d, b] = 1 \\ [c, b] = d^{k_1} & [d, c] = 1 \\ \alpha a = b^{-1}\alpha d^{k_2} & \alpha^2 = d^{k_4} \\ \alpha b = a^{-1}\alpha d^{k_2} & \alpha d = d\alpha \\ \alpha c = c^{-1}\alpha d^{k_3} & \end{matrix} \quad >$$

$$\lambda(\alpha) = \begin{pmatrix} 1 & k_2 & k_2 & k_3 & \frac{k_4}{2} \\ 0 & 0 & -1 & 0 & 0 \\ 0 & -1 & 0 & 0 & 0 \\ 0 & 0 & 0 & -1 & 0 \\ 0 & 0 & 0 & 0 & 1 \end{pmatrix}$$

$H^2(Q, \mathbb{Z}) = \mathbb{Z} \oplus (\mathbb{Z}_2)^2 = \mathbb{Z}^4/A,$

$A = \{(k_1, \ldots, k_4)\|k_1 = 0, \ k_3, k_4 \in 2\mathbb{Z}, \ k_2 \in \mathbb{Z}\}$

AB-groups:

$\forall k > 0, \ < (k, 0, 0, 1) >$

---

6. $Q = Pm$

$$E : < a, b, c, d, \alpha \ \| \quad \begin{matrix} [b, a] = 1 & [d, a] = 1 \\ [c, a] = d^{k_1} & [d, b] = 1 \\ [c, b] = 1 & [d, c] = 1 \\ \alpha a = a\alpha & \alpha^2 = d^{k_3} \\ \alpha b = b^{-1}\alpha d^{k_2} & \alpha d = d\alpha \\ \alpha c = c\alpha & \end{matrix} \quad >$$

$$\lambda(\alpha) = \begin{pmatrix} 1 & 0 & k_2 & 0 & \frac{k_3}{2} \\ 0 & 1 & 0 & 0 & 0 \\ 0 & 0 & -1 & 0 & 0 \\ 0 & 0 & 0 & 1 & 0 \\ 0 & 0 & 0 & 0 & 1 \end{pmatrix}$$

$H^2(Q, \mathbb{Z}) = \mathbb{Z} \oplus (\mathbb{Z}_2)^2 = \mathbb{Z}^3/A, \ A = \{(k_1, k_2, k_3) \| k_1 = 0, \ k_2, k_3 \in 2\mathbb{Z}\}$

AB-groups:
$\forall k > 0, \ < (k, 0, 1) >$

---

7. $Q = Pc$
Trivial action:

$$E: < a, b, c, d, \alpha \ \| \ \begin{array}{ll} [b, a] = 1 & [d, a] = 1 \\ [c, a] = d^{2k_1} & [d, b] = 1 \\ [c, b] = 1 & [d, c] = 1 \\ \alpha a = a\alpha d^{k_1} & \alpha^2 = cd^{k_3} \\ \alpha b = b^{-1}\alpha d^{k_2} & \alpha d = d\alpha \\ \alpha c = c\alpha \end{array} \ >$$

$$\lambda(\alpha) = \begin{pmatrix} 1 & \frac{k_1}{2} & k_2 & 0 & \frac{k_3}{2} \\ 0 & 1 & 0 & 0 & 0 \\ 0 & 0 & -1 & 0 & 0 \\ 0 & 0 & 0 & 1 & \frac{1}{2} \\ 0 & 0 & 0 & 0 & 1 \end{pmatrix}$$

$H^2(Q, \mathbb{Z}) = \mathbb{Z} \oplus \mathbb{Z}_2 = \mathbb{Z}^3/A, \ A = \{(k_1, k_2, k_3) \| k_1 = 0, \ k_2 \in 2\mathbb{Z}, \ k_3 \in \mathbb{Z}\}$

AB-groups:
$\forall k > 0, \ < (k, 0, 0) >$
$\forall k > 0, \ < (k, 1, 0) >$
Action $^\alpha d = d^{-1}$:

$$E: < a, b, c, d, \alpha \ \| \ \begin{array}{ll} [b, a] = d^{k_1} & [d, a] = 1 \\ [c, a] = 1 & [d, b] = 1 \\ [c, b] = d^{2k_2} & [d, c] = 1 \\ \alpha a = a\alpha d^{k_3} & \alpha^2 = cd^{k_4} \\ \alpha b = b^{-1}\alpha d^{k_2} & \alpha d = d^{-1}\alpha \\ \alpha c = c\alpha d^{-2k_4} \end{array} \ >$$

$$\lambda(\alpha) = \begin{pmatrix} -1 & -k_3 & -\frac{k_2}{2} & 2k_4 & 0 \\ 0 & 1 & 0 & 0 & 0 \\ 0 & 0 & -1 & 0 & 0 \\ 0 & 0 & 0 & 1 & \frac{1}{2} \\ 0 & 0 & 0 & 0 & 1 \end{pmatrix}$$

$H^2(Q, \mathbb{Z}) = \mathbb{Z}^2 \oplus \mathbb{Z}_2 = \mathbb{Z}^4/A,$

$$A = \{(k_1, \ldots, k_4) \| k_1 = k_2 = 0, \ k_3 \in 2\mathbb{Z}, \ k_4 \in \mathbb{Z}\}$$

AB-groups:
$\forall k > 0, \ < (k, 0, 0, 0) >$
$\forall k > 0, \ k \equiv 0 \bmod 2, \ < (k, 0, 1, 0) >$
$\forall k > 0, \ < (0, k, 0, 0) >$
$\forall k > 0, \ < (0, k, 1, 0) >$
Remark: If $k \not\equiv 0 \bmod 2 < (k, 0, 0, 0) > \ \cong \ < (k, 0, 1, 0) >$.

---

8. $Q = Cm$

$$E : < a, b, c, d, \alpha \ \| \begin{array}{ll} [b, a] = 1 & [d, a] = 1 \\ [c, a] = d^{k_1} & [d, b] = 1 \\ [c, b] = d^{k_1} & [d, c] = 1 \\ \alpha a = b\alpha d^{k_2} & \alpha^2 = d^{k_3} \\ \alpha b = a\alpha d^{-k_2} & \alpha d = d\alpha \\ \alpha c = c\alpha & \end{array} >$$

$$\lambda(\alpha) = \begin{pmatrix} 1 & k_2 & -k_2 & 0 & \frac{k_3}{2} \\ 0 & 0 & 1 & 0 & 0 \\ 0 & 1 & 0 & 0 & 0 \\ 0 & 0 & 0 & 1 & 0 \\ 0 & 0 & 0 & 0 & 1 \end{pmatrix}$$

$$H^2(Q, \mathbb{Z}) = \mathbb{Z} \oplus \mathbb{Z}_2 = \mathbb{Z}^3/A, \ A = \{(k_1, k_2, k_3) \| k_1 = 0, \ k_2 \in \mathbb{Z}, \ k_3 \in 2\mathbb{Z}\}$$

AB-groups:
$\forall k > 0, \ < (k, 0, 1) >$

---

9. $Q = Cc$
Trivial action:

$$E : < a, b, c, d, \alpha \ \| \begin{array}{ll} [b, a] = 1 & [d, a] = 1 \\ [c, a] = d^{k_1} & [d, b] = 1 \\ [c, b] = d^{k_1} & [d, c] = 1 \\ \alpha a = b\alpha d^{k_2} & \alpha^2 = cd^{k_3} \\ \alpha b = a\alpha d^{k_1 - k_2} & \alpha d = d\alpha \\ \alpha c = c\alpha & \end{array} >$$

$$\lambda(\alpha) = \begin{pmatrix} 1 & k_2 - \frac{k_1}{4} & \frac{3k_1}{4} - k_2 & 0 & \frac{k_3}{2} \\ 0 & 0 & 1 & 0 & 0 \\ 0 & 1 & 0 & 0 & 0 \\ 0 & 0 & 0 & 1 & \frac{1}{2} \\ 0 & 0 & 0 & 0 & 1 \end{pmatrix}$$

$$H^2(Q, \mathbb{Z}) = \mathbb{Z} = \mathbb{Z}^3/A, \ A = \{(k_1, k_2, k_3) \| k_1 = 0, \ k_2, k_3 \in \mathbb{Z}\}$$

AB-groups:
$\forall k > 0, \ <(k,0,0)>$
Action $^{\alpha}d = d^{-1}$:

$$E : <a,b,c,d,\alpha \ \| \quad \begin{array}{ll} [b,a] = d^{k_1} & [d,a] = 1 \\ [c,a] = d^{k_2} & [d,b] = 1 \\ [c,b] = d^{-k_2} & [d,c] = 1 \\ \alpha a = b\alpha d^{k_3} & \alpha^2 = cd^{k_4} \\ \alpha b = a\alpha d^{k_3 - k_2} & \alpha d = d^{-1}\alpha \\ \alpha c = c\alpha d^{-2k_4} & \end{array} \quad >$$

$$\lambda(\alpha) = \begin{pmatrix} -1 & \frac{k_2}{4} - k_3 & \frac{3k_2}{4} - k_3 & 2k_4 & 0 \\ 0 & 0 & 1 & 0 & 0 \\ 0 & 1 & 0 & 0 & 0 \\ 0 & 0 & 0 & 1 & \frac{1}{2} \\ 0 & 0 & 0 & 0 & 1 \end{pmatrix}$$

$$H^2(Q,\mathbb{Z}) = \mathbb{Z}^2 = \mathbb{Z}^4/A, \ A = \{(k_1,k_2,k_3,k_4)\|k_1 = k_2 = 0, \ k_3,k_4 \in \mathbb{Z}\}$$

AB-groups:
$\forall k > 0, \ <(k,0,0,0)>$
$\forall k > 0, \ <(0,k,0,0)>$

---

10. $Q = P2/m$

$$E : <a,b,c,d,\alpha,\beta \ \| \quad \begin{array}{ll} [b,a] = 1 & [d,a] = 1 \\ [c,a] = d^{k_1} & [d,b] = 1 \\ [c,b] = 1 & [d,c] = 1 \\ \alpha a = a^{-1}\alpha d^{k_2} & \alpha^2 = d^{k_4} \\ \alpha b = b\alpha & \alpha d = d\alpha \\ \alpha c = c^{-1}\alpha d^{k_3} & \\ \beta a = a^{-1}\beta d^{k_2} & \beta^2 = d^{k_6} \\ \beta b = b^{-1}\beta d^{k_5} & \beta d = d\beta \\ \beta c = c^{-1}\beta d^{k_3} & \alpha\beta = \beta\alpha \end{array} \quad >$$

$$\lambda(\alpha) = \begin{pmatrix} 1 & k_2 & 0 & k_3 & \frac{k_4}{2} \\ 0 & -1 & 0 & 0 & 0 \\ 0 & 0 & 1 & 0 & 0 \\ 0 & 0 & 0 & -1 & 0 \\ 0 & 0 & 0 & 0 & 1 \end{pmatrix} \quad \lambda(\beta) = \begin{pmatrix} 1 & k_2 & k_5 & k_3 & \frac{k_6}{2} \\ 0 & -1 & 0 & 0 & 0 \\ 0 & 0 & -1 & 0 & 0 \\ 0 & 0 & 0 & -1 & 0 \\ 0 & 0 & 0 & 0 & 1 \end{pmatrix}$$

$$H^2(Q,\mathbb{Z}) = \mathbb{Z} \oplus (\mathbb{Z}_2)^5 = \mathbb{Z}^6/A, \ A = \{(k_1,\ldots,k_6)\|k_1 = 0, \ k_2,\ldots,k_6 \in 2\mathbb{Z}\}$$

AB-groups:
None

13. $Q = P2/c$

$$E: <a,b,c,d,\alpha,\beta \parallel \quad [b,a] = 1 \qquad\qquad [d,a] = 1 \qquad\qquad >$$

$$[c,a] = d^{2k_1} \qquad\qquad [d,b] = 1$$
$$[c,b] = 1 \qquad\qquad [d,c] = 1$$
$$\alpha a = a^{-1}\alpha d^{k_2} \qquad\qquad \alpha^2 = d^{k_3}$$
$$\alpha b = b\alpha \qquad\qquad \alpha d = d\alpha$$
$$\alpha c = c^{-1}\alpha d^{-2k_6}$$
$$\beta a = a^{-1}\beta d^{k_1+k_2} \qquad \beta^2 = d^{k_5}$$
$$\beta b = b^{-1}\beta d^{k_4} \qquad\qquad \beta d = d\beta$$
$$\beta c = c^{-1}\beta d^{-2k_6} \qquad \alpha\beta = c\beta\alpha d^{k_6}$$

$$\lambda(\alpha) = \begin{pmatrix} 1 & \frac{k_1}{2}+k_2 & 0 & -2k_6 & \frac{k_3+k_6}{2} \\ 0 & -1 & 0 & 0 & 0 \\ 0 & 0 & 1 & 0 & 0 \\ 0 & 0 & 0 & -1 & \frac{1}{2} \\ 0 & 0 & 0 & 0 & 1 \end{pmatrix} \quad \lambda(\beta) = \begin{pmatrix} 1 & k_1+k_2 & k_4 & -2k_6 & \frac{k_5}{2} \\ 0 & -1 & 0 & 0 & 0 \\ 0 & 0 & -1 & 0 & 0 \\ 0 & 0 & 0 & -1 & 0 \\ 0 & 0 & 0 & 0 & 1 \end{pmatrix}$$

$$H^2(Q,\mathbb{Z}) = \mathbb{Z} \oplus (\mathbb{Z}_2)^4 = \mathbb{Z}^6/A,$$

$$A = \{(k_1,\ldots,k_6)\|k_1 = 0,\ k_2,\ldots,k_5 \in 2\mathbb{Z},\ k_6 \in \mathbb{Z}\}$$

AB-groups:
$$\forall k > 0,\ k \equiv 0 \bmod 2,\ <(k,0,1,0,1,0)>$$

---

11. $Q = P2_1/m$

$$E: <a,b,c,d,\alpha,\beta \parallel \quad [b,a] = 1 \qquad\qquad [d,a] = 1 \qquad\qquad >$$

$$[c,a] = d^{k_1} \qquad\qquad [d,b] = 1$$
$$[c,b] = 1 \qquad\qquad [d,c] = 1$$
$$\alpha a = a^{-1}\alpha d^{k_2} \qquad\qquad \alpha^2 = bd^{k_4}$$
$$\alpha b = b\alpha \qquad\qquad \alpha d = d\alpha$$
$$\alpha c = c^{-1}\alpha d^{k_3}$$
$$\beta a = a^{-1}\beta d^{k_2} \qquad\qquad \beta^2 = d^{k_5}$$
$$\beta b = b^{-1}\beta d^{-2k_6} \qquad \beta d = d\beta$$
$$\beta c = c^{-1}\beta d^{k_3} \qquad\qquad \alpha\beta = b\beta\alpha d^{k_6}$$

$$\lambda(\alpha) = \begin{pmatrix} 1 & k_2 & 0 & k_3 & \frac{k_4}{2} \\ 0 & -1 & 0 & 0 & 0 \\ 0 & 0 & 1 & 0 & \frac{1}{2} \\ 0 & 0 & 0 & -1 & 0 \\ 0 & 0 & 0 & 0 & 1 \end{pmatrix} \quad \lambda(\beta) = \begin{pmatrix} 1 & k_2 & -2k_6 & k_3 & \frac{k_5}{2} \\ 0 & -1 & 0 & 0 & 0 \\ 0 & 0 & -1 & 0 & 0 \\ 0 & 0 & 0 & -1 & 0 \\ 0 & 0 & 0 & 0 & 1 \end{pmatrix}$$

$$H^2(Q,\mathbb{Z}) = \mathbb{Z} \oplus (\mathbb{Z}_2)^4 = \mathbb{Z}^6/A,$$

$$A = \{(k_1,\ldots,k_6)\|k_1 = 0,\ k_2, k_3, k_5, (k_4 - k_6) \in 2\mathbb{Z}\}$$

AB-groups:
$$\forall k > 0,\ k \equiv 0 \bmod 2,\ <(k,0,0,0,1,0)>$$

14. $Q = P2_1/c$
Trivial Action:

$$E : < a, b, c, d, \alpha, \beta \ \| \quad \begin{aligned} [b,a] &= 1 \\ [c,a] &= d^{2k_1} \\ [c,b] &= 1 \\ \alpha a &= a^{-1}\alpha d^{k_2} \\ \alpha b &= b\alpha \\ \alpha c &= c^{-1}\alpha d^{k_3} \\ \beta a &= a^{-1}\beta d^{k_1+k_2} \\ \beta b &= b^{-1}\beta d^{-k_3-2k_6} \\ \beta c &= c^{-1}\beta d^{k_3} \end{aligned} \quad \begin{aligned} [d,a] &= 1 \\ [d,b] &= 1 \\ [d,c] &= 1 \\ \alpha^2 &= bd^{k_4} \\ \alpha d &= d\alpha \\ \\ \beta^2 &= d^{k_5} \\ \beta d &= d\beta \\ \alpha\beta &= bc\beta\alpha d^{k_6} \end{aligned} \quad >$$

$$\lambda(\alpha) = \begin{pmatrix} 1 & \frac{k_1}{2}+k_2 & 0 & k_3 & -\frac{k_3}{4}+\frac{k_4}{2} \\ 0 & -1 & 0 & 0 & 0 \\ 0 & 0 & 1 & 0 & \frac{1}{2} \\ 0 & 0 & 0 & -1 & \frac{1}{2} \\ 0 & 0 & 0 & 0 & 1 \end{pmatrix} \quad \lambda(\beta) = \begin{pmatrix} 1 & k_1+k_2 & -k_3-2k_6 & k_3 & \frac{k_5}{2} \\ 0 & -1 & 0 & 0 & 0 \\ 0 & 0 & -1 & 0 & 0 \\ 0 & 0 & 0 & -1 & 0 \\ 0 & 0 & 0 & 0 & 1 \end{pmatrix}$$

$$H^2(Q, \mathbb{Z}) = \mathbb{Z} \oplus (\mathbb{Z}_2)^2 \oplus \mathbb{Z}_4 = \mathbb{Z}^6/A,$$

$$A = \{(k_1, \dots, k_6) \| k_1 = 0, \ k_2, k_5 \in 2\mathbb{Z}, \ (-k_3 + 2k_4 - 2k_6) \in 4\mathbb{Z}\}$$

AB-groups:
$\forall k > 0, \ k \equiv 0 \bmod 2, \ < (k, 0, 0, 0, 1, 0) >$
$\forall k > 0, \ k \equiv 0 \bmod 2, \ < (k, 0, 0, 1, 1, 0) >$
$\forall k > 0, \ k \not\equiv 0 \bmod 2, \ < (k, 1, 0, 0, 1, 0) >$
Remark: $\forall k \not\equiv 0 \bmod 2, \ < (k, 1, 0, 0, 1, 0) > \equiv < (k, 1, 0, 1, 1, 0) >$
Action $^\alpha d = d^{-1}, \ ^\beta d = d$:

$$E : < a, b, c, d, \alpha, \beta \ \| \quad \begin{aligned} [b,a] &= d^{2k_1} \\ [c,a] &= 1 \\ [c,b] &= d^{2k_2} \\ \alpha a &= a^{-1}\alpha d^{k_1} \\ \alpha b &= b\alpha d^{-2k_3} \\ \alpha c &= c^{-1}\alpha d^{-k_2} \\ \beta a &= a^{-1}\beta d^{k_4} \\ \beta b &= b^{-1}\beta d^{k_2-2k_3} \\ \beta c &= c^{-1}\beta d^{k_2+2k_3-2k_5+2k_6} \end{aligned} \quad \begin{aligned} [d,a] &= 1 \\ [d,b] &= 1 \\ [d,c] &= 1 \\ \alpha^2 &= bd^{k_3} \\ \alpha d &= d^{-1}\alpha \\ \\ \beta^2 &= d^{k_5} \\ \beta d &= d\beta \\ \alpha\beta &= bc\beta\alpha d^{k_6} \end{aligned} \quad >$$

$$\lambda(\alpha) = \begin{pmatrix} -1 & -\frac{k_1}{2} & -\frac{k_2}{2}+2k_3 & \frac{k_2}{2} & 0 \\ 0 & -1 & 0 & 0 & 0 \\ 0 & 0 & 1 & 0 & \frac{1}{2} \\ 0 & 0 & 0 & -1 & \frac{1}{2} \\ 0 & 0 & 0 & 0 & 1 \end{pmatrix}$$

$$\lambda(\beta) = \begin{pmatrix} 1 & k_4 & k_2 - 2k_3 & k_2 + 2k_3 - 2k_5 + 2k_6 & \frac{k_5}{2} \\ 0 & -1 & 0 & 0 & 0 \\ 0 & 0 & -1 & 0 & 0 \\ 0 & 0 & 0 & -1 & 0 \\ 0 & 0 & 0 & 0 & 1 \end{pmatrix}$$

$$H^2(Q, \mathbb{Z}) = \mathbb{Z}^2 \oplus (\mathbb{Z}_2)^2 = \mathbb{Z}^6/A,$$

$$A = \{(k_1, \ldots, k_6) \| k_1 = k_2 = 0, k_4, k_5 \in 2\mathbb{Z}, k_3, k_6 \in \mathbb{Z}\}$$

AB-groups:

$\forall k > 0, \; < (k, 0, 0, 0, 1, 0) >$

$\forall k > 0, \; k \equiv 0 \bmod 2, \; < (0, k, 0, 0, 1, 0) >$

---

12. $Q = C2/m$

$$E : \; < a, b, c, d, \alpha, \beta \; \| \quad \begin{aligned} &[b, a] = 1 & &[d, a] = 1 \quad > \\ &[c, a] = d^{k_1} & &[d, b] = 1 \\ &[c, b] = d^{k_1} & &[d, c] = 1 \\ &\alpha a = b^{-1}\alpha d^{k_2} & &\alpha^2 = d^{k_4} \\ &\alpha b = a^{-1}\alpha d^{k_2} & &\alpha d = d\alpha \\ &\alpha c = c^{-1}\alpha d^{k_3} & & \\ &\beta a = a^{-1}\beta d^{k_5} & &\beta^2 = d^{k_6} \\ &\beta b = b^{-1}\beta d^{2k_2 - k_5} & &\beta d = d\beta \\ &\beta c = c^{-1}\beta d^{k_3} & &\alpha\beta = \beta\alpha \end{aligned}$$

$$\lambda(\alpha) = \begin{pmatrix} 1 & k_2 & k_2 & k_3 & \frac{k_4}{2} \\ 0 & 0 & -1 & 0 & 0 \\ 0 & -1 & 0 & 0 & 0 \\ 0 & 0 & 0 & -1 & 0 \\ 0 & 0 & 0 & 0 & 1 \end{pmatrix} \quad \lambda(\beta) = \begin{pmatrix} 1 & k_5 & 2k_2 - k_5 & k_3 & \frac{k_6}{2} \\ 0 & -1 & 0 & 0 & 0 \\ 0 & 0 & -1 & 0 & 0 \\ 0 & 0 & 0 & -1 & 0 \\ 0 & 0 & 0 & 0 & 1 \end{pmatrix}$$

$$H^2(Q, \mathbb{Z}) = \mathbb{Z} \oplus (\mathbb{Z}_2)^4 = \mathbb{Z}^6/A,$$

$$A = \{(k_1, \ldots, k_6) \| k_1 = 0, \; k_3, k_4, k_5, k_6 \in 2\mathbb{Z}, \; k_2 \in \mathbb{Z}\}$$

AB-groups:

None

---

15. $Q = C2/c$

$$E : \; < a, b, c, d, \alpha, \beta \; \| \quad \begin{aligned} &[b, a] = 1 & &[d, a] = 1 \quad > \\ &[c, a] = d^{k_1} & &[d, b] = 1 \\ &[c, b] = d^{k_1} & &[d, c] = 1 \\ &\alpha a = b^{-1}\alpha d^{k_2} & &\alpha^2 = d^{k_3} \\ &\alpha b = a^{-1}\alpha d^{k_2} & &\alpha d = d\alpha \\ &\alpha c = c^{-1}\alpha d^{-2k_6} & & \\ &\beta a = a^{-1}\beta d^{k_4} & &\beta^2 = d^{k_5} \\ &\beta b = b^{-1}\beta d^{k_1 + 2k_2 - k_4} & &\beta d = d\beta \\ &\beta c = c^{-1}\beta d^{-2k_6} & &\alpha\beta = c\beta\alpha d^{k_6} \end{aligned}$$

$$\lambda(\alpha) = \begin{pmatrix} 1 & \frac{k_1}{4}+k_2 & \frac{k_1}{4}+k_2 & -2k_6 & \frac{k_3}{2}+\frac{k_6}{2} \\ 0 & 0 & -1 & 0 & 0 \\ 0 & -1 & 0 & 0 & 0 \\ 0 & 0 & 0 & -1 & \frac{1}{2} \\ 0 & 0 & 0 & 0 & 1 \end{pmatrix}$$

$$\lambda(\beta) = \begin{pmatrix} 1 & k_4 & k_1+2k_2-k_4 & -2k_6 & \frac{k_5}{2} \\ 0 & -1 & 0 & 0 & 0 \\ 0 & 0 & -1 & 0 & 0 \\ 0 & 0 & 0 & -1 & 0 \\ 0 & 0 & 0 & 0 & 1 \end{pmatrix}$$

$$H^2(Q,\mathbb{Z}) = \mathbb{Z} \oplus (\mathbb{Z}_2)^3 = \mathbb{Z}^6/A,$$

$$A = \{(k_1,\ldots,k_6)\|k_1 = 0,\ k_3,k_4,k_5 \in 2\mathbb{Z},\ k_2,k_6 \in \mathbb{Z}\}$$

AB-groups:
$\forall k > 0\ k \equiv \bmod 2,\ <(k,0,1,0,1,0)>$

---

18. $Q = P2_12_12$

$E: <a,b,c,d,\alpha,\beta\ \|\ \begin{array}{ll} [b,a]=d^{2k_1} & [d,a]=1 \\ [c,a]=1 & [d,b]=1 \\ [c,b]=1 & [d,c]=1 \\ \alpha a = a^{-1}\alpha d^{-k_1-2k_3+2k_4+2k_2} & \alpha^2 = d^{k_2} \\ \alpha b = b^{-1}\alpha d^{k_1-2k_3} & \alpha d = d\alpha \\ \alpha c = c\alpha & \\ \beta a = a^{-1}\beta d^{k_1} & \beta^2 = bd^{k_3} \\ \beta b = b\beta d^{-2k_3} & \beta d = d^{-1}\beta \\ \beta c = c^{-1}\beta & \alpha\beta = ab^{-1}\beta\alpha d^{k_4} \end{array}\ >$

$$\lambda(\alpha) = \begin{pmatrix} 1 & -k_1+2(k_2-k_3+k_4) & \frac{k_1}{2}-2k_3 & 0 & \frac{k_1}{4}+\frac{k_3}{2}-\frac{k_4}{2} \\ 0 & -1 & 0 & 0 & \frac{1}{2} \\ 0 & 0 & -1 & 0 & 0 \\ 0 & 0 & 0 & 1 & 0 \\ 0 & 0 & 0 & 0 & 1 \end{pmatrix}$$

$$\lambda(\beta) = \begin{pmatrix} -1 & -\frac{k_1}{2} & 2k_3 & 0 & 0 \\ 0 & -1 & 0 & 0 & 0 \\ 0 & 0 & 1 & 0 & \frac{1}{2} \\ 0 & 0 & 0 & -1 & 0 \\ 0 & 0 & 0 & 0 & 1 \end{pmatrix}$$

$$H^2(Q,\mathbb{Z}) = \mathbb{Z} \oplus \mathbb{Z}_2 = \mathbb{Z}^4/A,\ A = \{(k_1,\ldots,k_4)\|k_1 = 0,\ k_2 \in 2\mathbb{Z},\ k_3,k_4 \in \mathbb{Z}\}$$

AB-groups:
$\forall k > 0\ k \equiv \bmod 2,\ <(k,1,0,0)>$

---

19. $Q = P2_12_12_1$
Action $^\alpha d = d^{-1}$, $^\beta d = d$:

$E: <a,b,c,d,\alpha,\beta \|$

$[b,a] = 1$	$[d,a] = 1$
$[c,a] = d^{2k_1}$	$[d,b] = 1$
$[c,b] = 1$	$[d,c] = 1$
$\alpha a = a^{-1}\alpha d^{k_1}$	$\alpha^2 = cd^{k_2}$
$\alpha b = b^{-1}\alpha$	$\alpha d = d^{-1}\alpha$
$\alpha c = c\alpha d^{-2k_2}$	
$\beta a = a^{-1}\beta d^{3k_1+2(k_2-k_3+k_4)}$	$\beta^2 = bd^{k_3}$
$\beta b = b\beta$	$\beta d = d\beta$
$\beta c = c^{-1}\beta d^{-k_1-2k_2}$	$\alpha\beta = ab^{-1}c\beta\alpha d^{k_4}$

$>$

$$\lambda(\alpha) = \begin{pmatrix} -1 & -\frac{k_1}{2} & 0 & \frac{k_1}{2}+2k_2 & 0 \\ 0 & -1 & 0 & 0 & \frac{1}{2} \\ 0 & 0 & -1 & 0 & 0 \\ 0 & 0 & 0 & 1 & \frac{1}{2} \\ 0 & 0 & 0 & 0 & 1 \end{pmatrix}$$

$$\lambda(\beta) = \begin{pmatrix} 1 & 3k_1+2(k_2-k_3+k_4) & 0 & -k_1-2k_2 & \frac{k_3}{2} \\ 0 & -1 & 0 & 0 & 0 \\ 0 & 0 & 1 & 0 & \frac{1}{2} \\ 0 & 0 & 0 & -1 & 0 \\ 0 & 0 & 0 & 0 & 1 \end{pmatrix}$$

$$H^2(Q,\mathbb{Z}) = \mathbb{Z} = \mathbb{Z}^4/A, \ A = \{(k_1,\ldots,k_4)\|k_1 = 0, \ k_2,k_3,k_4 \in \mathbb{Z}\}$$

AB-groups:
$\forall k > 0 \ <(k,0,0,0)>$
Action $^\alpha d = d$, $^\beta d = d^{-1}$:

$E: <a,b,c,d,\alpha,\beta \|$

$[b,a] = d^{2k_1}$	$[d,a] = 1$
$[c,a] = 1$	$[d,b] = 1$
$[c,b] = 1$	$[d,c] = 1$
$\alpha a = a^{-1}\alpha d^{-k_1-2k_3+2k_4+2k_2}$	$\alpha^2 = cd^{k_2}$
$\alpha b = b^{-1}\alpha d^{k_1-2k_3}$	$\alpha d = d\alpha$
$\alpha c = c\alpha$	
$\beta a = a^{-1}\beta d^{k_1}$	$\beta^2 = bd^{k_3}$
$\beta b = b\beta d^{-2k_3}$	$\beta d = d^{-1}\beta$
$\beta c = c^{-1}\beta$	$\alpha\beta = ab^{-1}c\beta\alpha d^{k_4}$

$>$

$$\lambda(\alpha) = \begin{pmatrix} 1 & -k_1+2(k_2-k_3+k_4) & \frac{k_1}{2}-2k_3 & 0 & \frac{k_1}{4}+\frac{k_3-k_4}{2} \\ 0 & -1 & 0 & 0 & \frac{1}{2} \\ 0 & 0 & -1 & 0 & 0 \\ 0 & 0 & 0 & 1 & \frac{1}{2} \\ 0 & 0 & 0 & 0 & 1 \end{pmatrix}$$

$$\lambda(\beta) = \begin{pmatrix} -1 & -\frac{k_1}{2} & 2k_3 & 0 & 0 \\ 0 & -1 & 0 & 0 & 0 \\ 0 & 0 & 1 & 0 & \frac{1}{2} \\ 0 & 0 & 0 & -1 & 0 \\ 0 & 0 & 0 & 0 & 1 \end{pmatrix}$$

$H^2(Q, \mathbb{Z}) = \mathbb{Z} = \mathbb{Z}^4/A, \ A = \{(k_1, \ldots, k_4) \| k_1 = 0, \ k_2, k_3, k_4 \in \mathbb{Z}\}$

AB-groups:

All groups are isomorphic to one of the previous case.

Action $^\alpha d = d^{-1}, \ ^\beta d = d^{-1}$:

$$E: <a, b, c, d, \alpha, \beta \ \| \quad \begin{aligned} [b, a] &= 1 \\ [c, a] &= 1 \\ [c, b] &= d^{2k_1} \\ \alpha a &= a^{-1}\alpha \\ \alpha b &= b^{-1}\alpha d^{k_1} \\ \alpha c &= c\alpha d^{-2k_2} \\ \beta a &= a^{-1}\beta \\ \beta b &= b\beta d^{-2k_3} \\ \beta c &= c^{-1}\beta d^{-k_1} \end{aligned} \qquad \begin{aligned} [d, a] &= 1 \\ [d, b] &= 1 \\ [d, c] &= 1 \\ \alpha^2 &= cd^{k_2} \\ \alpha d &= d^{-1}\alpha \\ \\ \beta^2 &= bd^{k_3} \\ \beta d &= d^{-1}\beta \\ \alpha\beta &= ab^{-1}c\beta\alpha d^{k_4} \end{aligned} \qquad >$$

$$\lambda(\alpha) = \begin{pmatrix} -1 & 0 & -\frac{k_1}{2} & 2k_2 & \frac{3k_1}{4} + \frac{k_4}{2} \\ 0 & -1 & 0 & 0 & \frac{1}{2} \\ 0 & 0 & -1 & 0 & 0 \\ 0 & 0 & 0 & 1 & \frac{1}{2} \\ 0 & 0 & 0 & 0 & 1 \end{pmatrix}$$

$$\lambda(\beta) = \begin{pmatrix} -1 & 0 & 2k_3 & \frac{k_1}{2} & 0 \\ 0 & -1 & 0 & 0 & 0 \\ 0 & 0 & 1 & 0 & \frac{1}{2} \\ 0 & 0 & 0 & -1 & 0 \\ 0 & 0 & 0 & 0 & 1 \end{pmatrix}$$

$H^2(Q, \mathbb{Z}) = \mathbb{Z} = \mathbb{Z}^4/A, \ A = \{(k_1, \ldots, k_4) \| k_1 = 0, \ k_2, k_3, k_4 \in \mathbb{Z}\}$

AB-groups:

All groups are isomorphic to one of the previous case.

---

27. $Q = Pcc2$

$$E: <a, b, c, d, \alpha, \beta \ \| \quad \begin{aligned} [b, a] &= d^{k_1} \\ [c, a] &= 1 \\ [c, b] &= 1 \\ \alpha a &= a^{-1}\alpha d^{k_2} \\ \alpha b &= b^{-1}\alpha d^{k_3} \\ \alpha c &= c\alpha \\ \beta a &= a\beta d^{k_2} \\ \beta b &= b^{-1}\beta \\ \beta c &= c\beta d^{-2k_5} \end{aligned} \qquad \begin{aligned} [d, a] &= 1 \\ [d, b] &= 1 \\ [d, c] &= 1 \\ \alpha^2 &= d^{k_4} \\ \alpha d &= d\alpha \\ \\ \beta^2 &= cd^{k_5} \\ \beta d &= d^{-1}\beta \\ \alpha\beta &= \beta\alpha d^{-k_4} \end{aligned} \qquad >$$

$$\lambda(\alpha) = \begin{pmatrix} 1 & k_2 & k_3 & 0 & \frac{k_4}{2} \\ 0 & -1 & 0 & 0 & 0 \\ 0 & 0 & -1 & 0 & 0 \\ 0 & 0 & 0 & 1 & 0 \\ 0 & 0 & 0 & 0 & 1 \end{pmatrix} \quad \lambda(\beta) = \begin{pmatrix} -1 & -k_2 & 0 & 2k_5 & 0 \\ 0 & 1 & 0 & 0 & 0 \\ 0 & 0 & -1 & 0 & 0 \\ 0 & 0 & 0 & 1 & \frac{1}{2} \\ 0 & 0 & 0 & 0 & 1 \end{pmatrix}$$

$$H^2(Q, \mathbb{Z}) = \mathbb{Z} \oplus (\mathbb{Z}_2)^3 = \mathbb{Z}^5/A,$$

$$A = \{(k_1, \ldots, k_5) \| k_1 = 0,\ k_2, k_3, k_4 \in 2\mathbb{Z},\ k_5 \in \mathbb{Z}\}$$

AB-groups:
$\forall k > 0\ k \equiv \mathrm{mod}\,2,\ < (k, 0, 0, 1, 0) >$

---

30. $Q = Pnc2$

$$E :\ < a, b, c, d, \alpha, \beta\ \|$$

$[b, a] = d^{2k_1}$	$[d, a] = 1$
$[c, a] = 1$	$[d, b] = 1$
$[c, b] = 1$	$[d, c] = 1$
$\alpha a = a^{-1}\alpha d^{k_2}$	$\alpha^2 = d^{k_3}$
$\alpha b = b^{-1}\alpha d^{2k_3 + 2k_5}$	$\alpha d = d\alpha$
$\alpha c = c\alpha$	
$\beta a = a\beta d^{k_1 + k_2}$	$\beta^2 = cd^{k_4}$
$\beta b = b^{-1}\beta$	$\beta d = d^{-1}\beta$
$\beta c = c\beta d^{-2k_4}$	$\alpha\beta = b\beta\alpha d^{k_5}$

$\quad >$

$$\lambda(\alpha) = \begin{pmatrix} 1 & \frac{k_1}{2} + k_2 & 2(k_3 + k_5) & 0 & -\frac{k_5}{2} \\ 0 & -1 & 0 & 0 & 0 \\ 0 & 0 & -1 & 0 & \frac{1}{2} \\ 0 & 0 & 0 & 1 & 0 \\ 0 & 0 & 0 & 0 & 1 \end{pmatrix} \quad \lambda(\beta) = \begin{pmatrix} -1 & -k_1 - k_2 & 0 & 2k_4 & 0 \\ 0 & 1 & 0 & 0 & 0 \\ 0 & 0 & -1 & 0 & 0 \\ 0 & 0 & 0 & 1 & \frac{1}{2} \\ 0 & 0 & 0 & 0 & 1 \end{pmatrix}$$

$$H^2(Q, \mathbb{Z}) = \mathbb{Z} \oplus (\mathbb{Z}_2)^2 = \mathbb{Z}^5/A,$$

$$A = \{(k_1, \ldots, k_5) \| k_1 = 0,\ k_2, k_3 \in 2\mathbb{Z},\ k_4, k_5 \in \mathbb{Z}\}$$

AB-groups:
$\forall k > 0\ < (k, 0, 1, 0, 0) >$

---

32. $Q = Pba2$

$$E :\ < a, b, c, d, \alpha, \beta\ \|$$

$[b, a] = d^{2k_1}$	$[d, a] = 1$
$[c, a] = 1$	$[d, b] = 1$
$[c, b] = 1$	$[d, c] = 1$
$\alpha a = a^{-1}\alpha d^{-k_1 - 2k_4}$	$\alpha^2 = d^{k_2}$
$\alpha b = b^{-1}\alpha d^{-3k_1 - 2k_4 + 2k_5 + 2k_2}$	$\alpha d = d\alpha$
$\alpha c = c\alpha$	
$\beta a = a\beta d^{-2k_4}$	$\beta^2 = ad^{k_4}$
$\beta b = b^{-1}\beta d^{-k_1}$	$\beta d = d^{-1}\beta$
$\beta c = c\beta d^{k_3}$	$\alpha\beta = a^{-1}b\beta\alpha d^{k_5}$

$\quad >$

$$\lambda(\alpha) = \begin{pmatrix} 1 & -\frac{k_1}{2} - 2k_4 & -3k_1 + 2(k_2 - k_4 + k_5) & 0 & \frac{3k_1}{4} + \frac{k_4}{2} - \frac{k_5}{2} \\ 0 & -1 & 0 & 0 & 0 \\ 0 & 0 & -1 & 0 & \frac{1}{2} \\ 0 & 0 & 0 & 1 & 0 \\ 0 & 0 & 0 & 0 & 1 \end{pmatrix}$$

$$\lambda(\beta) = \begin{pmatrix} -1 & 2k_4 & \frac{k_1}{2} & -k_3 & 0 \\ 0 & 1 & 0 & 0 & \frac{1}{2} \\ 0 & 0 & -1 & 0 & 0 \\ 0 & 0 & 0 & 1 & 0 \\ 0 & 0 & 0 & 0 & 1 \end{pmatrix}$$

$$H^2(Q, \mathbb{Z}) = \mathbb{Z} \oplus (\mathbb{Z}_2)^2 = \mathbb{Z}^5 / A,$$

$$A = \{(k_1, \ldots, k_5) \| k_1 = 0, \ k_2, k_3 \in 2\mathbb{Z}, \ k_4, k_5 \in \mathbb{Z}\}$$

AB-groups:

$\forall k > 0 \ k \equiv 0 \bmod 2, \ < (k, 1, 0, 0, 0) >$

$\forall k > 0 \ k \equiv 0 \bmod 2, \ < (k, 1, 1, 0, 0) >$

---

34. $Q = Pnn2$

$E : \; < a, b, c, d, \alpha, \beta \, \| \quad$

$[b, a] = d^{2k_1}$	$[d, a] = 1$
$[c, a] = 1$	$[d, b] = 1$
$[c, b] = 1$	$[d, c] = 1$
$\alpha a = a^{-1} \alpha d^{k_2}$	$\alpha^2 = d^{k_3}$
$\alpha b = b^{-1} \alpha d^{-2k_1 + 2k_3 + 2k_5 + k_2}$	$\alpha d = d\alpha$
$\alpha c = c\alpha$	
$\beta a = a\beta d^{k_1 + k_2}$	$\beta^2 = acd^{k_4}$
$\beta b = b^{-1} \beta d^{-k_1}$	$\beta d = d^{-1} \beta$
$\beta c = c\beta d^{-k_1 - k_2 - 2k_4}$	$\alpha\beta = a^{-1} b\beta \alpha d^{k_5}$

$>$

$$\lambda(\alpha) = \begin{pmatrix} 1 & \frac{k_1}{2} + k_2 & k_2 + 2(k_3 - k_1 + k_5) & 0 & \frac{k_1}{2} - \frac{k_2}{4} - \frac{k_5}{2} \\ 0 & -1 & 0 & 0 & 0 \\ 0 & 0 & -1 & 0 & \frac{1}{2} \\ 0 & 0 & 0 & 1 & 0 \\ 0 & 0 & 0 & 0 & 1 \end{pmatrix}$$

$$\lambda(\beta) = \begin{pmatrix} -1 & -k_1 - k_2 & \frac{k_1}{2} & k_1 + k_2 + 2k_4 & 0 \\ 0 & 1 & 0 & 0 & \frac{1}{2} \\ 0 & 0 & -1 & 0 & 0 \\ 0 & 0 & 0 & 1 & \frac{1}{2} \\ 0 & 0 & 0 & 0 & 1 \end{pmatrix}$$

$$H^2(Q, \mathbb{Z}) = \mathbb{Z} \oplus (\mathbb{Z}_2)^2 = \mathbb{Z}^5 / A,$$

$$A = \{(k_1, \ldots, k_5) \| k_1 = 0, \ k_2, k_3 \in 2\mathbb{Z}, \ k_4, k_5 \in \mathbb{Z}\}$$

AB-groups:

$\forall k > 0 \ < (k, 0, 1, 0, 0) >$

26. $Q = Pmc2_1$

$$E : \; < a, b, c, d, \alpha, \beta \; || \quad \begin{array}{ll} [b,a] = 1 & [d,a] = 1 \\ [c,a] = d^{2k_1} & [d,b] = 1 \\ [c,b] = 1 & [d,c] = 1 \\ \alpha a = a^{-1}\alpha d^{k_1} & \alpha^2 = cd^{k_2} \\ \alpha b = b^{-1}\alpha & \alpha d = d^{-1}\alpha \\ \alpha c = c\alpha d^{-2k_2} & \\ \beta a = a\beta & \beta^2 = d^{k_4} \\ \beta b = b^{-1}\beta d^{k_3} & \beta d = d\beta \\ \beta c = c\beta & \alpha\beta = \beta\alpha d^{k_4} \end{array} \quad >$$

$$\lambda(\alpha) = \begin{pmatrix} -1 & -\frac{k_1}{2} & 0 & 2k_2 & 0 \\ 0 & -1 & 0 & 0 & 0 \\ 0 & 0 & -1 & 0 & 0 \\ 0 & 0 & 0 & 1 & \frac{1}{2} \\ 0 & 0 & 0 & 0 & 1 \end{pmatrix} \qquad \lambda(\beta) = \begin{pmatrix} 1 & 0 & k_3 & 0 & \frac{k_4}{2} \\ 0 & 1 & 0 & 0 & 0 \\ 0 & 0 & -1 & 0 & 0 \\ 0 & 0 & 0 & 1 & 0 \\ 0 & 0 & 0 & 0 & 1 \end{pmatrix}$$

$$H^2(Q, \mathbb{Z}) = \mathbb{Z} \oplus (\mathbb{Z}_2)^2 = \mathbb{Z}^4/A$$

$$A = \{(k_1, \ldots, k_4) \| k_1 = 0, \; k_3, k_4 \in 2\mathbb{Z}, \; k_2 \in \mathbb{Z}\}$$

AB-groups:
$\forall k > 0 \; < (k, 0, 0, 1) >$

---

31. $Q = Pmn2_1$

$$E : \; < a, b, c, d, \alpha, \beta \; || \quad \begin{array}{ll} [b,a] = 1 & [d,a] = 1 \\ [c,a] = d^{2k_1} & [d,b] = 1 \\ [c,b] = 1 & [d,c] = 1 \\ \alpha a = a^{-1}\alpha d^{k_1} & \alpha^2 = cd^{k_2} \\ \alpha b = b^{-1}\alpha & \alpha d = d^{-1}\alpha \\ \alpha c = c\alpha d^{-2k_2} & \\ \beta a = a\beta & \beta^2 = d^{k_3} \\ \beta b = b^{-1}\beta d^{-2k_3+2k_4} & \beta d = d\beta \\ \beta c = c\beta & \alpha\beta = b\beta\alpha d^{k_4} \end{array} \quad >$$

$$\lambda(\alpha) = \begin{pmatrix} -1 & -\frac{k_1}{2} & 0 & 2k_2 & 0 \\ 0 & -1 & 0 & 0 & 0 \\ 0 & 0 & -1 & 0 & \frac{1}{2} \\ 0 & 0 & 0 & 1 & \frac{1}{2} \\ 0 & 0 & 0 & 0 & 1 \end{pmatrix} \qquad \lambda(\beta) = \begin{pmatrix} 1 & 0 & -2k_3+2k_4 & 0 & \frac{k_3}{2} \\ 0 & 1 & 0 & 0 & 0 \\ 0 & 0 & -1 & 0 & 0 \\ 0 & 0 & 0 & 1 & 0 \\ 0 & 0 & 0 & 0 & 1 \end{pmatrix}$$

$$H^2(Q, \mathbb{Z}) = \mathbb{Z} \oplus \mathbb{Z}_2 = \mathbb{Z}^4/A, \quad A = \{(k_1, \ldots, k_4) \| k_1 = 0, \; k_3 \in 2\mathbb{Z}, \; k_2, k_4 \in \mathbb{Z}\}$$

AB-groups:
$\forall k > 0 \; < (k, 0, 1, 0) >$

29. $Q = Pca2_1$

Action $^\alpha d = d^{-1}$, $^\beta d = d$:

$$E : < a, b, c, d, \alpha, \beta \ \|$$

$[b, a] = 1$	$[d, a] = 1$
$[c, a] = d^{2k_1}$	$[d, b] = 1$
$[c, b] = 1$	$[d, c] = 1$
$\alpha a = a^{-1}\alpha d^{k_1}$	$\alpha^2 = cd^{k_2}$
$\alpha b = b^{-1}\alpha$	$\alpha d = d^{-1}\alpha$
$\alpha c = c\alpha d^{-2k_2}$	
$\beta a = a\beta d^{k_1}$	$\beta^2 = cd^{k_3}$
$\beta b = b^{-1}\beta d^{2k_2 - 2k_3 + 2k_4}$	$\beta d = d\beta$
$\beta c = c\beta$	$\alpha\beta = b\beta\alpha d^{k_4}$

$> $

$$\lambda(\alpha) = \begin{pmatrix} -1 & -\frac{k_1}{2} & 0 & 2k_2 & 0 \\ 0 & -1 & 0 & 0 & 0 \\ 0 & 0 & -1 & 0 & \frac{1}{2} \\ 0 & 0 & 0 & 1 & \frac{1}{2} \\ 0 & 0 & 0 & 0 & 1 \end{pmatrix} \quad \lambda(\beta) = \begin{pmatrix} 1 & \frac{k_1}{2} & 2k_2 - 2k_3 + 2k_4 & 0 & \frac{k_3}{2} \\ 0 & 1 & 0 & 0 & 0 \\ 0 & 0 & -1 & 0 & 0 \\ 0 & 0 & 0 & 1 & \frac{1}{2} \\ 0 & 0 & 0 & 0 & 1 \end{pmatrix}$$

$H^2(Q, \mathbb{Z}) = \mathbb{Z} \oplus \mathbb{Z}_2 = \mathbb{Z}^4/A, \ A = \{(k_1, \ldots, k_4) \| k_1 = 0, \ (k_3 - k_2) \in 2\mathbb{Z}, \ k_4 \in \mathbb{Z}\}$

AB-groups:

$\forall k > 0 \ < (k, 0, 0, 0) >$

$\forall k > 0 \ k \equiv 0 \mod 2 \ < (k, 1, 0, 0) >$

Remark: If $k \not\equiv 0 \mod 2 \ < (k, 0, 0, 0) > \ \cong < (k, 1, 0, 0) >$

Action $^\alpha d = d$, $^\beta d = d^{-1}$:

$$E : < a, b, c, d, \alpha, \beta \ \|$$

$[b, a] = d^{2k_1}$	$[d, a] = 1$
$[c, a] = 1$	$[d, b] = 1$
$[c, b] = 1$	$[d, c] = 1$
$\alpha a = a^{-1}\alpha d^{k_2}$	$\alpha^2 = cd^{k_3}$
$\alpha b = b^{-1}\alpha d^{2k_3 - 2k_4 + 2k_5}$	$\alpha d = d\alpha$
$\alpha c = c\alpha$	
$\beta a = a\beta d^{k_1 + k_2}$	$\beta^2 = cd^{k_4}$
$\beta b = b^{-1}\beta$	$\beta d = d^{-1}\beta$
$\beta c = c\beta d^{-2k_4}$	$\alpha\beta = b\beta\alpha d^{k_5}$

$> $

$$\lambda(\alpha) = \begin{pmatrix} 1 & \frac{k_1}{2} + k_2 & 2(k_3 - k_4 + k_5) & 0 & \frac{k_4 - k_5}{2} \\ 0 & -1 & 0 & 0 & 0 \\ 0 & 0 & -1 & 0 & \frac{1}{2} \\ 0 & 0 & 0 & 1 & \frac{1}{2} \\ 0 & 0 & 0 & 0 & 1 \end{pmatrix}$$

$$\lambda(\beta) = \begin{pmatrix} -1 & -k_1 - k_2 & 0 & 2k_4 & 0 \\ 0 & 1 & 0 & 0 & 0 \\ 0 & 0 & -1 & 0 & 0 \\ 0 & 0 & 0 & 1 & \frac{1}{2} \\ 0 & 0 & 0 & 0 & 1 \end{pmatrix}$$

$H^2(Q, \mathbb{Z}) = \mathbb{Z} \oplus (\mathbb{Z}_2)^2 = \mathbb{Z}^5/A, \, A = \{(k_1, \ldots, k_5) \| k_1 = 0, (k_3 - k_4), k_2 \in 2\mathbb{Z}, k_5 \in \mathbb{Z}\}$

AB-groups:

$\forall k > 0 \, < (k, 1, 0, 0, 0) >$

$\forall k > 0 \, < (k, 0, 0, 0, 0) >$

$\forall k > 0 \, < (k, 0, 1, 0, 0) >$

Remark: $< (k, 1, 0, 0, 0) > \cong < (k, 1, 1, 0, 0) >$.

Action $^\alpha d = d^{-1}$, $^\beta d = d^{-1}$:

$$E : < a, b, c, d, \alpha, \beta \, \| \quad \begin{array}{ll} [b, a] = 1 & [d, a] = 1 \qquad > \\ [c, a] = 1 & [d, b] = 1 \\ [c, b] = d^{4k_1} & [d, c] = 1 \\ \alpha a = a^{-1}\alpha & \alpha^2 = cd^{k_2} \\ \alpha b = b^{-1}\alpha d^{2k_1} & \alpha d = d^{-1}\alpha \\ \alpha c = c\alpha d^{-2k_2} & \\ \beta a = a\beta d^{k_3} & \beta^2 = cd^{k_1+k_2} \\ \beta b = b^{-1}\beta d^{2k_1} & \beta d = d^{-1}\beta \\ \beta c = c\beta d^{-2k_1-2k_2} & \alpha\beta = b\beta\alpha d^{k_4} \end{array}$$

$$\lambda(\alpha) = \begin{pmatrix} -1 & 0 & -k_1 & k_1 + 2k_2 & -k_1 \\ 0 & -1 & 0 & 0 & 0 \\ 0 & 0 & -1 & 0 & \frac{1}{2} \\ 0 & 0 & 0 & 1 & \frac{1}{2} \\ 0 & 0 & 0 & 0 & 1 \end{pmatrix}$$

$$\lambda(\beta) = \begin{pmatrix} -1 & -k_3 & -k_1 & 2(k_1 + k_2) & -\frac{k_4}{2} \\ 0 & 1 & 0 & 0 & 0 \\ 0 & 0 & -1 & 0 & 0 \\ 0 & 0 & 0 & 1 & \frac{1}{2} \\ 0 & 0 & 0 & 0 & 1 \end{pmatrix}$$

$H^2(Q, \mathbb{Z}) = \mathbb{Z} \oplus \mathbb{Z}_2 = \mathbb{Z}^4/A, \; A = \{(k_1, \ldots, k_4) \| k_1 = 0, \; k_3 \in 2\mathbb{Z}, \; k_2, k_4 \in \mathbb{Z}\}$

AB-groups:

$\forall k > 0 \, < (k, 0, 0, 0) >$

$\forall k > 0 \, < (k, 0, 1, 0) >$

---

33. $Q = Pna2_1$

Action $^\alpha d = d^{-1}$, $^\beta d = d$:

$$E : < a, b, c, d, \alpha, \beta \, \| \quad \begin{array}{ll} [b, a] = 1 & [d, a] = 1 \qquad > \\ [c, a] = d^{2k_1} & [d, b] = 1 \\ [c, b] = 1 & [d, c] = 1 \\ \alpha a = a^{-1}\alpha d^{k_1} & \alpha^2 = cd^{k_2} \\ \alpha b = b^{-1}\alpha & \alpha d = d^{-1}\alpha \\ \alpha c = c\alpha d^{-2k_2} & \\ \beta a = a\beta & \beta^2 = ad^{k_3} \\ \beta b = b^{-1}\beta d^{-k_1-2k_3+2k_4} & \beta d = d\beta \\ \beta c = c\beta d^{-k_1} & \alpha\beta = a^{-1}b\beta\alpha d^{k_4} \end{array}$$

$$\lambda(\alpha)=\begin{pmatrix} -1 & -\frac{k_1}{2} & 0 & 2k_2 & 0 \\ 0 & -1 & 0 & 0 & 0 \\ 0 & 0 & -1 & 0 & \frac{1}{2} \\ 0 & 0 & 0 & 1 & \frac{1}{2} \\ 0 & 0 & 0 & 0 & 1 \end{pmatrix} \quad \lambda(\beta)=\begin{pmatrix} 1 & 0 & 2(k_4-k_3)-k_1 & -\frac{k_1}{2} & \frac{k_3}{2} \\ 0 & 1 & 0 & 0 & \frac{1}{2} \\ 0 & 0 & -1 & 0 & 0 \\ 0 & 0 & 0 & 1 & 0 \\ 0 & 0 & 0 & 0 & 1 \end{pmatrix}$$

$$H^2(Q,\mathbb{Z})=\mathbb{Z}=\mathbb{Z}^4/A, \quad A=\{(k_1,\dots,k_4)||k_1=0, \ k_2,k_3,k_4\in\mathbb{Z}\}$$

AB-groups:
$\forall k>0 \ <(k,0,0,0)>$
Action $^\alpha d=d, \ ^\beta d=d^{-1}$:

$$E: <a,b,c,d,\alpha,\beta \ || \quad \begin{array}{ll} [b,a]=d^{2k_1} & [d,a]=1 \\ [c,a]=1 & [d,b]=1 \\ [c,b]=1 & [d,c]=1 \\ \alpha a=a^{-1}\alpha d^{-k_1-2k_4} & \alpha^2=cd^{k_2} \\ \alpha b=b^{-1}\alpha d^{-3k_1+2k_2+k_3-2k_4+2k_5} & \alpha d=d\alpha \\ \alpha c=c\alpha & \\ \beta a=a\beta d^{-2k_4} & \beta^2=ad^{k_4} \\ \beta b=b^{-1}\beta d^{-k_1} & \beta d=d^{-1}\beta \\ \beta c=c\beta d^{k_3} & \alpha\beta=a^{-1}b\beta\alpha d^{k_5} \end{array} \quad >$$

$$\lambda(\alpha)=\begin{pmatrix} 1 & -\frac{k_1}{2}-2k_4 & -3k_1+2(k_2-k_4+k_5)+k_3 & 0 & \frac{3k_1-k_3+2k_4-2k_5}{4} \\ 0 & -1 & 0 & 0 & 0 \\ 0 & 0 & -1 & 0 & \frac{1}{2} \\ 0 & 0 & 0 & 1 & \frac{1}{2} \\ 0 & 0 & 0 & 0 & 1 \end{pmatrix}$$

$$\lambda(\beta)=\begin{pmatrix} -1 & 2k_4 & \frac{k_1}{2} & -k_3 & 0 \\ 0 & 1 & 0 & 0 & \frac{1}{2} \\ 0 & 0 & -1 & 0 & 0 \\ 0 & 0 & 0 & 1 & 0 \\ 0 & 0 & 0 & 0 & 1 \end{pmatrix}$$

$$H^2(Q,\mathbb{Z})=\mathbb{Z}\oplus\mathbb{Z}_4=\mathbb{Z}^5/A,$$

$$A=\{(k_1,\dots,k_5)||k_1=0, \ (k_3+2k_2)\in 4\mathbb{Z}, \ k_4,k_5\in\mathbb{Z}\}$$

AB-groups:
$\forall k>0 \ <(k,0,0,0,0)>$
$\forall k>0, \ k\equiv 0 \bmod 2, \ <(k,1,0,0,0)>$
$\forall k>0 \ <(k,0,1,0,0)>$
Remark: If $k\not\equiv 0 \bmod 2 <(k,0,0,0,0)>\cong<(k,1,0,0,0)>$.
$<(k,0,1,0,0)>\cong<(k,1,1,0,0)>$

Action $^\alpha d = d^{-1}$, $^\beta d = d^{-1}$:

$$E : \;< a,b,c,d,\alpha,\beta \;\|\quad \begin{array}{ll} [b,a]=1 & [d,a]=1 \\ [c,a]=1 & [d,b]=1 \\ [c,b]=d^{2k_1} & [d,c]=1 \\ \alpha a = a^{-1}\alpha & \alpha^2 = cd^{k_2} \\ \alpha b = b^{-1}\alpha d^{k_1} & \alpha d = d^{-1}\alpha \\ \alpha c = c\alpha d^{-2k_2} & \\ \beta a = a\beta d^{-2k_3} & \beta^2 = ad^{k_3} \\ \beta b = b^{-1}\beta & \beta d = d^{-1}\beta \\ \beta c = c\beta d^{-k_1-2k_2} & \alpha\beta = a^{-1}b\beta\alpha d^{k_4} \end{array} \quad >$$

$$\lambda(\alpha)=\begin{pmatrix} -1 & 0 & \frac{k_1}{2} & \frac{k_1}{2}+2k_2 & \frac{k_2}{2} \\ 0 & -1 & 0 & 0 & 0 \\ 0 & 0 & -1 & 0 & \frac{1}{2} \\ 0 & 0 & 0 & 1 & \frac{1}{2} \\ 0 & 0 & 0 & 0 & 1 \end{pmatrix} \quad \lambda(\beta)=\begin{pmatrix} -1 & 2k_3 & 0 & k_1+2k_2 & -\frac{k_4}{2} \\ 0 & 1 & 0 & 0 & \frac{1}{2} \\ 0 & 0 & -1 & 0 & 0 \\ 0 & 0 & 0 & 1 & 0 \\ 0 & 0 & 0 & 0 & 1 \end{pmatrix}$$

$$H^2(Q,\mathbb{Z}) = \mathbb{Z} = \mathbb{Z}^4/A, \; A = \{(k_1,\dots,k_4)\|k_1 = 0, \; k_2, k_3, k_4 \in \mathbb{Z}\}$$

AB-groups:
$$\forall k > 0 \; < (k,0,0,0) >$$

---

37. $Q = Ccc2$

$$E : \;< a,b,c,d,\alpha,\beta \;\|\quad \begin{array}{ll} [b,a]=d^{k_1} & [d,a]=1 \\ [c,a]=1 & [d,b]=1 \\ [c,b]=1 & [d,c]=1 \\ \alpha a = a^{-1}\alpha d^{k_2} & \alpha^2 = d^{k_3} \\ \alpha b = b^{-1}\alpha d^{-k_2+2k_4} & \alpha d = d\alpha \\ \alpha c = c\alpha & \\ \beta a = b\beta d^{k_4} & \beta^2 = cd^{k_5} \\ \beta b = a\beta d^{k_4} & \beta d = d^{-1}\beta \\ \beta c = c\beta d^{-2k_5} & \alpha\beta = \beta\alpha d^{-k_3} \end{array} \quad >$$

$$\lambda(\alpha) = \begin{pmatrix} 1 & k_2 & -k_2+2k_4 & 0 & \frac{k_3}{2} \\ 0 & -1 & 0 & 0 & 0 \\ 0 & 0 & -1 & 0 & 0 \\ 0 & 0 & 0 & 1 & 0 \\ 0 & 0 & 0 & 0 & 1 \end{pmatrix} \quad \lambda(\beta) = \begin{pmatrix} -1 & -k_4 & -k_4 & 2k_5 & 0 \\ 0 & 0 & 1 & 0 & 0 \\ 0 & 1 & 0 & 0 & 0 \\ 0 & 0 & 0 & 1 & \frac{1}{2} \\ 0 & 0 & 0 & 0 & 1 \end{pmatrix}$$

$$H^2(Q,\mathbb{Z}) = \mathbb{Z} \oplus (\mathbb{Z}_2)^2 = \mathbb{Z}^5/A,$$

$$A = \{(k_1,\dots,k_5)\|k_1 = 0, \; k_2, k_3 \in 2\mathbb{Z}, \; k_4, k_5 \in \mathbb{Z}\}$$

AB-groups:
$$\forall k > 0, k \equiv 0 \bmod 2 \; < (k,0,1,0,0) >$$

36. $Q = Cmc2_1$

$$E : < a,b,c,d,\alpha,\beta \| \quad [b,a] = 1 \qquad [d,a] = 1 \qquad >$$

$$[c,a] = d^{2k_1} \qquad [d,b] = 1$$
$$[c,b] = d^{2k_1} \qquad [d,c] = 1$$
$$\alpha a = a^{-1}\alpha d^{k_1} \quad \alpha^2 = cd^{k_2}$$
$$\alpha b = b^{-1}\alpha d^{k_1} \quad \alpha d = d^{-1}\alpha$$
$$\alpha c = c\alpha d^{-2k_2}$$
$$\beta a = b\beta d^{k_3} \qquad \beta^2 = d^{k_4}$$
$$\beta b = a\beta d^{-k_3} \qquad \beta d = d\beta$$
$$\beta c = c\beta \qquad \alpha\beta = \beta\alpha d^{k_4}$$

$$\lambda(\alpha) = \begin{pmatrix} -1 & -\frac{k_1}{2} & -\frac{k_1}{2} & 2k_2 & 0 \\ 0 & -1 & 0 & 0 & 0 \\ 0 & 0 & -1 & 0 & 0 \\ 0 & 0 & 0 & 1 & \frac{1}{2} \\ 0 & 0 & 0 & 0 & 1 \end{pmatrix} \quad \lambda(\beta) = \begin{pmatrix} 1 & k_3 & -k_3 & 0 & \frac{k_4}{2} \\ 0 & 0 & 1 & 0 & 0 \\ 0 & 1 & 0 & 0 & 0 \\ 0 & 0 & 0 & 1 & 0 \\ 0 & 0 & 0 & 0 & 1 \end{pmatrix}$$

$$H^2(Q, \mathbb{Z}) = \mathbb{Z} \oplus \mathbb{Z}_2 = \mathbb{Z}^4/A, \; A = \{(k_1, \ldots, k_4) \| k_1 = 0, \; k_4 \in 2\mathbb{Z}, \; k_2, k_3 \in \mathbb{Z}\}$$

AB-groups:

$\forall k > 0 \; < (k,0,0,1) >$

---

41. $Q = Aba2$

$$E : < a,b,c,d,\alpha,\beta \| \quad [b,a] = 1 \qquad [d,a] = 1 \qquad >$$

$$[c,a] = d^{2k_1} \qquad [d,b] = 1$$
$$[c,b] = d^{2k_1} \qquad [d,c] = 1$$
$$\alpha a = b^{-1}\alpha d^{k_2} \qquad \alpha^2 = d^{k_3}$$
$$\alpha b = a^{-1}\alpha d^{k_2} \qquad \alpha d = d\alpha$$
$$\alpha c = c^{-1}\alpha d^{2k_1 - 2k_5}$$
$$\beta a = b^{-1}\beta d^{k_4} \qquad \beta^2 = cd^{k_5}$$
$$\beta b = a^{-1}\beta d^{2k_1 - k_4} \qquad \beta d = d^{-1}\beta$$
$$\beta c = c\beta d^{-2k_5} \qquad \alpha\beta = abc^{-1}\beta\alpha d^{k_1+k_2-k_3+k_5}$$

$$\lambda(\alpha) = \begin{pmatrix} 1 & k_2 & k_2 & k_1-!k_5 & \frac{k_3-k_2}{2} \\ 0 & 0 & -1 & 0 & \frac{1}{2} \\ 0 & -1 & 0 & 0 & \frac{1}{2} \\ 0 & 0 & 0 & -1 & 0 \\ 0 & 0 & 0 & 0 & 1 \end{pmatrix} \quad \lambda(\beta) = \begin{pmatrix} -1 & \frac{k_1}{2}-k_4 & \frac{-3k_1}{2}+k_4 & 2k_5 & 0 \\ 0 & 0 & -1 & 0 & 0 \\ 0 & -1 & 0 & 0 & 0 \\ 0 & 0 & 0 & 1 & \frac{1}{2} \\ 0 & 0 & 0 & 0 & 1 \end{pmatrix}$$

$$H^2(Q, \mathbb{Z}) = \mathbb{Z} \oplus (\mathbb{Z}_2)^2 = \mathbb{Z}^5/A,$$

$$A = \{(k_1, \ldots, k_5) \| k_1 = 0, \; k_3, (k_2 + k_4) \in 2\mathbb{Z}, \; k_5 \in \mathbb{Z}\}$$

AB-groups:

$\forall k > 0 \; < (k,0,1,0,0) >$

$\forall k > 0 \; < (k,1,1,0,0) >$

43. $Q = Fdd2$

$$E : < a, b, c, d, \alpha, \beta \;\|\; \begin{array}{ll} [b,a] = d^{k_1} & [d,a] = 1 \\ [c,a] = d^{k_1} & [d,b] = 1 \\ [c,b] = d^{-k_1} & [d,c] = 1 \\ \alpha a = bc^{-1}\alpha d^{k_2 + 2k_3 + 2k_5} & \alpha^2 = d^{k_3} \\ \alpha b = ac^{-1}\alpha d^{k_2} & \alpha d = d\alpha \\ \alpha c = c^{-1}\alpha d^{2(k_2 + k_3 + k_5)} & \\ \beta a = bc^{-1}\beta d^{-k_2 - 2k_4} & \beta^2 = bd^{k_4} \\ \beta b = b\beta d^{-2k_4} & \beta d = d^{-1}\beta \\ \beta c = a^{-1}b\beta d^{k_1 + k_2} & \alpha\beta = ab^{-1}\beta\alpha d^{k_5} \end{array} >$$

$$\lambda(\alpha) = \begin{pmatrix} 1 & -\frac{k_1}{4} + k_2 + 2(k_3 + k_5) & \frac{k_1}{4} + k_2 & 2(k_2 + k_3 + k_5) & -\frac{k_2 + k_5}{2} \\ 0 & 0 & 1 & 0 & 0 \\ 0 & 1 & 0 & 0 & 0 \\ 0 & -1 & -1 & -1 & \frac{1}{2} \\ 0 & 0 & 0 & 0 & 1 \end{pmatrix},$$

$$\lambda(\beta) = \begin{pmatrix} -1 & -\frac{k_1}{4} + k_2 + 2k_4 & 2k_4 & -\frac{k_1}{4} - k_2 & 0 \\ 0 & 0 & 0 & -1 & 0 \\ 0 & 1 & 1 & 1 & \frac{1}{2} \\ 0 & -1 & 0 & 0 & 0 \\ 0 & 0 & 0 & 0 & 1 \end{pmatrix}$$

$H^2(Q, \mathbb{Z}) = \mathbb{Z} \oplus \mathbb{Z}_2 = \mathbb{Z}^5/A,\ A = \{(k_1, \ldots, k_5)\|k_1 = 0,\ k_3 \in 2\mathbb{Z},\ k_2, k_4, k_5 \in \mathbb{Z}\}$

AB-groups:
$\forall k > 0\ \ < (k, 0, 1, 0, 0) >$

---

45. $Q = Iba2$

$$E : < a, b, c, d, \alpha, \beta \;\|\; \begin{array}{ll} [b,a] = 1 & [d,a] = 1 \\ [c,a] = d^{k_1} & [d,b] = 1 \\ [c,b] = d^{-k_1} & [d,c] = 1 \\ \alpha a = ba\alpha d^{k_2} & \alpha^2 = d^{k_3} \\ \alpha b = a\alpha d^{-k_2} & \alpha d = d\alpha \\ \alpha c = a^{-1}b^{-1}c^{-1}\alpha d^{k_2 - 2k_4 - 2k_5} & \\ \beta a = c^{-1}\beta d^{k_4} & \beta^2 = abd^{k_5} \\ \beta b = abc\beta d^{-k_4 - 2k_5} & \beta d = d^{-1}\beta \\ \beta c = a^{-1}\beta d^{-k_4} & \alpha\beta = \beta\alpha d^{-k_3} \end{array} >$$

$$\lambda(\alpha) = \begin{pmatrix} 1 & k_2 & -k_2 & k_2 - 2k_4 - 2k_5 & \frac{k_3}{2} \\ 0 & 0 & 1 & -1 & 0 \\ 0 & 1 & 0 & -1 & 0 \\ 0 & 0 & 0 & -1 & 0 \\ 0 & 0 & 0 & 0 & 1 \end{pmatrix} \quad \lambda(\beta) = \begin{pmatrix} -1 & -k_4 & k_4 + 2k_5 & k_4 & 0 \\ 0 & 0 & 1 & -1 & \frac{1}{2} \\ 0 & 0 & 1 & 0 & \frac{1}{2} \\ 0 & -1 & 1 & 0 & 0 \\ 0 & 0 & 0 & 0 & 1 \end{pmatrix}$$

$$H^2(Q, \mathbb{Z}) = \mathbb{Z} \oplus (\mathbb{Z}_2)^2 = \mathbb{Z}^5/A,$$

$$A = \{(k_1, \ldots, k_5) \| k_1 = 0, \ k_3, (k_2 + k_5) \in 2\mathbb{Z}, \ k_4 \in \mathbb{Z}\}$$

AB-groups:

$\forall k > 0, \ k \equiv 0 \bmod 2, \ < (k, 0, 1, 0, 0) >$

$\forall k > 0, \ k \equiv 0 \bmod 2, \ < (k, 1, 1, 0, 0) >$

---

60. $Q = P\frac{2_1}{b}\frac{2}{c}\frac{2_1}{n}$

$E : < a, b, c, d, \alpha, \beta, \gamma \ \|$

$[b, a] = d^{4k_1}$	$[d, a] = 1$
$[c, a] = 1$	$[d, b] = 1$
$[c, b] = 1$	$[d, c] = 1$
$\alpha a = a^{-1} \alpha d^{-2(k_1 - k_2 + k_3 - k_4)}$	$\alpha^2 = d^{k_2}$
$\alpha b = b^{-1} \alpha d^{2(k_1 - k_3)}$	$\alpha d = d\alpha$
$\alpha \gamma = a\gamma \alpha d^{k_1 - k_2 + k_3 - k_4}$	$\alpha c = c\alpha$
$\beta a = a^{-1} \beta d^{2k_1}$	$\beta^2 = bd^{k_3}$
$\beta b = b\beta d^{-2k_3}$	$\beta d = d^{-1}\beta$
$\beta c = c^{-1}\beta$	$\alpha \beta = ab^{-1}\beta \alpha d^{k_4}$
$\gamma a = a^{-1}\gamma d^{-2(k_1 - k_2 + k_3 - k_4)}$	$\gamma^2 = d^{k_6}$
$\gamma b = b^{-1}\gamma d^{-2k_3}$	$\gamma d = d\gamma$
$\gamma c = c^{-1}\gamma d^{2(k_3 + k_5 - k_6)}$	$\beta \gamma = bc\gamma \beta d^{k_5}$

$>$

$$\lambda(\alpha) = \begin{pmatrix} 1 & 2(k_2 - k_1 + k_4 - k_3) & k_1 - 2k_3 & 0 & \frac{k_1 + k_3 - k_4}{2} \\ 0 & -1 & 0 & 0 & \frac{1}{2} \\ 0 & 0 & -1 & 0 & 0 \\ 0 & 0 & 0 & 1 & 0 \\ 0 & 0 & 0 & 0 & 1 \end{pmatrix}$$

$$\lambda(\beta) = \begin{pmatrix} -1 & -k_1 & 2k_3 & 0 & 0 \\ 0 & -1 & 0 & 0 & 0 \\ 0 & 0 & 1 & 0 & \frac{1}{2} \\ 0 & 0 & 0 & -1 & \frac{1}{2} \\ 0 & 0 & 0 & 0 & 1 \end{pmatrix}$$

$$\lambda(\gamma) = \begin{pmatrix} 1 & -2(k_1 - k_2 + k_3 - k_4) & -2k_3 & 2k_3 + 2k_5 - 2k_6 & \frac{k_6}{2} \\ 0 & -1 & 0 & 0 & 0 \\ 0 & 0 & -1 & 0 & 0 \\ 0 & 0 & 0 & -1 & 0 \\ 0 & 0 & 0 & 0 & 1 \end{pmatrix}$$

$$H^2(Q, \mathbb{Z}) = \mathbb{Z} \oplus (\mathbb{Z}_2)^2 = \mathbb{Z}^6 / A,$$

$$A = \{(k_1, \ldots, k_6) \| k_1 = 0, \ k_2, k_6 \in 2\mathbb{Z}, \ k_3, k_4, k_5 \in \mathbb{Z}\}$$

AB-groups:

$\forall k > 0 \ < (k, 1, 0, 0, 0, 1) >$

---

56. $Q = P\frac{2_1}{c}\frac{2_1}{c}\frac{2}{n}$

$E : \,< a, b, c, d, \alpha, \beta, \gamma \, \|$

$[b, a] = d^{2k_1}$	$[d, a] = 1$
$[c, a] = 1$	$[d, b] = 1$
$[c, b] = 1$	$[d, c] = 1$
$\alpha a = a^{-1}\alpha d^{k_1 + 2k_3 - 2k_4}$	$\alpha^2 = d^{k_2}$
$\alpha b = b^{-1}\alpha d^{k_1 - 2k_3}$	$\alpha d = d\alpha$
$\alpha c = c\alpha$	$\alpha\gamma = ab\gamma\alpha d^{k_4}$
$\beta a = a^{-1}\beta d^{k_1}$	$\beta^2 = bd^{k_3}$
$\beta b = b\beta d^{-2k_3}$	$\beta d = d^{-1}\beta$
$\alpha\beta = ab^{-1}\beta\alpha d^{k_1 - k_2 + 2k_3 - k_4}$	$\beta c = c^{-1}\beta$
$\gamma a = a^{-1}\gamma d^{2(k_1 + k_3 - k_4)}$	$\gamma^2 = d^{k_6}$
$\gamma b = b^{-1}\gamma d^{-2k_3}$	$\gamma d = d\gamma$
$\gamma c = c^{-1}\gamma d^{2(k_3 + k_5 - k_6)}$	$\beta\gamma = bc\gamma\beta d^{k_5}$

$> $

$$\lambda(\alpha) = \begin{pmatrix} 1 & \frac{3k_1}{2} + 2(k_3 - k_4) & \frac{k_1}{2} - 2k_3 & 0 & \frac{k_2 + k_4 - k_1}{2} \\ 0 & -1 & 0 & 0 & \frac{1}{2} \\ 0 & 0 & -1 & 0 & \frac{1}{2} \\ 0 & 0 & 0 & 1 & 0 \\ 0 & 0 & 0 & 0 & 1 \end{pmatrix}$$

$$\lambda(\beta) = \begin{pmatrix} -1 & -\frac{k_1}{2} & 2k_3 & 0 & 0 \\ 0 & -1 & 0 & 0 & 0 \\ 0 & 0 & 1 & 0 & \frac{1}{2} \\ 0 & 0 & 0 & -1 & \frac{1}{2} \\ 0 & 0 & 0 & 0 & 1 \end{pmatrix}$$

$$\lambda(\gamma) = \begin{pmatrix} 1 & 2(k_1 + k_3 - k_4) & -2k_3 & 2(k_3 + k_5 - k_6) & \frac{k_6}{2} \\ 0 & -1 & 0 & 0 & 0 \\ 0 & 0 & -1 & 0 & 0 \\ 0 & 0 & 0 & -1 & 0 \\ 0 & 0 & 0 & 0 & 1 \end{pmatrix}$$

$$H^2(Q, \mathbb{Z}) = \mathbb{Z} \oplus (\mathbb{Z}_2)^2 = \mathbb{Z}^6/A,$$

$$A = \{(k_1, \ldots, k_6) \| k_1 = 0, \ k_2, k_6 \in 2\mathbb{Z}, \ k_3, k_4, k_5 \in \mathbb{Z}\}$$

AB-groups:
$\forall k > 0, k \equiv 0 \bmod 2 \ < (k, 1, 0, 0, 0, 1) >$

55. $Q = P\frac{2_1}{b}\frac{2_1}{a}\frac{2}{m}$

$E : < a,b,c,d,\alpha,\beta,\gamma \parallel$

$$[b,a] = d^{2k_1} \qquad\qquad [d,a] = 1$$
$$[c,a] = 1 \qquad\qquad [d,b] = 1$$
$$[c,b] = 1 \qquad\qquad [d,c] = 1$$
$$\alpha a = a^{-1}\alpha d^{-k_1-2k_2+2k_3-2k_4} \quad \alpha^2 = d^{k_2}$$
$$\alpha b = b^{-1}\alpha d^{-k_1-2k_3} \qquad \alpha d = d\alpha$$
$$\alpha c = c\alpha \qquad\qquad \alpha\gamma = \gamma\alpha$$
$$\beta a = a^{-1}\beta d^{k_1} \qquad\qquad \beta^2 = bd^{k_3}$$
$$\beta b = b\beta d^{-2k_3} \qquad\qquad \beta d = d^{-1}\beta$$
$$\beta c = c^{-1}\beta \qquad\qquad \alpha\beta = a^{-1}b^{-1}\beta\alpha d^{k_4}$$
$$\gamma a = a^{-1}\gamma d^{-k_1-2(k_2-k_3+k_4)} \quad \gamma^2 = d^{2k_1+k_2+k_4+k_6}$$
$$\gamma b = b^{-1}\gamma d^{-k_1-2k_3} \qquad \gamma d = d\gamma$$
$$\gamma c = c^{-1}\gamma d^{k_5} \qquad\qquad \beta\gamma = ab\gamma\beta d^{k_6}$$

$>$

$$\lambda(\alpha) = \begin{pmatrix} 1 & -k_1-2(k_2-k_3+k_4) & -k_1-2k_3 & 0 & \frac{k_2}{2} \\ 0 & -1 & 0 & 0 & 0 \\ 0 & 0 & -1 & 0 & 0 \\ 0 & 0 & 0 & 1 & 0 \\ 0 & 0 & 0 & 0 & 1 \end{pmatrix}$$

$$\lambda(\beta) = \begin{pmatrix} -1 & -\frac{k_1}{2} & \frac{k_1}{2}+2k_3 & 0 & 0 \\ 0 & -1 & 0 & 0 & \frac{1}{2} \\ 0 & 0 & 1 & 0 & \frac{1}{2} \\ 0 & 0 & 0 & -1 & 0 \\ 0 & 0 & 0 & 0 & 1 \end{pmatrix}$$

$$\lambda(\gamma) = \begin{pmatrix} 1 & -k_1-2(k_2-k_3+k_4) & -k_1-2k_3 & k_5 & \frac{2k_1+k_2+k_4+k_6}{2} \\ 0 & -1 & 0 & 0 & 0 \\ 0 & 0 & -1 & 0 & 0 \\ 0 & 0 & 0 & -1 & 0 \\ 0 & 0 & 0 & 0 & 1 \end{pmatrix}$$

$$H^2(Q,\mathbb{Z}) = \mathbb{Z} \oplus (\mathbb{Z}_2)^3 = \mathbb{Z}^6/A,$$

$$A = \{(k_1,\ldots,k_6) \| k_1 = 0, k_2, k_5, (k_4+k_6) \in 2\mathbb{Z}, \ k_3 \in \mathbb{Z}\}$$

AB-groups:
None

58. $Q = P\frac{2_1}{n}\frac{2_1}{n}\frac{2}{m}$

$E :\; <a,b,c,d,\alpha,\beta,\gamma \parallel$

$[b,a] = d^{2k_1}$ $\qquad\qquad\qquad$ $[d,a] = 1$ $\qquad\qquad$ >

$[c,a] = 1$ $\qquad\qquad\qquad\qquad\;\;$ $[d,b] = 1$

$[c,b] = 1$ $\qquad\qquad\qquad\qquad\;\;$ $[d,c] = 1$

$\alpha a = a^{-1}\alpha d^{-k_1-2(k_2-k_3+k_4)}$ $\qquad$ $\alpha^2 = d^{k_2}$

$\alpha b = b^{-1}\alpha d^{-k_1-2k_3}$ $\qquad\qquad\quad$ $\alpha d = d\alpha$

$\alpha c = c\alpha$ $\qquad\qquad\qquad\qquad\quad\;$ $\alpha\gamma = \gamma\alpha$

$\beta a = a^{-1}\beta d^{k_1}$ $\qquad\qquad\qquad\;\;$ $\beta^2 = bd^{k_3}$

$\beta b = b\beta d^{-2k_3}$ $\qquad\qquad\qquad\;$ $\beta d = d^{-1}\beta$

$\beta c = c^{-1}\beta$ $\qquad\qquad\qquad\qquad$ $\alpha\beta = a^{-1}b^{-1}\beta\alpha d^{k_4}$

$\gamma a = a^{-1}\gamma d^{-k_1-2(k_2-k_3+k_4)}$ $\qquad$ $\gamma^2 = d^{k_6}$

$\gamma b = b^{-1}\gamma d^{-k_1-2k_3}$ $\qquad\qquad\;$ $\gamma d = d\gamma$

$\gamma c = c^{-1}\gamma d^{2(2k_1+k_2+k_4+k_5-k_6)}$ $\quad$ $\beta\gamma = abc\gamma\beta d^{k_5}$

$$\lambda(\alpha) = \begin{pmatrix} 1 & -k_1-2(k_2-k_3+k_4) & -k_1-2k_3 & 0 & \frac{k_2}{2} \\ 0 & -1 & 0 & 0 & 0 \\ 0 & 0 & -1 & 0 & 0 \\ 0 & 0 & 0 & 1 & 0 \\ 0 & 0 & 0 & 0 & 1 \end{pmatrix}$$

$$\lambda(\beta) = \begin{pmatrix} -1 & -\frac{k_1}{2} & \frac{k_1}{2}+2k_3 & 0 & 0 \\ 0 & -1 & 0 & 0 & \frac{1}{2} \\ 0 & 0 & 1 & 0 & \frac{1}{2} \\ 0 & 0 & 0 & -1 & \frac{1}{2} \\ 0 & 0 & 0 & 0 & 1 \end{pmatrix}$$

$$\lambda(\gamma) = \begin{pmatrix} 1 & -k_1-2(k_2-k_3+k_4) & -k_1-2k_3 & 2(2k_1+k_2+k_4+k_5-k_6) & \frac{k_6}{2} \\ 0 & -1 & 0 & 0 & 0 \\ 0 & 0 & -1 & 0 & 0 \\ 0 & 0 & 0 & -1 & 0 \\ 0 & 0 & 0 & 0 & 1 \end{pmatrix}$$

$$H^2(Q,\mathbb{Z}) = \mathbb{Z} \oplus (\mathbb{Z}_2)^2 = \mathbb{Z}^6/A,$$

$$A = \{(k_1,\dots,k_6)\| k_1 = 0,\; k_2,k_6 \in 2\mathbb{Z},\; k_3,k_4,k_5 \in \mathbb{Z}\}$$

AB-groups:
None

62. $Q = P\frac{2_1}{n}\frac{2_1}{m}\frac{2_1}{a}$

$E: < a, b, c, d, \alpha, \beta, \gamma \parallel$

$[b, a] = 1$	$[d, a] = 1$
$[c, a] = d^{2k_1}$	$[d, b] = 1$
$[c, b] = 1$	$[d, c] = 1$
$\alpha a = a^{-1}\alpha d^{k_1}$	$\alpha^2 = cd^{k_2}$
$\alpha b = b^{-1}\alpha$	$\alpha d = d^{-1}\alpha$
$\alpha c = c\alpha d^{-2k_2}$	$\alpha\gamma = ac\gamma\alpha d^{k_3}$
$\beta a = a^{-1}\beta d^{3k_1 + 2(k_2 - k_3 + k_4)}$	$\beta^2 = bd^{k_3}$
$\beta b = b\beta$	$\beta d = d\beta$
$\beta c = c^{-1}\beta d^{-k_1 - 2k_2}$	$\alpha\beta = ab^{-1}c\beta\alpha d^{k_4}$
$\gamma a = a^{-1}\gamma d^{3k_1 + 2(k_2 - k_3 + k_4)}$	$\gamma^2 = d^{k_3 - k_4 + k_5}$
$\gamma b = b^{-1}\gamma d^{-2k_6}$	$\gamma d = d\gamma$
$\gamma c = c^{-1}\gamma d^{-k_1 - 2k_2}$	$\beta\gamma = b\gamma\beta d^{k_6}$

$>$

$$\lambda(\alpha) = \begin{pmatrix} -1 & \frac{-k_1}{2} & 0 & \frac{k_1}{2} + 2k_2 & 0 \\ 0 & -1 & 0 & 0 & \frac{1}{2} \\ 0 & 0 & -1 & 0 & 0 \\ 0 & 0 & 0 & 1 & \frac{1}{2} \\ 0 & 0 & 0 & 0 & 1 \end{pmatrix}$$

$$\lambda(\beta) = \begin{pmatrix} 1 & 3k_1 + 2(k_2 - k_3 + k_4) & 0 & -k_1 - 2k_2 & \frac{k_3}{2} \\ 0 & -1 & 0 & 0 & 0 \\ 0 & 0 & 1 & 0 & \frac{1}{2} \\ 0 & 0 & 0 & -1 & 0 \\ 0 & 0 & 0 & 0 & 1 \end{pmatrix}$$

$$\lambda(\gamma) = \begin{pmatrix} 1 & 3k_1 + 2(k_2 - k_3 + k_4) & -2k_6 & -k_1 - 2k_2 & \frac{k_3 - k_4 + k_5}{2} \\ 0 & -1 & 0 & 0 & 0 \\ 0 & 0 & -1 & 0 & 0 \\ 0 & 0 & 0 & -1 & 0 \\ 0 & 0 & 0 & 0 & 1 \end{pmatrix}$$

$$H^2(Q, \mathbb{Z}) = \mathbb{Z} \oplus (\mathbb{Z}_2)^2 = \mathbb{Z}^6 / A,$$

$$A = \{(k_1, \ldots, k_6) \| k_1 = 0, \ (k_3 - k_6), \ (k_3 - k_4 + k_5) \in 2\mathbb{Z}, \ k_2 \in \mathbb{Z}\}$$

AB-groups:
$\forall k > 0, \ k \equiv 0 \bmod 2, \ < (k, 0, 0, 0, 1, 0) >$

61. $Q = P\frac{2_1}{b}\frac{2_1}{c}\frac{2_1}{a}$

<u>Action $\alpha d = d^{-1}\alpha$, $\beta d = d\beta$:</u>

$E :\ < a, b, c, d, \alpha, \beta, \gamma\ ||\ $

$[b, a] = 1$	$[d, a] = 1$	>
$[c, a] = d^{4k_1}$	$[d, b] = 1$	
$[c, b] = 1$	$[d, c] = 1$	
$\alpha a = a^{-1}\alpha d^{2k_1}$	$\alpha^2 = cd^{k_2}$	
$\alpha b = b^{-1}\alpha$	$\alpha d = d^{-1}\alpha$	
$\alpha c = c\alpha d^{-2k_2}$	$\alpha\gamma = ac\gamma\alpha d^{k_5}$	
$\beta a = a^{-1}\beta d^{6k_1 + 2(k_2 - k_3 + k_4)}$	$\beta^2 = bd^{k_3}$	
$\beta b = b\beta$	$\beta d = d\beta$	
$\beta c = c^{-1}\beta d^{-2(k_1 + k_2)}$	$\alpha\beta = ab^{-1}c\beta\alpha d^{k_4}$	
$\gamma a = a^{-1}\gamma d^{8k_1 + 2(k_2 - k_3 + k_4)}$	$\gamma^2 = d^{-k_1 + k_3 - k_4 + k_5}$	
$\gamma b = b^{-1}\gamma d^{2(k_1 + k_2 - k_6)}$	$\gamma d = d\gamma$	
$\gamma c = c^{-1}\gamma d^{-2(k_1 + k_2)}$	$\beta\gamma = bc\gamma\beta d^{k_6}$	

$$\lambda(\alpha) = \begin{pmatrix} -1 & -k_1 & 0 & k_1 + 2k_2 & 0 \\ 0 & -1 & 0 & 0 & \frac{1}{2} \\ 0 & 0 & -1 & 0 & 0 \\ 0 & 0 & 0 & 1 & \frac{1}{2} \\ 0 & 0 & 0 & 0 & 1 \end{pmatrix}$$

$$\lambda(\beta) = \begin{pmatrix} 1 & 7k_1 + k_2 - 2k_3 + 2k_4 & 0 & -2(k_1 + k_2) & \frac{k_1 + k_2 + k_3}{2} \\ 0 & -1 & 0 & 0 & 0 \\ 0 & 0 & 1 & 0 & \frac{1}{2} \\ 0 & 0 & 0 & -1 & \frac{1}{2} \\ 0 & 0 & 0 & 0 & 1 \end{pmatrix}$$

$$\lambda(\gamma) = \begin{pmatrix} 1 & 2(4k_1 + k_2 - k_3 + k_4) & 2(k_1 + k_2 - k_6) & -2(k_1 + k_2) & \frac{-k_1 + k_3 - k_4 + k_5}{2} \\ 0 & -1 & 0 & 0 & 0 \\ 0 & 0 & -1 & 0 & 0 \\ 0 & 0 & 0 & -1 & 0 \\ 0 & 0 & 0 & 0 & 1 \end{pmatrix}$$

$$H^2(Q, \mathbb{Z}) = \mathbb{Z} \oplus (\mathbb{Z}_2)^2 = \mathbb{Z}^6/A,$$

$$A = \{(k_1, \ldots, k_6) || k_1 = 0,\ (k_2 + k_3 - k_6),\ (k_3 - k_4 + k_5) \in 2\mathbb{Z}\}$$

AB-groups:

$\forall k > 0,\ k \equiv 0 \bmod 2,\ < (k, 0, 0, 0, 1, 0) >$

$\forall k > 0,\ k \equiv 0 \bmod 2,\ < (k, 0, 0, 0, 1, 1) >$

$\forall k > 0,\ k \not\equiv 0 \bmod 2,\ < (k, 0, 0, 0, 0, 0) >$

$\forall k > 0,\ k \not\equiv 0 \bmod 2,\ < (k, 0, 0, 0, 0, 1) >$

Action $\alpha d = d\alpha$, $\beta d = d^{-1}\beta$:

$E : < a, b, c, d, \alpha, \beta, \gamma \;\|\;$

$[b, a] = d^{4k_1}$	$[d, a] = 1$
$[c, a] = 1$	$[d, b] = 1$
$[c, b] = 1$	$[d, c] = 1$
$\alpha a = a^{-1}\alpha d^{2(-k_1+k_2-k_3+k_4)}$	$\alpha^2 = cd^{k_2}$
$\alpha b = b^{-1}\alpha d^{2(k_1-k_3)}$	$\alpha d = d\alpha$
$\alpha c = c\alpha$	$\alpha\gamma = ac\gamma\alpha d^{k_5}$
$\beta a = a^{-1}\beta d^{2k_1}$	$\beta^2 = bd^{k_3}$
$\beta b = b\beta d^{-2k_3}$	$\beta d = d^{-1}\beta$
$\gamma^2 = d^{-k_1+k_2+k_4+k_5+k_6}$	$\beta c = c^{-1}\beta$
$\gamma a = a^{-1}\gamma d^{2(-k_1+k_2-k_3+k_4)}$	$\alpha\beta = ab^{-1}c\beta\alpha d^{k_4}$
$\gamma b = b^{-1}\gamma d^{-2k_3}$	$\gamma d = d\gamma$
$\gamma c = c^{-1}\gamma d^{2(k_1-k_2+k_3-k_4-k_5)}$	$\beta\gamma = bc\gamma\beta d^{k_6}$

$>$

$$\lambda(\alpha) = \begin{pmatrix} 1 & 2(-k_1+k_2-k_3+k_4) & k_1-2k_3 & 0 & \frac{k_1+k_3-k_4}{2} \\ 0 & -1 & 0 & 0 & \frac{1}{2} \\ 0 & 0 & -1 & 0 & 0 \\ 0 & 0 & 0 & 1 & \frac{1}{2} \\ 0 & 0 & 0 & 0 & 1 \end{pmatrix}$$

$$\lambda(\beta) = \begin{pmatrix} -1 & -k_1 & 2k_3 & 0 & 0 \\ 0 & -1 & 0 & 0 & 0 \\ 0 & 0 & 1 & 0 & \frac{1}{2} \\ 0 & 0 & 0 & -1 & \frac{1}{2} \\ 0 & 0 & 0 & 0 & 1 \end{pmatrix} \quad \text{and} \quad \lambda(\gamma) =$$

$$\begin{pmatrix} 1 & 2(-k_1+k_2-k_3+k_4) & -2k_3 & 2(k_1-k_2+k_3-k_4-k_5) & \frac{-k_1+k_2+k_4+k_5+k_6}{2} \\ 0 & -1 & 0 & 0 & 0 \\ 0 & 0 & -1 & 0 & 0 \\ 0 & 0 & 0 & -1 & 0 \\ 0 & 0 & 0 & 0 & 1 \end{pmatrix}$$

$$H^2(Q, \mathbb{Z}) = \mathbb{Z} \oplus (\mathbb{Z}_2)^2 = \mathbb{Z}^6/A,$$

$$A = \{(k_1, \ldots, k_6) \| k_1 = 0, \ (k_2 + k_4 + k_5 + k_6), (k_3 - k_4 - k_5) \in 2\mathbb{Z}\}$$

AB-groups:

All groups are isomorphic to one of the previous case.

Action $\alpha d = d^{-1}\alpha$, $\beta d = d^{-1}\beta$:

$E : < a,b,c,d,\alpha,\beta,\gamma \parallel$

$[b,a] = 1$	$[d,a] = 1$
$[c,a] = 1$	$[d,b] = 1$
$[c,b] = d^{4k_1}$	$[d,c] = 1$
$\alpha a = a^{-1}\alpha$	$\alpha^2 = cd^{k_2}$
$\alpha b = b^{-1}\alpha d^{2k_1}$	$\alpha d = d^{-1}\alpha$
$\alpha c = c\alpha d^{-2k_2}$	$\alpha\gamma = ac\gamma\alpha d^{k_5}$
$\beta a = a^{-1}\beta$	$\beta^2 = bd^{k_3}$
$\beta b = b\beta d^{-2k_3}$	$\beta d = d^{-1}\beta$
$\beta c = c^{-1}\beta d^{-2k_1}$	$\alpha\beta = ab^{-1}c\beta\alpha d^{k_4}$
$\gamma a = a^{-1}\gamma d^{2(-k_1-k_3+k_5-k_6)}$	$\gamma^2 = d^{k_1+k_2+k_3+k_6}$
$\gamma b = b^{-1}\gamma d^{2(k_1-k_3)}$	$\gamma d = d\gamma$
$\gamma c = c^{-1}\gamma d^{-2k_2}$	$\beta\gamma = bc\gamma\beta d^{k_6}$

$>$

$$\lambda(\alpha) = \begin{pmatrix} -1 & 0 & -k_1 & 2k_2 & 2k_1 - \frac{k_2}{2} \\ 0 & -1 & 0 & 0 & \frac{1}{2} \\ 0 & 0 & -1 & 0 & 0 \\ 0 & 0 & 0 & 1 & \frac{1}{2} \\ 0 & 0 & 0 & 0 & 1 \end{pmatrix} \quad \lambda(\beta) = \begin{pmatrix} -1 & 0 & -k_1 + 2k_3 & k_1 & -\frac{k_4}{2} \\ 0 & -1 & 0 & 0 & 0 \\ 0 & 0 & 1 & 0 & \frac{1}{2} \\ 0 & 0 & 0 & -1 & \frac{1}{2} \\ 0 & 0 & 0 & 0 & 1 \end{pmatrix}$$

$$\lambda(\gamma) = \begin{pmatrix} 1 & 2(-k_1 - k_3 + k_5 - k_6) & 2(k_1 - k_3) & -2k_2 & \frac{k_1+k_2+k_3+k_6}{2} \\ 0 & -1 & 0 & 0 & 0 \\ 0 & 0 & -1 & 0 & 0 \\ 0 & 0 & 0 & -1 & 0 \\ 0 & 0 & 0 & 0 & 1 \end{pmatrix}$$

$$H^2(Q,\mathbb{Z}) = \mathbb{Z} \oplus (\mathbb{Z}_2)^2 = \mathbb{Z}^6/A,$$

$$A = \{(k_1,\ldots,k_6) \| k_1 = 0,\ (k_2 - k_4 - k_5 + k_6),\ (k_2 + k_3 + k_6) \in 2\mathbb{Z}\}$$

AB-groups:
All groups are isomorphic to one of the previous case.

---

75. $Q = P4$

$E : < a,b,c,d,\alpha \parallel$

$[b,a] = d^{k_1}$	$[d,a] = 1$
$[c,a] = 1$	$[d,b] = 1$
$[c,b] = 1$	$[d,c] = 1$
$\alpha a = b\alpha d^{k_2}$	$\alpha^4 = d^{k_4}$
$\alpha b = a^{-1}\alpha d^{k_3}$	$\alpha d = d\alpha$
$\alpha c = c\alpha$	

$>$

$$\lambda(\alpha) = \begin{pmatrix} 1 & k_2 & k_3 & 0 & \frac{k_4}{4} \\ 0 & 0 & -1 & 0 & 0 \\ 0 & 1 & 0 & 0 & 0 \\ 0 & 0 & 0 & 1 & 0 \\ 0 & 0 & 0 & 0 & 1 \end{pmatrix}$$

$$H^2(Q, \mathbb{Z}) = \mathbb{Z} \oplus \mathbb{Z}_2 \oplus \mathbb{Z}_4 = \mathbb{Z}^4/A,$$

$$A = \{(k_1, \ldots, k_4) \| k_1 = 0, \ (k_2 + k_3) \in 2\mathbb{Z}, \ k_4 \in 4\mathbb{Z}\}$$

AB-groups:
$\forall k > 0, \ k \equiv 0 \bmod 2, \ <(k, 0, 0, 1)>$
$\forall k > 0, \ k \equiv 0 \bmod 4, \ <(k, 0, 0, 3)>$
Remark: $<(k, 0, 0, 1)> \cong <(k, 0, 0, 3)> \ \forall k = 4l + 2, \ l \in \mathbb{Z}$.

---

76. $Q = P4_1$

$$E : <a, b, c, d, \alpha \ \| \quad \begin{array}{ll} [b,a] = d^{k_1} & [d,a] = 1 \\ [c,a] = 1 & [d,b] = 1 \\ [c,b] = 1 & [d,c] = 1 \\ \alpha a = b\alpha d^{k_2} & \alpha^4 = cd^{k_4} \\ \alpha b = a^{-1}\alpha d^{k_3} & \alpha d = d\alpha \\ \alpha c = c\alpha & \end{array} \quad >$$

$$\lambda(\alpha) = \begin{pmatrix} 1 & k_2 & k_3 & 0 & \frac{k_4}{4} \\ 0 & 0 & -1 & 0 & 0 \\ 0 & 1 & 0 & 0 & 0 \\ 0 & 0 & 0 & 1 & \frac{1}{4} \\ 0 & 0 & 0 & 0 & 1 \end{pmatrix}$$

$$H^2(Q, \mathbb{Z}) = \mathbb{Z} \oplus \mathbb{Z}_2 = \mathbb{Z}^4/A,$$

$$A = \{(k_1, \ldots, k_4) \| k_1 = 0, \ (k_2 + k_3) \in 2\mathbb{Z}, \ k_4 \in \mathbb{Z}\}$$

AB-groups:
$\forall k > 0, \ <(k, 0, 0, 0)>$
$\forall k > 0, \ k \equiv 0 \bmod 2, \ <(k, 1, 0, 0)>$
Remark: $<(k, 0, 0, 0)> \cong <(k, 1, 0, 0)> \ \forall k, \ k \not\equiv 0 \bmod 2$

---

77. $Q = P4_2$

$$E : <a, b, c, d, \alpha \ \| \quad \begin{array}{ll} [b,a] = d^{k_1} & [d,a] = 1 \\ [c,a] = 1 & [d,b] = 1 \\ [c,b] = 1 & [d,c] = 1 \\ \alpha a = b\alpha d^{k_2} & \alpha^4 = c^2 d^{k_4} \\ \alpha b = a^{-1}\alpha d^{k_3} & \alpha d = d\alpha \\ \alpha c = c\alpha & \end{array} \quad >$$

$$\lambda(\alpha) = \begin{pmatrix} 1 & k_2 & k_3 & 0 & \frac{k_4}{4} \\ 0 & 0 & -1 & 0 & 0 \\ 0 & 1 & 0 & 0 & 0 \\ 0 & 0 & 0 & 1 & \frac{1}{2} \\ 0 & 0 & 0 & 0 & 1 \end{pmatrix}$$

$$H^2(Q, \mathbb{Z}) = \mathbb{Z} \oplus (\mathbb{Z}_2)^2 = \mathbb{Z}^4/A, \ A = \{(k_1, \ldots, k_4) \| k_1 = 0, \ (k_2 + k_3), k_4 \in 2\mathbb{Z}\}$$

AB-groups:
$\forall k > 0, \; k \equiv 0 \bmod 2, \; < (k, 0, 0, 1) >$

---

79. $Q = I_4$

$$E : < a, b, c, d, \alpha \; \| \quad \begin{array}{ll} [b, a] = 1 & [d, a] = 1 \\ [c, a] = d^{k_1} & [d, b] = 1 \\ [c, b] = d^{-k_1} & [d, c] = 1 \\ \alpha a = c^{-1} \alpha d^{k_2} & \alpha^4 = d^{k_4} \\ \alpha b = abc \alpha d^{-k_2} & \alpha d = d\alpha \\ \alpha c = b^{-1} \alpha d^{k_3} \end{array} \quad >$$

$$\lambda(\alpha) = \begin{pmatrix} 1 & k_2 & -k_2 & k_3 & \frac{k_4}{4} \\ 0 & 0 & 1 & 0 & 0 \\ 0 & 0 & 1 & -1 & 0 \\ 0 & -1 & 1 & 0 & 0 \\ 0 & 0 & 0 & 0 & 1 \end{pmatrix}$$

$H^2(Q, \mathbb{Z}) = \mathbb{Z} \oplus \mathbb{Z}_4 = \mathbb{Z}^4 / A, \; A = \{(k_1, \ldots, k_4) \| k_1 = 0, \; k_2, k_3 \in \mathbb{Z}, \; k_4 \in 4\mathbb{Z}\}$

AB-groups:
$\forall k > 0, \; k \equiv 0 \bmod 2, \; < (k, 0, 0, 1) >$
$\forall k > 0, \; k \equiv 0 \bmod 2, \; < (k, 0, 0, 3) >$

---

80. $Q = I4_1$

$$E : < a, b, c, d, \alpha \; \| \quad \begin{array}{ll} [b, a] = 1 & [d, a] = 1 \\ [c, a] = d^{k_1} & [d, b] = 1 \\ [c, b] = d^{-k_1} & [d, c] = 1 \\ \alpha a = c^{-1} \alpha d^{k_2} & \alpha^4 = abd^{k_4} \\ \alpha b = abc \alpha d^{-k_2} & \alpha d = d\alpha \\ \alpha c = b^{-1} \alpha d^{k_3} \end{array} \quad >$$

$$\lambda(\alpha) = \begin{pmatrix} 1 & \frac{-k_1}{4} + k_2 & \frac{k_1}{4} - k_2 & k_3 & \frac{k_1}{16} + \frac{k_4 + k_3 - k_2}{4} \\ 0 & 0 & 1 & 0 & \frac{1}{2} \\ 0 & 0 & 1 & -1 & 0 \\ 0 & -1 & 1 & 0 & 0 \\ 0 & 0 & 0 & 0 & 1 \end{pmatrix}$$

$H^2(Q, \mathbb{Z}) = \mathbb{Z} \oplus \mathbb{Z}_2 = \mathbb{Z}^4 / A, \; A = \{(k_1, \ldots, k_4) \| k_1 = 0, \; (k_2 + k_3 + k_4) \in 2\mathbb{Z}\}$

AB-groups:
$\forall k > 0, \; < (k, 0, 0, 1) >$

---

81. $Q = P\bar{4}$

$$E : <a,b,c,d,\alpha \parallel \quad [b,a] = d^{k_1} \qquad [d,a] = 1 \quad >$$

$$\begin{array}{ll} [c,a] = 1 & [d,b] = 1 \\ [c,b] = 1 & [d,c] = 1 \\ \alpha a = b^{-1}\alpha d^{k_2} & \alpha^4 = d^{k_5} \\ \alpha b = a\alpha d^{k_3} & \alpha d = d\alpha \\ \alpha c = c^{-1}\alpha d^{k_4} & \end{array}$$

$$\lambda(\alpha) = \begin{pmatrix} 1 & k_2 & k_3 & k_4 & \frac{k_5}{4} \\ 0 & 0 & 1 & 0 & 0 \\ 0 & -1 & 0 & 0 & 0 \\ 0 & 0 & 0 & -1 & 0 \\ 0 & 0 & 0 & 0 & 1 \end{pmatrix}$$

$$H^2(Q,\mathbb{Z}) = \mathbb{Z} \oplus (\mathbb{Z}_2)^2 \oplus \mathbb{Z}_4 = \mathbb{Z}^5/A,$$

$$A = \{(k_1,\dots,k_5) \parallel k_1 = 0, \ (k_2 + k_3), k_4 \in 2\mathbb{Z}, \ k_5 \in 4\mathbb{Z}\}$$

AB-groups:

$\forall k > 0, \ k \equiv 0 \bmod 2 < (k,0,0,0,1) >$

$\forall k > 0, \ k \equiv 0 \bmod 2 < (k,0,0,1,1) >$

$\forall k > 0, \ k \equiv 0 \bmod 4 < (k,0,0,0,3) >$

Remark: $\forall k > 0, k \equiv 0 \bmod 2, \ < (k,0,0,1,1) > \ \cong \ < (k,0,0,1,3) >$

$\forall k = 4l+2, \ l \in \mathbb{Z}, \ < (k,0,0,0,1) > \ \cong \ < (k,0,0,0,3) >$

---

82. $Q = I\bar{4}$

$$E : <a,b,c,d,\alpha \parallel \quad [b,a] = 1 \qquad\qquad [d,a] = 1 \quad >$$

$$\begin{array}{ll} [c,a] = d^{k_1} & [d,b] = 1 \\ [c,b] = d^{-k_1} & [d,c] = 1 \\ \alpha a = c\alpha d^{k_2} & \alpha^4 = d^{k_5} \\ \alpha b = a^{-1}b^{-1}c^{-1}\alpha d^{k_3} & \alpha d = d\alpha \\ \alpha c = b\alpha d^{k_4} & \end{array}$$

$$\lambda(\alpha) = \begin{pmatrix} 1 & k_2 & k_3 & k_4 & \frac{k_5}{4} \\ 0 & 0 & -1 & 0 & 0 \\ 0 & 0 & -1 & 1 & 0 \\ 0 & 1 & -1 & 0 & 0 \\ 0 & 0 & 0 & 0 & 1 \end{pmatrix}$$

$$H^2(Q,\mathbb{Z}) = \mathbb{Z} \oplus (\mathbb{Z}_4)^2 = \mathbb{Z}^5/A,$$

$$A = \{(k_1,\dots,k_5) \parallel k_1 = 0, \ (2k_4 - k_2 + k_3), k_5 \in 4\mathbb{Z}\}$$

AB-groups:

$\forall k > 0, \ k \equiv 0 \bmod 2 < (k,0,0,0,1) >$

$\forall k > 0, \ k \equiv 0 \bmod 4 < (k,0,0,0,3) >$

$\forall k > 0, \ k \equiv 0 \bmod 2 < (k,0,0,1,1) >$

$\forall k > 0, \ k = 4l+2, \ l \in \mathbb{Z}, \ < (k,0,0,1,3) >$

$\forall k > 0,\ k \not\equiv 0 \bmod 2 < (k,1,0,0,1) >$

Remark: $\forall k = 4l+2,\ l \in \mathbb{Z},\ < (k,0,0,0,1) > \cong < (k,0,0,0,3) >$

$\forall k = 4l,\ l \in \mathbb{Z},\ < (k,0,0,1,1) > \cong < (k,0,0,1,3) >$

$\forall k \not\equiv \bmod 0,\ < (k,1,0,0,1) > \cong < (k,1,0,0,3) >$

$\cong < (k,1,0,1,1) > \cong < (k,1,0,1,3) >$

---

83. $Q = P4/m$

$$E :\ < a,b,c,d,\alpha,\beta\ \|\quad \begin{array}{ll} [b,a] = d^{k_1} & [d,a] = 1 \quad > \\ [c,a] = 1 & [d,b] = 1 \\ [c,b] = 1 & [d,c] = 1 \\ \alpha a = b\alpha d^{k_2} & \alpha^4 = d^{k_4} \\ \alpha b = a^{-1}\alpha d^{k_3} & \alpha d = d\alpha \\ \alpha c = c\alpha & \\ \beta a = a^{-1}\beta d^{k_2+k_3} & \beta^2 = d^{k_6} \\ \beta b = b^{-1}\beta d^{k_3-k_2} & \beta d = d\beta \\ \beta c = c^{-1}\beta d^{k_5} & \alpha\beta = \beta\alpha \end{array}$$

$$\lambda(\alpha) = \begin{pmatrix} 1 & k_2 & k_3 & 0 & \frac{k_4}{4} \\ 0 & 0 & -1 & 0 & 0 \\ 0 & 1 & 0 & 0 & 0 \\ 0 & 0 & 0 & 1 & 0 \\ 0 & 0 & 0 & 0 & 1 \end{pmatrix} \quad \lambda(\beta) = \begin{pmatrix} 1 & k_2+k_3 & k_3-k_2 & k_5 & \frac{k_6}{2} \\ 0 & -1 & 0 & 0 & 0 \\ 0 & 0 & -1 & 0 & 0 \\ 0 & 0 & 0 & -1 & 0 \\ 0 & 0 & 0 & 0 & 1 \end{pmatrix}$$

$$H^2(Q,\mathbb{Z}) = \mathbb{Z} \oplus (\mathbb{Z}_2)^3 \oplus \mathbb{Z}_4 = \mathbb{Z}^6/A,$$

$$A = \{(k_1,\ldots,k_6)\|k_1 = 0,\ (k_2+k_3),\ k_5,\ k_6 \in 2\mathbb{Z},\ k_4 \in 4\mathbb{Z}\}$$

AB–groups:

None

---

84. $Q = P4_2/m$

$$E :\ < a,b,c,d,\alpha,\beta\ \|\quad \begin{array}{ll} [b,a] = d^{k_1} & [d,a] = 1 \quad > \\ [c,a] = 1 & [d,b] = 1 \\ [c,b] = 1 & [d,c] = 1 \\ \alpha a = b\alpha d^{k_2} & \alpha^4 = c^2 d^{k_4} \\ \alpha b = a^{-1}\alpha d^{k_3} & \alpha d = d\alpha \\ \alpha c = c\alpha & \\ \beta a = a^{-1}\beta d^{k_2+k_3} & \beta^2 = d^{k_5} \\ \beta b = b^{-1}\beta d^{k_3-k_2} & \beta d = d\beta \\ \beta c = c^{-1}\beta d^{-2k_6} & \alpha\beta = c\beta\alpha d^{k_6} \end{array}$$

$$\lambda(\alpha) = \begin{pmatrix} 1 & k_2 & k_3 & 0 & \frac{k_4}{4} \\ 0 & 0 & -1 & 0 & 0 \\ 0 & 1 & 0 & 0 & 0 \\ 0 & 0 & 0 & 1 & \frac{1}{2} \\ 0 & 0 & 0 & 0 & 1 \end{pmatrix} \quad \lambda(\beta) = \begin{pmatrix} 1 & k_2+k_3 & k_3-k_2 & -2k_6 & \frac{k_5}{2} \\ 0 & -1 & 0 & 0 & 0 \\ 0 & 0 & -1 & 0 & 0 \\ 0 & 0 & 0 & -1 & 0 \\ 0 & 0 & 0 & 0 & 1 \end{pmatrix}$$

$$H^2(Q, \mathbb{Z}) = \mathbb{Z} \oplus (\mathbb{Z}_2)^2 \oplus \mathbb{Z}_4 = \mathbb{Z}^6/A,$$

$$A = \{(k_1, \ldots, k_6) \| k_1 = 0, \ (k_2 + k_3), k_5 \in 2\mathbb{Z}, \ (k_4 - 2k_6) \in 4\mathbb{Z}\}$$

AB-groups:
None

---

85. $Q = P4/n$

$$E : \ <a, b, c, d, \alpha, \beta \ \|$$

$[b, a] = d^{2k_1}$	$[d, a] = 1$
$[c, a] = 1$	$[d, b] = 1$
$[c, b] = 1$	$[d, c] = 1$
$\alpha a = b \alpha d^{k_2}$	$\alpha^4 = d^{k_3}$
$\alpha b = a^{-1} \alpha d^{-k_1 + k_2 - 2k_6}$	$\alpha d = d\alpha$
$\alpha c = c\alpha$	
$\beta a = a^{-1}\beta d^{2(k_2 - k_6)}$	$\beta^2 = d^{k_5}$
$\beta b = b^{-1}\beta d^{-2k_6}$	$\beta d = d\beta$
$\beta c = c^{-1}\beta d^{k_4}$	$\alpha\beta = b\beta\alpha d^{k_6}$

$\ \ >$

$$\lambda(\alpha) = \begin{pmatrix} 1 & k_2 & -\frac{k_1}{2} + k_2 - 2k_6 & 0 & \frac{k_1 + 2k_3 + 4k_6}{8} \\ 0 & 0 & -1 & 0 & 0 \\ 0 & 1 & 0 & 0 & \frac{1}{2} \\ 0 & 0 & 0 & 1 & 0 \\ 0 & 0 & 0 & 0 & 1 \end{pmatrix}$$

$$\lambda(\beta) = \begin{pmatrix} 1 & 2(k_2 - k_6) & -2k_6 & k_4 & \frac{k_5}{2} \\ 0 & -1 & 0 & 0 & 0 \\ 0 & 0 & -1 & 0 & 0 \\ 0 & 0 & 0 & -1 & 0 \\ 0 & 0 & 0 & 0 & 1 \end{pmatrix}$$

$$H^2(Q, \mathbb{Z}) = \mathbb{Z} \oplus (\mathbb{Z}_2)^2 \oplus \mathbb{Z}_4 = \mathbb{Z}^6/A,$$

$$A = \{(k_1, \ldots, k_6) \| k_1 = 0, \ k_4, k_5 \in 2\mathbb{Z}, \ k_3 \in 4\mathbb{Z}, \ k_2, k_6 \in \mathbb{Z}\}$$

AB-groups:
$\forall k > 0, \ k \equiv 0 \bmod 2, \ < (k, 0, 1, 0, 1, 0) >$
$\forall k > 0, \ k \equiv 0 \bmod 2, \ < (k, 0, 3, 0, 1, 0) >$

---

86. $Q = P4_2/n$

$$E : \ <a, b, c, d, \alpha, \beta \ \|$$

$[b, a] = d^{2k_1}$	$[d, a] = 1$
$[c, a] = 1$	$[d, b] = 1$
$[c, b] = 1$	$[d, c] = 1$
$\alpha a = b \alpha d^{k_2}$	$\alpha^4 = c^2 d^{k_4}$
$\alpha b = a^{-1} \alpha d^{k_3}$	$\alpha d = d\alpha$
$\alpha c = c\alpha$	
$\beta a = a^{-1}\beta d^{k_1 + k_2 + k_3}$	$\beta^2 = d^{k_5}$
$\beta b = b^{-1}\beta d^{k_1 - k_2 + k_3}$	$\beta d = d\beta$
$\beta c = c^{-1}\beta d^{-k_1 + k_2 - k_3 - 2k_6}$	$\alpha\beta = bc\beta\alpha d^{k_6}$

$\ \ >$

$$\lambda(\alpha) = \begin{pmatrix} 1 & k_2 & \frac{k_1}{2} + k_3 & 0 & \frac{-k_1 + 2(k_2 - k_3 + k_4)}{8} \\ 0 & 0 & -1 & 0 & 0 \\ 0 & 1 & 0 & 0 & \frac{1}{2} \\ 0 & 0 & 0 & 1 & \frac{1}{2} \\ 0 & 0 & 0 & 0 & 1 \end{pmatrix}$$

$$\lambda(\beta) = \begin{pmatrix} 1 & k_1 + k_2 + k_3 & k_1 - k_2 + k_3 & -k_1 + k_2 - k_3 - 2k_6 & \frac{k_5}{2} \\ 0 & -1 & 0 & 0 & 0 \\ 0 & 0 & -1 & 0 & 0 \\ 0 & 0 & 0 & -1 & 0 \\ 0 & 0 & 0 & 0 & 1 \end{pmatrix}$$

$$H^2(Q, \mathbb{Z}) = \mathbb{Z} \oplus (\mathbb{Z}_2)^2 \oplus \mathbb{Z}_4 = \mathbb{Z}^6 / A,$$

$$A = \{(k_1, \ldots, k_6) \| k_1 = 0, \ (k_2 + k_3), k_5 \in 2\mathbb{Z}, \ (k_2 - k_3 + k_4 - 2k_6) \in 4\mathbb{Z}\}$$

AB-groups:
$\forall k > 0, \ k \equiv 0 \bmod 2, \ < (k, 0, 0, 1, 1, 0) >$
$\forall k > 0, \ k \equiv 0 \bmod 2, \ < (k, 0, 0, 1, 1, 1) >$

---

87. $Q = I4/m$

$$E : \ < a, b, c, d, \alpha, \beta \ \|$$

$[b, a] = 1$	$[d, a] = 1 \quad >$
$[c, a] = d^{k_1}$	$[d, b] = 1$
$[c, b] = d^{-k_1}$	$[d, c] = 1$
$\alpha a = c^{-1} \alpha d^{k_2}$	$\alpha^4 = d^{k_4}$
$\alpha b = abc\alpha d^{-k_2}$	$\alpha d = d\alpha$
$\alpha c = b^{-1} \alpha d^{k_3}$	
$\beta a = a^{-1} \beta d^{k_5}$	$\beta^2 = d^{k_6}$
$\beta b = b^{-1} \beta d^{-2k_2 + 2k_3 + k_5}$	$\beta d = d\beta$
$\beta c = c^{-1} \beta d^{2k_2 - k_5}$	$\alpha\beta = \beta\alpha$

$$\lambda(\alpha) = \begin{pmatrix} 1 & k_2 & -k_2 & k_3 & \frac{k_4}{2} \\ 0 & 0 & 1 & 0 & 0 \\ 0 & 0 & 1 & -1 & 0 \\ 0 & -1 & 1 & 0 & 0 \\ 0 & 0 & 0 & 0 & 1 \end{pmatrix} \qquad \lambda(\beta) = \begin{pmatrix} 1 & k_5 & -2k_2 + 2k_3 + k_5 & 2k_2 - k_5 & \frac{k_6}{2} \\ 0 & -1 & 0 & 0 & 0 \\ 0 & 0 & -1 & 0 & 0 \\ 0 & 0 & 0 & -1 & 0 \\ 0 & 0 & 0 & 0 & 1 \end{pmatrix}$$

$$H^2(Q, \mathbb{Z}) = \mathbb{Z} \oplus (\mathbb{Z}_2)^2 \oplus \mathbb{Z}_4 = \mathbb{Z}^6 / A,$$

$$A = \{(k_1, \ldots, k_6) \| k_1 = 0, \ k_5, k_6 \in 2\mathbb{Z}, \ k_4 \in 4\mathbb{Z}, \ k_2, k_3 \in \mathbb{Z}\}$$

AB-groups:
None

88. $Q = I4_1/a$

$E : \; < a, b, c, d, \alpha, \beta \; \| \quad [b,a] = 1 \qquad\qquad [d,a] = 1 \qquad\qquad >$
$\qquad\qquad\qquad\qquad\qquad [c,a] = d^{k_1} \qquad\qquad [d,b] = 1$
$\qquad\qquad\qquad\qquad\qquad [c,b] = d^{-k_1} \qquad\qquad [d,c] = 1$
$\qquad\qquad\qquad\qquad\qquad \alpha a = c^{-1}\alpha d^{k_2} \qquad\quad \alpha^4 = a^{-1}b^{-1}d^{k_4}$
$\qquad\qquad\qquad\qquad\qquad \alpha b = abc\alpha d^{-k_2} \qquad\quad \alpha d = d\alpha$
$\qquad\qquad\qquad\qquad\qquad \alpha c = b^{-1}\alpha d^{k_3}$
$\qquad\qquad\qquad\qquad\qquad \beta a = a^{-1}\beta d^{2k_2+2k_6} \qquad \beta^2 = d^{k_5}$
$\qquad\qquad\qquad\qquad\qquad \beta b = b^{-1}\beta d^{-k_1+2k_3+2k_6} \qquad \beta d = d\beta$
$\qquad\qquad\qquad\qquad\qquad \beta c = c^{-1}\beta d^{-2k_6} \qquad\qquad \alpha\beta = c\beta\alpha d^{k_6}$

$$\lambda(\alpha) = \begin{pmatrix} 1 & k_2 & -k_2 & \frac{-k_1}{4}+k_3 & \frac{k_1-4(k_2+k_3-k_4)}{16} \\ 0 & 0 & 1 & 0 & 0 \\ 0 & 0 & 1 & -1 & 0 \\ 0 & -1 & 1 & 0 & \frac{1}{2} \\ 0 & 0 & 0 & 0 & 1 \end{pmatrix}$$

$$\lambda(\beta) = \begin{pmatrix} 1 & 2(k_2+k_6) & -k_1+2k_3+2k_6 & -2k_6 & \frac{k_5}{2} \\ 0 & -1 & 0 & 0 & 0 \\ 0 & 0 & -1 & 0 & 0 \\ 0 & 0 & 0 & -1 & 0 \\ 0 & 0 & 0 & 0 & 1 \end{pmatrix}.$$

$$H^2(Q, \mathbb{Z}) = \mathbb{Z} \oplus \mathbb{Z}_2 \oplus \mathbb{Z}_4 = \mathbb{Z}^6/A,$$

$$A = \{(k_1, \ldots, k_6) \| k_1 = 0,\; k_5 \in 2\mathbb{Z},\; (-k_2 - k_3 + k_4 - 2k_6) \in 4\mathbb{Z}\}$$

AB-groups:
$\forall k > 0,\; k \equiv 0 \bmod 2,\; < (k,0,0,1,1,0) >$
$\forall k > 0,\; k \equiv 0 \bmod 2,\; < (k,0,0,1,1,1) >$

---

103. $Q = P4cc$

$E : \; < a, b, c, d, \alpha, \beta \; \| \quad [b,a] = d^{k_1} \qquad\qquad [d,a] = 1 \qquad\qquad >$
$\qquad\qquad\qquad\qquad\qquad [c,a] = 1 \qquad\qquad\qquad [d,b] = 1$
$\qquad\qquad\qquad\qquad\qquad [c,b] = 1 \qquad\qquad\qquad [d,c] = 1$
$\qquad\qquad\qquad\qquad\qquad \alpha a = b\alpha d^{k_2} \qquad\qquad \alpha^4 = d^{k_4}$
$\qquad\qquad\qquad\qquad\qquad \alpha b = a^{-1}\alpha d^{k_3} \qquad\quad \alpha d = d\alpha$
$\qquad\qquad\qquad\qquad\qquad \alpha c = c\alpha$
$\qquad\qquad\qquad\qquad\qquad \beta a = a\beta d^{k_2+k_3} \qquad \beta^2 = cd^{k_5}$
$\qquad\qquad\qquad\qquad\qquad \beta b = b^{-1}\beta \qquad\qquad \beta d = d^{-1}\beta$
$\qquad\qquad\qquad\qquad\qquad \beta c = c\beta d^{-2k_5} \qquad\quad \alpha\beta = \beta\alpha^3 d^{-k_4}$

$$\lambda(\alpha) = \begin{pmatrix} 1 & k_2 & k_3 & 0 & \frac{k_4}{4} \\ 0 & 0 & -1 & 0 & 0 \\ 0 & 1 & 0 & 0 & 0 \\ 0 & 0 & 0 & 1 & 0 \\ 0 & 0 & 0 & 0 & 1 \end{pmatrix} \qquad \lambda(\beta) = \begin{pmatrix} -1 & -k_2-k_3 & 0 & 2k_5 & 0 \\ 0 & 1 & 0 & 0 & 0 \\ 0 & 0 & -1 & 0 & 0 \\ 0 & 0 & 0 & 1 & \frac{1}{2} \\ 0 & 0 & 0 & 0 & 1 \end{pmatrix}$$

$$H^2(Q, \mathbb{Z}) = \mathbb{Z} \oplus \mathbb{Z}_2 \oplus \mathbb{Z}_4 = \mathbb{Z}^5/A,$$
$$A = \{(k_1, \ldots, k_5) \| k_1 = 0,\ (k_2 + k_3) \in 2\mathbb{Z},\ k_4 \in 4\mathbb{Z},\ k_5 \in \mathbb{Z}\}$$

AB-groups:

$\forall k > 0,\ k \equiv 0 \bmod 2,\ < (k, 0, 0, 1, 0) >$

$\forall k > 0,\ k \equiv 0 \bmod 4,\ < (k, 0, 0, 3, 0) >$

Remark: If $k = 4l + 2$ then $< (k, 0, 0, 1, 0) > \cong < (k, 0, 0, 3, 0) >$.

---

106. $Q = P4_2bc$

$E :\ < a, b, c, d, \alpha, \beta \ \|$

$[b, a] = d^{2k_1}$	$[d, a] = 1$
$[c, a] = 1$	$[d, b] = 1$
$[c, b] = 1$	$[d, c] = 1$
$\alpha a = b\alpha d^{k_2}$	$\alpha^4 = c^2 d^{k_3}$
$\alpha b = a^{-1}\alpha d^{-k_1-k_2-2k_5}$	$\alpha d = d\alpha$
$\alpha c = c\alpha$	
$\beta a = a\beta d^{-2k_5}$	$\beta^2 = ad^{k_5}$
$\beta b = b^{-1}\beta d^{-k_1}$	$\beta d = d^{-1}\beta$
$\beta c = c\beta d^{2k_4}$	$\alpha\beta = bc^{-1}\beta\alpha^3 d^{-k_2-k_3-3k_4-k_5}$

$>$

$$\lambda(\alpha) = \begin{pmatrix} 1 & \frac{k_1}{2}+k_2 & -k_1-k_2-2k_5 & 0 & \frac{k_1}{8}+\frac{k_3}{4}+\frac{k_5}{2} \\ 0 & 0 & -1 & 0 & \frac{1}{2} \\ 0 & 1 & 0 & 0 & 0 \\ 0 & 0 & 0 & 1 & \frac{1}{2} \\ 0 & 0 & 0 & 0 & 1 \end{pmatrix}$$

$$\lambda(\beta) = \begin{pmatrix} -1 & 2k_5 & \frac{k_1}{2} & -2k_4 & 0 \\ 0 & 1 & 0 & 0 & \frac{1}{2} \\ 0 & 0 & -1 & 0 & 0 \\ 0 & 0 & 0 & 1 & 0 \\ 0 & 0 & 0 & 0 & 1 \end{pmatrix}$$

$H^2(Q, \mathbb{Z}) = \mathbb{Z} \oplus \mathbb{Z}_4 = \mathbb{Z}^5/A,\ A = \{(k_1, \ldots, k_5) \| k_1 = 0, (k_3+2k_4) \in 4\mathbb{Z}, k_2, k_5 \in \mathbb{Z}\}$

AB-groups:

$\forall k > 0,\ k \equiv 0 \bmod 2,\ < (k, 0, 1, 0, 0) >$

$\forall k > 0,\ k \equiv 0 \bmod 2,\ < (k, 0, 1, 1, 0) >$

---

104. $Q = P4nc$

$E :\ < a, b, c, d, \alpha, \beta \ \|$

$[b, a] = d^{2k_1}$	$[d, a] = 1$
$[c, a] = 1$	$[d, b] = 1$
$[c, b] = 1$	$[d, c] = 1$
$\alpha a = b\alpha d^{k_2}$	$\alpha^4 = d^{k_3}$
$\alpha b = a^{-1}\alpha d^{-k_1+k_2+2k_3+2k_5}$	$\alpha d = d\alpha$
$\alpha c = c\alpha$	
$\beta a = a\beta d^{2k_2+2k_3+2k_5}$	$\beta^2 = acd^{k_4}$
$\beta b = b^{-1}\beta d^{-k_1}$	$\beta d = d^{-1}\beta$
$\beta c = c\beta d^{-2k_2-2k_3-2k_4-2k_5}$	$\alpha\beta = b\beta\alpha^3 d^{k_5}$

$>$

$$\lambda(\alpha) = \begin{pmatrix} 1 & \frac{k_1}{2} + k_2 & -k_1 + k_2 + 2k_3 + 2k_5 & 0 & \frac{k_1 - 4k_2 - 2k_3 - 4k_5}{8} \\ 0 & 0 & -1 & 0 & \frac{1}{2} \\ 0 & 1 & 0 & 0 & 0 \\ 0 & 0 & 0 & 1 & 0 \\ 0 & 0 & 0 & 0 & 1 \end{pmatrix}$$

$$\lambda(\beta) = \begin{pmatrix} -1 & -2(k_2 + k_3 + k_5) & \frac{k_1}{2} & 2(k_2 + k_3 + k_4 + k_5) & 0 \\ 0 & 1 & 0 & 0 & \frac{1}{2} \\ 0 & 0 & -1 & 0 & 0 \\ 0 & 0 & 0 & 1 & \frac{1}{2} \\ 0 & 0 & 0 & 0 & 1 \end{pmatrix}$$

$$H^2(Q, \mathbb{Z}) = \mathbb{Z} \oplus \mathbb{Z}_4 = \mathbb{Z}^5 / A,$$

$$A = \{(k_1, \ldots, k_5) \| k_1 = 0, \ k_3 \in 4\mathbb{Z}, \ k_2, k_4, k_5 \in \mathbb{Z}\}$$

AB-groups:
$\forall k > 0, \ k \equiv 0 \bmod 2, \ < (k, 0, 1, 0, 0) >$
$\forall k > 0, \ k \equiv 0 \bmod 2, \ < (k, 0, 3, 0, 0) >$

---

110. $Q = I4_1cd$

$E : < a, b, c, d, \alpha, \beta \|$

$[b, a] = 1$		$[d, a] = 1$
$[c, a] = d^{2k_1}$		$[d, b] = 1$
$[c, b] = d^{-2k_1}$		$[d, c] = 1$
$\alpha a = c^{-1} \alpha d^{4k_1 - 4k_2 + 2k_3 - 3k_4 + 2k_5}$		$\alpha^4 = a^{-1}b^{-1}d^{k_3}$
$\alpha b = abc\alpha d^{-4k_1 + 4k_2 - 2k_3 + 3k_4 - 2k_5}$		$\alpha d = d\alpha$
$\alpha c = b^{-1}\alpha d^{k_2}$		
$\beta a = c^{-1}\beta d^{k_4}$		$\beta^2 = abd^{k_1 - k_2 - k_4}$
$\beta b = abc\beta d^{-2k_1 + 2k_2 + k_4}$		$\beta d = d^{-1}\beta$
$\beta c = a^{-1}\beta d^{-k_4}$		$\alpha\beta = abc\beta\alpha^3 d^{k_5}$

$>$

$$\lambda(\alpha) =$$

$$\begin{pmatrix} 1 & 4k_1 - 4k_2 + 2k_3 + 2k_5 - 3k_4 & 4k_2 - 2k_3 - 2k_5 + 3k_4 - 4k_1 & \frac{-k_1}{2} + k_2 & \frac{6k_2 - 2k_3 + 6k_4 - 4k_5 - 7k_1}{8} \\ 0 & 0 & 1 & 0 & 0 \\ 0 & 0 & 1 & -1 & 0 \\ 0 & -1 & 1 & 0 & \frac{1}{2} \\ 0 & 0 & 0 & 0 & 1 \end{pmatrix}$$

$$\lambda(\beta) = \begin{pmatrix} -1 & -k_4 & 2(k1 - k2) - k_4 & k_4 & 0 \\ 0 & 0 & 1 & -1 & \frac{1}{2} \\ 0 & 0 & 1 & 0 & \frac{1}{2} \\ 0 & -1 & 1 & 0 & 0 \\ 0 & 0 & 0 & 0 & 1 \end{pmatrix}$$

$$H^2(Q, \mathbb{Z}) = \mathbb{Z} \oplus \mathbb{Z}_4 = \mathbb{Z}^5 / A,$$

$$A = \{(k_1, \ldots, k_5) \| k_1 = 0, \ (k_2 - k_3 + k_4) \in 4\mathbb{Z}, \ k_5 \in \mathbb{Z}\}$$

AB-groups:
$\forall k > 0, \ < (k, 0, 1, 0, 0) >$

$\forall k > 0, \ < (k, 0, 3, 0, 0) >$

---

114. $Q = P\bar{4}2_1c$

$$E : \ < a, b, c, d, \alpha, \beta \ \| \quad \begin{array}{ll} [b, a] = d^{2k_1} & [d, a] = 1 \\ [c, a] = 1 & [d, b] = 1 \\ [c, b] = 1 & [d, c] = 1 \\ \alpha a = b^{-1}\alpha d^{k_2} & \alpha^4 = d^{k_3} \\ \alpha b = a\alpha d^{k_1 - k_2 - 2k_4} & \alpha d = d\alpha \\ \alpha c = c^{-1}\alpha d^{2(k_1 - k_2 + k_3 - k_4 + k_5)} & \\ \beta a = a^{-1}\beta d^{k_1} & \beta^2 = bd^{k_4} \\ \beta b = b\beta d^{-2k_4} & \beta d = d^{-1}\beta \\ \beta c = c^{-1}\beta & \alpha\beta = ac\beta\alpha^3 d^{k_5} \end{array} \ >$$

$$\lambda(\alpha) = \begin{pmatrix} 1 & \frac{-k_1}{2} + k_2 & k_1 - k_2 - 2k_4 & 2(k_1 - k_2 + k_3 - k_4 + k_5) & -\frac{k_1 + 2k_3 + 4k_5}{8} \\ 0 & 0 & 1 & 0 & \frac{1}{2} \\ 0 & -1 & 0 & 0 & 0 \\ 0 & 0 & 0 & -1 & \frac{1}{2} \\ 0 & 0 & 0 & 0 & 1 \end{pmatrix}$$

$$\lambda(\beta) = \begin{pmatrix} -1 & -\frac{k_1}{2} & 2k_4 & 0 & 0 \\ 0 & -1 & 0 & 0 & 0 \\ 0 & 0 & 1 & 0 & \frac{1}{2} \\ 0 & 0 & 0 & -1 & 0 \\ 0 & 0 & 0 & 0 & 1 \end{pmatrix}$$

$H^2(Q, \mathbb{Z}) = \mathbb{Z} \oplus \mathbb{Z}_4 = \mathbb{Z}^5/A, \ A = \{(k_1, \ldots, k_5) \| k_1 = 0, \ k_3 \in 4\mathbb{Z}, \ k_2, k_4, k_5 \in \mathbb{Z}\}$

AB–groups:
$\forall k > 0, \ k \equiv 0 \bmod 2, \ < (k, 0, 1, 0, 0) >$
$\forall k > 0, \ k \equiv 0 \bmod 2, \ < (k, 0, 3, 0, 0) >$

---

146. $Q = R3$

$$E : \ < a, b, c, d, \alpha \ \| \quad \begin{array}{ll} [b, a] = d^{k_1} & [d, a] = 1 \\ [c, a] = d^{-k_1} & [d, b] = 1 \\ [c, b] = d^{k_1} & [d, c] = 1 \\ \alpha a = b\alpha d^{k_2} & \alpha^3 = d^{k_4} \\ \alpha b = c\alpha d^{k_3} & \alpha d = d\alpha \\ \alpha c = a\alpha d^{-k_2 - k_3} & \end{array} \ >$$

$$\lambda(\alpha) = \begin{pmatrix} 1 & k_2 & k_3 & -k_2 - k_3 & \frac{k_4}{3} \\ 0 & 0 & 0 & 1 & 0 \\ 0 & 1 & 0 & 0 & 0 \\ 0 & 0 & 1 & 0 & 0 \\ 0 & 0 & 0 & 0 & 1 \end{pmatrix}$$

$H^2(Q, \mathbb{Z}) = \mathbb{Z} \oplus \mathbb{Z}_3 = \mathbb{Z}^4/A, \ A = \{(k_1, k_2, k_3, k_4) \| k_1 = 0, \ k_2, k_3 \in \mathbb{Z}, \ k_4 \in 3\mathbb{Z}\}$

AB-groups:
$\forall k > 0, \; < (k,0,0,1) >$
$\forall k > 0, \; < (k,0,0,2) >$

---

144. $Q = P3_1$

$$E : \; < a,b,c,d,\alpha \; \| \quad \begin{aligned} &[b,a] = d^{k_1} && [d,a] = 1 \quad > \\ &[c,a] = 1 && [d,b] = 1 \\ &[c,b] = 1 && [d,c] = 1 \\ &\alpha a = b\alpha d^{k_2} && \alpha^3 = cd^{k_4} \\ &\alpha b = a^{-1}b^{-1}\alpha d^{k_3} && \alpha d = d\alpha \\ &\alpha c = c\alpha \end{aligned}$$

$$\lambda(\alpha) = \begin{pmatrix} 1 & k_2 & \frac{-k_1}{2}+k_3 & 0 & \frac{k_4}{3} \\ 0 & 0 & -1 & 0 & 0 \\ 0 & 1 & -1 & 0 & 0 \\ 0 & 0 & 0 & 1 & \frac{1}{3} \\ 0 & 0 & 0 & 0 & 1 \end{pmatrix}$$

$$H^2(Q,\mathbb{Z}) = \mathbb{Z} \oplus \mathbb{Z}_3 = \mathbb{Z}^4/A,$$

$$A = \{(k_1,k_2,k_3,k_4)\|k_1 = 0, \; (k_3 - k_2) \in 3\mathbb{Z}, \; k_4 \in \mathbb{Z}\}$$

AB-groups:
$\forall k > 0, \; < (k,0,0,0) >$
$\forall k > 0, \; k \equiv 0 \bmod 3, \; < (k,1,0,0) >$
Remark: $\quad \forall k \; < (k,1,0,0) > \; \cong \; < (k,2,0,0) >$
$\qquad\qquad \forall k \not\equiv 0 \bmod 3 : \; < (k,0,0,0) > \; \cong \; < (k,1,0,0) >$

---

148. $Q = R\bar{3}$

$$E : \; < a,b,c,d,\alpha \; \| \quad \begin{aligned} &[b,a] = d^{k_1} && [d,a] = 1 \quad > \\ &[c,a] = d^{-k_1} && [d,b] = 1 \\ &[c,b] = d^{k_1} && [d,c] = 1 \\ &\alpha a = b^{-1}\alpha d^{k_2} && \alpha^6 = d^{k_5} \\ &\alpha b = c^{-1}\alpha d^{k_3} && \alpha d = d\alpha \\ &\alpha c = a^{-1}\alpha d^{k_4} \end{aligned}$$

$$\lambda(\alpha) = \begin{pmatrix} 1 & k_2 & k_3 & k_4 & \frac{k_5}{6} \\ 0 & 0 & 0 & -1 & 0 \\ 0 & -1 & 0 & 0 & 0 \\ 0 & 0 & -1 & 0 & 0 \\ 0 & 0 & 0 & 0 & 1 \end{pmatrix}$$

$$H^2(Q,\mathbb{Z}) = \mathbb{Z} \oplus \mathbb{Z}_2 \oplus \mathbb{Z}_6 = \mathbb{Z}^5/A,$$

$$A = \{(k_1,\ldots,k_5)\|k_1 = 0, \; (k_2 + k_3 - k_4) \in 2\mathbb{Z}, \; k_5 \in 6\mathbb{Z}\}$$

AB-groups:
$\forall k > 0, \; k \equiv 0 \bmod 2 \; < (k,0,0,0,1) >$

$\forall k > 0, k \equiv 0 \bmod 2 \quad < (k,0,0,0,5) >$

---

147. $Q = P\bar{3}$

$$E : \; < a,b,c,d,\alpha \; \| \quad
\begin{array}{ll}
[b,a] = d^{k_1} & [d,a] = 1 \\
[c,a] = 1 & [d,b] = 1 \\
[c,b] = 1 & [d,c] = 1 \\
\alpha a = b^{-1}\alpha d^{k_2} & \alpha^6 = d^{k_5} \\
\alpha b = ab\alpha d^{k_3} & \alpha d = d\alpha \\
\alpha c = c^{-1}\alpha d^{k_4} &
\end{array}
\quad >$$

$$\lambda(\alpha) = \begin{pmatrix}
1 & k_2 & -\frac{k_1}{2}+k_3 & k_4 & \frac{k_5}{6} \\
0 & 0 & 1 & 0 & 0 \\
0 & -1 & 1 & 0 & 0 \\
0 & 0 & 0 & -1 & 0 \\
0 & 0 & 0 & 0 & 1
\end{pmatrix}$$

$$H^2(Q,\mathbb{Z}) = \mathbb{Z} \oplus \mathbb{Z}_2 \oplus \mathbb{Z}_6 = \mathbb{Z}^5/A,$$

$$A = \{(k_1,\ldots,k_5)\| k_1 = 0, \; k_4 \in 2\mathbb{Z}, \; k_5 \in 6\mathbb{Z}, \; k_2, k_3 \in \mathbb{Z}\}$$

AB-groups:

$\forall k > 0, k \equiv 0 \bmod 6 \quad < (k,0,0,0,1) >$
$\forall k > 0, k \equiv 0 \bmod 6 \quad < (k,0,0,0,5) >$
$\forall k > 0, k \equiv 2 \bmod 6 \quad < (k,0,0,0,1) >$
$\forall k > 0, k \equiv 4 \bmod 6 \quad < (k,0,0,0,5) >$

---

161. $Q = R3c$

$$E : \; < a,b,c,d,\alpha,\beta \; \| \quad
\begin{array}{ll}
[b,a] = d^{k_1} & [d,a] = 1 \\
[c,a] = d^{-k_1} & [d,b] = 1 \\
[c,b] = d^{k_1} & [d,c] = 1 \\
\alpha a = b\alpha d^{k_2} & \alpha^3 = abcd^{k_4} \\
\alpha b = c\alpha d^{k_3} & \alpha d = d\alpha \\
\alpha c = a\alpha d^{-k_2-k_3} & \\
\beta a = b\beta d^{k_2+2k_3-2k_5} & \beta^2 = cd^{k_5} \\
\beta b = a\beta d^{k_1+k_2+2k_3-2k_5} & \beta d = d^{-1}\beta \\
\beta c = c\beta d^{-2k_5} & \beta\alpha = b^{-1}\alpha^2\beta d^{k_2+k_3+k_4-2k_5}
\end{array}
\quad >$$

$$\lambda(\alpha) = \begin{pmatrix}
1 & \frac{k_1}{4}+k_2 & k_3 & -\frac{k_1}{4}-k_2-k_3 & -\frac{5k_1}{24}-\frac{k_2}{6}-\frac{k_3}{3}+\frac{k_4}{4} \\
0 & 0 & 0 & 1 & \frac{1}{2} \\
0 & 1 & 0 & 0 & \frac{1}{2} \\
0 & 0 & 1 & 0 & 0 \\
0 & 0 & 0 & 0 & 1
\end{pmatrix}$$

$$\lambda(\beta) = \begin{pmatrix}
-1 & -\frac{k_1}{4}-k_2-2k_3+2k_5 & -\frac{3k_1}{4}-k_2-2k_3+2k_5 & 2k_5 & 0 \\
0 & 0 & 1 & 0 & 0 \\
0 & 1 & 0 & 0 & 0 \\
0 & 0 & 0 & 1 & \frac{1}{2} \\
0 & 0 & 0 & 0 & 1
\end{pmatrix}$$

$$H^2(Q, \mathbb{Z}) = \mathbb{Z} \oplus \mathbb{Z}_3 = \mathbb{Z}^5/A,$$

$$A = \{(k_1, \ldots, k_5) \| k_1 = 0, \ (k_2 + 2k_3 + k_4) \in 3\mathbb{Z}, \ k_5 \in \mathbb{Z}\}$$

AB-groups:

$\forall k > 0, \ k \equiv 0 \bmod 3, \ < (k, 1, 0, 0, 0) >$

$\forall k > 0, \ k \equiv 0 \bmod 3, \ < (k, 2, 0, 0, 0) >$

$\forall k > 0, \ k \equiv 1 \bmod 3, \ < (k, 0, 0, 0, 0) >$

$\forall k > 0, \ k \equiv 1 \bmod 3, \ < (k, 2, 0, 0, 0) >$

$\forall k > 0, \ k \equiv 2 \bmod 3, \ < (k, 0, 0, 0, 0) >$

$\forall k > 0, \ k \equiv 2 \bmod 3, \ < (k, 1, 0, 0, 0) >$

---

158. $Q = P3c1$

$$E : \ < a, b, c, d, \alpha, \beta \ \|$$

$[b, a] = d^{k_1}$	$[d, a] = 1$
$[c, a] = 1$	$[d, b] = 1$
$[c, b] = 1$	$[d, c] = 1$
$\alpha a = b\alpha d^{k_2}$	$\alpha^3 = d^{k_4}$
$\alpha b = a^{-1}b^{-1}\alpha d^{k_3}$	$\alpha d = d\alpha$
$\alpha c = c\alpha$	
$\beta a = b^{-1}\beta d^{k_2}$	$\beta^2 = cd^{k_5}$
$\beta b = a^{-1}\beta d^{-k_2}$	$\beta d = d^{-1}\beta$
$\beta c = c\beta d^{-2k_5}$	$\beta\alpha = \alpha^2\beta d^{k_4}$

$$>$$

$$\lambda(\alpha) = \begin{pmatrix} 1 & k_2 & -\frac{k_1}{2} + k_3 & 0 & \frac{k_4}{3} \\ 0 & 0 & -1 & 0 & 0 \\ 0 & 1 & -1 & 0 & 0 \\ 0 & 0 & 0 & 1 & 0 \\ 0 & 0 & 0 & 0 & 1 \end{pmatrix} \quad \lambda(\beta) = \begin{pmatrix} -1 & -k_2 & k_2 & 2k_5 & 0 \\ 0 & 0 & -1 & 0 & 0 \\ 0 & -1 & 0 & 0 & 0 \\ 0 & 0 & 0 & 1 & \frac{1}{2} \\ 0 & 0 & 0 & 0 & 1 \end{pmatrix}$$

$$H^2(Q, \mathbb{Z}) = \mathbb{Z} \oplus (\mathbb{Z}_3)^2 = \mathbb{Z}^5/A,$$

$$A = \{(k_1, \ldots, k_5) \| k_1 = 0, \ (k_3 - k_2), k_4 \in 3\mathbb{Z}, \ k_5 \in \mathbb{Z}\}$$

AB-groups:

$\forall k > 0, \ k \equiv 0 \bmod 3, \ < (k, 0, 0, 1, 0) >$

$\forall k > 0, \ k \equiv 0 \bmod 3, \ < (k, 0, 0, 2, 0) >$

$\forall k > 0, \ k \not\equiv 0 \bmod 3, \ < (k, 1, 0, 1, 0) >$

Remark:

$\forall k \equiv 1 \bmod 3, \ < (k, 0, 0, 2, 0) > \cong < (k, 1, 0, 1, 0) > \cong < (k, 2, 0, 2, 0) >$

$\forall k \equiv 2 \bmod 3, \ < (k, 0, 0, 1, 0) > \cong < (k, 1, 0, 1, 0) > \cong < (k, 2, 0, 2, 0) >$

159. $Q = P31c$

$$E : \langle a, b, c, d, \alpha, \beta \, \| \quad [b, a] = d^{k_1} \qquad\qquad [d, a] = 1 \qquad \rangle$$

$$
\begin{array}{ll}
[c, a] = 1 & [d, b] = 1 \\
[c, b] = 1 & [d, c] = 1 \\
\alpha a = b\alpha d^{k_1 - 2k_2 + 3k_4} & \alpha^3 = d^{k_3} \\
\alpha b = a^{-1}b^{-1}\alpha d^{k_2} & \alpha d = d\alpha \\
\alpha c = c\alpha & \\
\beta a = b\beta d^{k_4} & \beta^2 = cd^{k_5} \\
\beta b = a\beta d^{k_4} & \beta d = d^{-1}\beta \\
\beta c = c\beta d^{-2k_5} & \beta\alpha = \alpha^2\beta d^{k_3}
\end{array}
$$

$$
\lambda(\alpha) = \begin{pmatrix}
1 & k_1 - 2k_2 + 3k_4 & -\frac{k_1}{2} + k_2 & 0 & \frac{k_3}{3} \\
0 & 0 & -1 & 0 & 0 \\
0 & 1 & -1 & 0 & 0 \\
0 & 0 & 0 & 1 & 0 \\
0 & 0 & 0 & 0 & 1
\end{pmatrix}
\quad
\lambda(\beta) = \begin{pmatrix}
-1 & -k_4 & -k_4 & 2k_5 & 0 \\
0 & 0 & 1 & 0 & 0 \\
0 & 1 & 0 & 0 & 0 \\
0 & 0 & 0 & 1 & \frac{1}{2} \\
0 & 0 & 0 & 0 & 1
\end{pmatrix}
$$

$H^2(Q, \mathbb{Z}) = \mathbb{Z} \oplus \mathbb{Z}_3 = \mathbb{Z}^5/A,\ A = \{(k_1, \ldots, k_5) \| k_1 = 0, k_3 \in 3\mathbb{Z}, k_2, k_4, k_5 \in \mathbb{Z}\}$

AB-groups:
$\forall k > 0,\ k \equiv 0 \bmod 3,\ \langle (k, 0, 1, 0, 0) \rangle$
$\forall k > 0,\ k \equiv 0 \bmod 3,\ \langle (k, 0, 2, 0, 0) \rangle$
$\forall k > 0,\ k \equiv 1 \bmod 3,\ \langle (k, 0, 1, 0, 0) \rangle$
$\forall k > 0,\ k \equiv 2 \bmod 3,\ \langle (k, 0, 2, 0, 0) \rangle$

---

168. $Q = P6$

$$E : \langle a, b, c, d, \alpha \, \| \quad [b, a] = d^{k_1} \qquad [d, a] = 1 \qquad \rangle$$

$$
\begin{array}{ll}
[c, a] = 1 & [d, b] = 1 \\
[c, b] = 1 & [d, c] = 1 \\
\alpha a = b^{-1}\alpha d^{k_2} & \alpha^6 = d^{k_4} \\
\alpha b = ab\alpha d^{k_3} & \alpha d = d\alpha \\
\alpha c = c\alpha &
\end{array}
$$

$$
\lambda(\alpha) = \begin{pmatrix}
1 & k_2 & -\frac{k_1}{2} + k_3 & 0 & \frac{k_4}{6} \\
0 & 0 & 1 & 0 & 0 \\
0 & -1 & 1 & 0 & 0 \\
0 & 0 & 0 & 1 & 0 \\
0 & 0 & 0 & 0 & 1
\end{pmatrix}
$$

$H^2(Q, \mathbb{Z}) = \mathbb{Z} \oplus \mathbb{Z}_6 = \mathbb{Z}^4/A,\ A = \{(k_1, \ldots, k_4) \| k_1 = 0,\ k_4 \in 6\mathbb{Z},\ k_2, k_3 \in \mathbb{Z}\}$

AB-groups:
$\forall k > 0, k \equiv 0 \bmod 6\ \langle (k, 0, 0, 1) \rangle$
$\forall k > 0, k \equiv 0 \bmod 6\ \langle (k, 0, 0, 5) \rangle$
$\forall k > 0, k \equiv 2 \bmod 6\ \langle (k, 0, 0, 1) \rangle$
$\forall k > 0, k \equiv 4 \bmod 6\ \langle (k, 0, 0, 5) \rangle$

172. $Q = P6_4$

$$E : \; < a, b, c, d, \alpha \; \| \quad [b,a] = d^{k_1} \qquad [d,a] = 1 \qquad >$$

$$\begin{aligned}
[c,a] &= 1 & [d,b] &= 1 \\
[c,b] &= 1 & [d,c] &= 1 \\
\alpha a &= b^{-1}\alpha d^{k_2} & \alpha^6 &= c^2 d^{k_4} \\
\alpha b &= ab\alpha d^{k_3} & \alpha d &= d\alpha \\
\alpha c &= c\alpha
\end{aligned}$$

$$\lambda(\alpha) = \begin{pmatrix}
1 & k_2 & -\frac{k_1}{2} + k_3 & 0 & \frac{k_4}{6} \\
0 & 0 & 1 & 0 & 0 \\
0 & -1 & 1 & 0 & 0 \\
0 & 0 & 0 & 1 & \frac{1}{3} \\
0 & 0 & 0 & 0 & 1
\end{pmatrix}$$

$H^2(Q,\mathbb{Z}) = \mathbb{Z} \oplus \mathbb{Z}_2 = \mathbb{Z}^4/A, \; A = \{(k_1,\ldots,k_4)\|k_1 = 0, \; k_4 \in 2\mathbb{Z}, \; k_2, k_3 \in \mathbb{Z}\}$

AB-groups:

$\forall k > 0, k \equiv 0 \bmod 2 \; < (k,0,0,1) >$

---

173. $Q = P6_3$

$$E : \; < a, b, c, d, \alpha \; \| \quad [b,a] = d^{k_1} \qquad [d,a] = 1 \qquad >$$

$$\begin{aligned}
[c,a] &= 1 & [d,b] &= 1 \\
[c,b] &= 1 & [d,c] &= 1 \\
\alpha a &= b^{-1}\alpha d^{k_2} & \alpha^6 &= c^3 d^{k_4} \\
\alpha b &= ab\alpha d^{k_3} & \alpha d &= d\alpha \\
\alpha c &= c\alpha
\end{aligned}$$

$$\lambda(\alpha) = \begin{pmatrix}
1 & k_2 & -\frac{k_1}{2} + k_3 & 0 & \frac{k_4}{6} \\
0 & 0 & 1 & 0 & 0 \\
0 & -1 & 1 & 0 & 0 \\
0 & 0 & 0 & 1 & \frac{1}{2} \\
0 & 0 & 0 & 0 & 1
\end{pmatrix}$$

$H^2(Q,\mathbb{Z}) = \mathbb{Z} \oplus \mathbb{Z}_3 = \mathbb{Z}^4/A, \; A = \{(k_1,\ldots,k_4)\|k_1 = 0, \; k_4 \in 3\mathbb{Z}, \; k_2, k_3 \in \mathbb{Z}\}$

AB-groups:

$\forall k > 0, k \equiv 0 \bmod 3 \; < (k,0,0,1) >$

$\forall k > 0, k \equiv 2 \bmod 3 \; < (k,0,0,1) >$

$\forall k > 0, k \equiv 0 \bmod 3 \; < (k,0,0,2) >$

$\forall k > 0, k \equiv 1 \bmod 3 \; < (k,0,0,2) >$

---

169. $Q = P6_1$

$$E : \; < a, b, c, d, \alpha \; \| \quad [b,a] = d^{k_1} \qquad [d,a] = 1 \qquad >$$

$$\begin{aligned}
[c,a] &= 1 & [d,b] &= 1 \\
[c,b] &= 1 & [d,c] &= 1 \\
\alpha a &= b^{-1}\alpha d^{k_2} & \alpha^6 &= c^5 d^{k_4} \\
\alpha b &= ab\alpha d^{k_3} & \alpha d &= d\alpha \\
\alpha c &= c\alpha
\end{aligned}$$

$$\lambda(\alpha) = \begin{pmatrix} 1 & k_2 & -\frac{k_1}{2} + k_3 & 0 & \frac{k_4}{6} \\ 0 & 0 & 1 & 0 & 0 \\ 0 & -1 & 1 & 0 & 0 \\ 0 & 0 & 0 & 1 & \frac{5}{6} \\ 0 & 0 & 0 & 0 & 1 \end{pmatrix}$$

$$H^2(Q, \mathbb{Z}) = \mathbb{Z} = \mathbb{Z}^4/A, \quad A = \{(k_1, \ldots, k_4) \| k_1 = 0, \ k_2, k_3, k_4 \in \mathbb{Z}\}$$

AB-groups:

$\forall k > 0, \ < (k, 0, 0, 0) >$

---

174. $Q = P\bar{6}$

$$E : \ < a, b, c, d, \alpha \ \| \quad \begin{aligned}[b, a] &= d^{k_1} & [d, a] &= 1 \quad >\\ [c, a] &= 1 & [d, b] &= 1 \\ [c, b] &= 1 & [d, c] &= 1 \\ \alpha a &= b\alpha d^{k_2} & \alpha^6 &= d^{k_5} \\ \alpha b &= a^{-1}b^{-1}\alpha d^{k_3} & \alpha d &= d\alpha \\ \alpha c &= c^{-1}\alpha d^{k_4} \end{aligned}$$

$$\lambda(\alpha) = \begin{pmatrix} 1 & k_2 & -\frac{k_1}{2} + k_3 & k_4 & \frac{k_5}{6} \\ 0 & 0 & -1 & 0 & 0 \\ 0 & 1 & -1 & 0 & 0 \\ 0 & 0 & 0 & -1 & 0 \\ 0 & 0 & 0 & 0 & 1 \end{pmatrix}$$

$$H^2(Q, \mathbb{Z}) = \mathbb{Z} \oplus \mathbb{Z}_2 \oplus \mathbb{Z}_3 \oplus \mathbb{Z}_6 = \mathbb{Z}^5/A,$$

$$A = \{(k_1, \ldots, k_5) \| k_1 = 0, \ (k_3 - k_2) \in 3\mathbb{Z}, \ k_4 \in 2\mathbb{Z}, \ k_5 \in 6\mathbb{Z}\}$$

AB-groups:

$\forall k > 0, \ k \equiv 0 \bmod 3, \ < (k, 0, 0, 0, 1) >$
$\forall k > 0, \ k \equiv 0 \bmod 3, \ < (k, 0, 0, 0, 5) >$
$\forall k > 0, \ k \not\equiv 0 \bmod 3, \ < (k, 1, 0, 0, 5) >$

Remark:

$\forall k \equiv 1 \bmod 3 \ < (k, 0, 0, 0, 1) > \cong < (k, 1, 0, 0, 5) > \cong < (k, 2, 0, 0, 1) >$
$\forall k \equiv 2 \bmod 3 \ < (k, 0, 0, 0, 5) > \cong < (k, 1, 0, 0, 5) > \cong < (k, 2, 0, 0, 1) >$

---

175. $Q = P6/m$

$$E : \ < a, b, c, d, \alpha, \beta \ \| \quad \begin{aligned}[b, a] &= d^{k_1} & [d, a] &= 1 \quad >\\ [c, a] &= 1 & [d, b] &= 1 \\ [c, b] &= 1 & [d, c] &= 1 \\ \alpha a &= b^{-1}\alpha d^{k_2} & \alpha^6 &= d^{k_4} \\ \alpha b &= ab\alpha d^{k_3} & \alpha d &= d\alpha \\ \alpha c &= c\alpha \\ \beta a &= a^{-1}\beta d^{k_1 - 2k_3} & \beta^2 &= d^{k_6} \\ \beta b &= b^{-1}\beta d^{-k_1 + 2k_2 + 2k_3} & \beta d &= d\beta \\ \beta c &= c^{-1}\beta d^{k_5} & \beta\alpha &= \alpha\beta \end{aligned}$$

$$\lambda(\alpha) = \begin{pmatrix} 1 & k_2 & -\frac{k_1}{2}+k_3 & 0 & \frac{k_4}{6} \\ 0 & 0 & 1 & 0 & 0 \\ 0 & -1 & 1 & 0 & 0 \\ 0 & 0 & 0 & 1 & 0 \\ 0 & 0 & 0 & 0 & 1 \end{pmatrix} \quad \lambda(\beta) = \begin{pmatrix} 1 & k_1-2k_3 & 2k_2+2k_3-k_1 & k_5 & \frac{k_6}{2} \\ 0 & -1 & 0 & 0 & 0 \\ 0 & 0 & -1 & 0 & 0 \\ 0 & 0 & 0 & -1 & 0 \\ 0 & 0 & 0 & 0 & 1 \end{pmatrix}$$

$$H^2(Q,\mathbb{Z}) = \mathbb{Z} \oplus (\mathbb{Z}_2)^2 \oplus \mathbb{Z}_6 = \mathbb{Z}^6/A,$$

$$A = \{(k_1,\dots,k_6)\|k_1 = 0,\ k_5, k_6 \in 2\mathbb{Z},\ k_4 \in 6\mathbb{Z},\ k_2, k_3 \in \mathbb{Z}\}$$

AB-groups:
None

---

176. $Q = P6_3/m$

$$E: <a,b,c,d,\alpha,\beta \|\ \begin{array}{ll} [b,a] = d^{k_1} & [d,a] = 1 \\ [c,a] = 1 & [d,b] = 1 \\ [c,b] = 1 & [d,c] = 1 \\ \alpha a = b^{-1}\alpha d^{k_2} & \alpha^6 = c^3 d^{k_4} \\ \alpha b = ab\alpha d^{k_3} & \alpha d = d\alpha \\ \alpha c = c\alpha & \\ \beta a = a^{-1}\beta d^{k_1-2k_3} & \beta^2 = d^{k_5} \\ \beta b = b^{-1}\beta d^{-k_1+2k_2+2k_3} & \beta d = d\beta \\ \beta c = c^{-1}\beta d^{2k_6} & \beta\alpha = c^{-1}\alpha\beta d^{k_6} \end{array} >$$

$$\lambda(\alpha) = \begin{pmatrix} 1 & k_2 & -\frac{k_1}{2}+k_3 & 0 & \frac{k_4}{6} \\ 0 & 0 & 1 & 0 & 0 \\ 0 & -1 & 1 & 0 & 0 \\ 0 & 0 & 0 & 1 & \frac{1}{2} \\ 0 & 0 & 0 & 0 & 1 \end{pmatrix} \quad \lambda(\beta) = \begin{pmatrix} 1 & k_1-2k_3 & 2k_2+2k_3-k_1 & 2k_6 & \frac{k_5}{2} \\ 0 & -1 & 0 & 0 & 0 \\ 0 & 0 & -1 & 0 & 0 \\ 0 & 0 & 0 & -1 & 0 \\ 0 & 0 & 0 & 0 & 1 \end{pmatrix}$$

$$H^2(Q,\mathbb{Z}) = \mathbb{Z} \oplus \mathbb{Z}_2 \oplus \mathbb{Z}_6 = \mathbb{Z}^6/A,$$

$$A = \{(k_1,\dots,k_6)\|k_1 = 0,\ k_5 \in 2\mathbb{Z},\ (k_4+3k_6) \in 6\mathbb{Z},\ k_2, k_3 \in \mathbb{Z}\}$$

AB-groups:
$\forall k > 0,\ k \equiv 0 \bmod 6,\ <(k,0,0,2,1,0)>$
$\forall k > 0,\ k \equiv 0 \bmod 6,\ <(k,0,0,4,1,0)>$
$\forall k > 0,\ k \equiv 2 \bmod 6,\ <(k,0,0,4,1,0)>$
$\forall k > 0,\ k \equiv 4 \bmod 6,\ <(k,0,0,2,1,0)>$

184. $Q = P6cc$

$$E : \; < a, b, c, d, \alpha, \beta \; \| \quad \begin{array}{ll} [b, a] = d^{k_1} & [d, a] = 1 \\ [c, a] = 1 & [d, b] = 1 \\ [c, b] = 1 & [d, c] = 1 \\ \alpha a = b^{-1} \alpha d^{k_2} & \alpha^6 = d^{k_4} \\ \alpha b = ab \alpha d^{k_3} & \alpha d = d \alpha \\ \alpha c = c \alpha & \\ \beta a = b^{-1} \beta d^{k_1 - k_2 - 2k_3} & \beta^2 = cd^{k_5} \\ \beta b = a^{-1} \beta d^{-k_1 + k_2 + 2k_3} & \beta d = d^{-1} \beta \\ \beta c = c \beta d^{-2k_5} & \beta \alpha = \alpha^5 \beta d^{k_4} \end{array} \quad >$$

$$\lambda(\alpha) = \begin{pmatrix} 1 & k_2 & -\frac{k_1}{2} + k_3 & 0 & \frac{k_4}{6} \\ 0 & 0 & 1 & 0 & 0 \\ 0 & -1 & 1 & 0 & 0 \\ 0 & 0 & 0 & 1 & 0 \\ 0 & 0 & 0 & 0 & 1 \end{pmatrix}$$

$$\lambda(\beta) = \begin{pmatrix} -1 & -k_1 + k_2 + 2k_3 & k_1 - k_2 - 2k_3 & 2k_5 & 0 \\ 0 & 0 & 0 & -1 & 0 & 0 \\ 0 & -1 & 0 & 0 & 0 \\ 0 & 0 & 0 & 1 & \frac{1}{2} \\ 0 & 0 & 0 & 0 & 1 \end{pmatrix}$$

$$H^2(Q, \mathbb{Z}) = \mathbb{Z} \oplus \mathbb{Z}_6 = \mathbb{Z}^5 / A, \; A = \{ (k_1, \ldots, k_5) \| k_1 = 0, k_4 \in 6\mathbb{Z}, k_2, k_3, k_5 \in \mathbb{Z} \}$$

AB–groups:

$\forall k > 0, \; k \equiv 0 \bmod 6, \; < (k, 0, 0, 1, 0) >$

$\forall k > 0, \; k \equiv 0 \bmod 6, \; < (k, 0, 0, 5, 0) >$

$\forall k > 0, \; k \equiv 2 \bmod 6, \; < (k, 0, 0, 1, 0) >$

$\forall k > 0, \; k \equiv 4 \bmod 6, \; < (k, 0, 0, 5, 0) >$

Comments and proofs

For the first groups we give detailed information on the choices of the actions and on the search for isomorphism types. Since the methods used are the same in all cases we do not give explicit information for most of the cases.

1. See classification of rank 4 nilpotent groups.
2. Since $\alpha^2 = 1$ in Q we only have to investigate the trivial action of Q on $\mathbb{Z}$.
The only torsion free groups E are parametrized as $(2k, 2l, 2m, 0, 0, 0, 1)$. One can see that, by doing the analogous changes of generators as we did during the classification of rank 4 nilpotent groups, any extension $(k_1, k_2, k_3, k_4, k_5, k_6, k_7)$ is isomorphic to the extension with parameters $((k_1, k_2, k_3), 0, 0, k'_4, k'_5, k'_6, k'_7)$. Therefore $(2k, 2l, 2m, 0, 0, 0, 1)$ is isomorphic to $(2(k, l, m), 0, 0, k'_4, k'_5, k'_6, k'_7)$ which has to be torsion free and so is equivalent with $(2(k, l, m), 0, 0, 0, 0, 0, 1)$.
3. $\alpha^2 = 1$ in Q $\Rightarrow$ only the trivial action of Q on $\mathbb{Z}$ has to be investigated.
4. Here we consider both the trivial and the non-trivial action. It is obvious that two extensions inducing a different action are not isomorphic. (e.g. In one group the center$= Z(N)$ while in the other extension the center will be trivial.)
Trivial action:
Isomorphism types:
Suppose $k \not\equiv 0 \bmod 2$, so $k = 2l+1$ for some $l \in \mathbb{Z}$. We have the following isomorphisms:

$\varphi_1 :< (k, 1, 0, 0) > \rightarrow < (k, 0, 0, 0) >,$
  with $\varphi_1(a) = ad^{-l}$, $\varphi_1(b) = b$, $\varphi_1(c) = c$, $\varphi_1(d) = d$, $\varphi_1(\alpha) = c^{-1}\alpha$
$\varphi_2 :< (k, 0, 1, 0) > \rightarrow < (k, 0, 0, 0) >,$
  with $\varphi_2(a) = a$, $\varphi_2(b) = b$, $\varphi_2(c) = cd^{-l}$, $\varphi_2(d) = d$, $\varphi_2(\alpha) = a\alpha$
$\varphi_3 :< (k, 1, 1, 0) > \rightarrow < (k, 0, 0, 0) >,$
  with $\varphi_3(a) = ad^{1+l}$, $\varphi_3(b) = bd^{-2l-1}$, $\varphi_3(c) = cd^{-l}$,
    $\varphi_3(d) = d$, $\varphi_3(\alpha) = aca$.

Suppose $k \equiv 0 \bmod 2$, so $k = 2l$ for some $l \in \mathbb{Z}$. We have the following isomorphisms:
$\varphi_1 :< (k, 0, 1, 0) > \rightarrow < (k, 1, 0, 0) >,$
  with $\varphi_1(a) = c$, $\varphi_1(b) = b$, $\varphi_1(c) = ad^{-1}$, $\varphi_1(d) = d^{-1}$, $\varphi_1(\alpha) = \alpha$
$\varphi_2 :< (k, 1, 1, 0) > \rightarrow < (k, 1, 0, 0) >,$

with $\varphi_2(a) = a$, $\varphi_2(b) = bd$, $\varphi_2(c) = ac$, $\varphi_2(d) = d$, $\varphi_2(\alpha) = a\alpha$.

There is no isomorphism between $< (2l, 0, 0, 0) >$ and $< (2l, 1, 0, 0) >$.
Suppose $\varphi :< (2l, 1, 0, 0) > \to < (2l, 0, 0, 0) >$ is an isomorphism of groups,
then

$$\varphi(a) = a^{\alpha_1} b^{\alpha_2} c^{\alpha_3} d^{\alpha_4}, \tag{7.1}$$
$$\varphi(b) = a^{\beta_1} b^{\beta_2} c^{\beta_3} d^{\beta_4}$$
$$\varphi(c) = a^{\gamma_1} b^{\gamma_2} c^{\gamma_3} d^{\gamma_4}$$
$$\varphi(d) = d^{\delta}, \quad \delta = \pm 1$$
$$\varphi(\alpha) = a^{m_1} b^{m_2} c^{m_3} d^{m_4} \alpha$$

now one has to see if this φ is compatible with the given relations, e.g.
the following should be satisfied:

$$\varphi(\alpha a) = \varphi(a^{-1} \alpha d)$$

or

$$a^{m_1} b^{m_2} c^{m_3} d^{m_4} \alpha a^{\alpha_1} b^{\alpha_2} c^{\alpha_2} d^{\alpha_4} = (a^{\alpha_1} b^{\alpha_2} c^{\alpha_3} d^{\alpha_4})^{-1} a^{m_1} b^{m_2} c^{m_3} d^{m_4} \alpha d^{\delta}$$

inducing $\delta \equiv 0 \bmod 2$ which is impossible.

Non-trivial action:
Isomorphism types:
Consider the group $< (k_1, k_2, k_3) >$. Suppose $(k_1, k_2) = pk_1 + qk_2$ for
some $p, q \in \mathbb{Z}$. There is an isomorphism $\varphi :< ((k_1, k_2), 0, k_3) > \to <
(k_1, k_2, k_3) >$ given by

$$\varphi(a) = a^p c^{-q}, \quad \varphi(b) = b, \quad \varphi(c) = a^{k_2/(k_1, k_2)} c^{k_1/(k_1, k_2)}, \quad \varphi(d) = d, \quad \varphi(\alpha) = \alpha.$$

7. Trivial action:
The groups $< (k, 0, 0) >$ and $< (k, 1, 0) >$ are not isomorphic:
Suppose $\varphi :< (k, 1, 0) > \to < (k, 0, 0) >$ is an isomorphism, then φ satisfies
(7.1) and one checks that the relation

$$\varphi(\alpha b) = \varphi(b^{-1} \alpha d)$$

can not be satisfied.

Non-trivial action:
Consider any extension $< (k_1, k_2, k_3, k_4) >$. By a new choice of generators
we can transform the presentation for $< (k_1, k_2, k_3, k_4) >$ to show that
this group is isomorphic to another one. Some situations:

1. $a' = a^{-1}$, $b' = b$, $c' = c$, $d' = d$, $\alpha' = \alpha$
   $< (k_1, k_2, k_3, k_4) > \longrightarrow < (-k_1, k_2, k'_3, k'_4) >$.

2. $a' = a$, $b' = b$, $c' = c$, $d' = d^{-1}$, $\alpha' = \alpha$
   $< (k_1, k_2, k_3, k_4) > \longrightarrow < (-k_1, -k_2, -k_3, -k_4) >$.

3. $a' = a$, $b' = b$, $c' = a^{2m}c$, $d' = d$, $\alpha' = a^m\alpha$
   $< (k_1, k_2, k_3, k_4) > \longrightarrow < (k_1, k_2 - mk_1, k'_3, k'_4) >$.

4. $a' = ac^m$, $b' = b$, $c' = c$, $d' = d$, $\alpha' = \alpha$
   $< (k_1, k_2, k_3, k_4) > \longrightarrow < (k_1 - 2mk2, k_2, k'_3, k'_4) >$.

This shows that one can reduce k_1 to $k_1 \bmod (2k_2)$ or $-k_1 \bmod (2k2)$. One of these two values will be $\leq k_2/2$. One can also reduce $k_2 \bmod k_1$. This shows that after a finite number of reductions of these two kinds one finds:
Situation 1: (If $(k_1, k_2) = (k_1, 2k_2)$)
$< (k_1, k_2, k_3, k_4) > \cong < ((k_1, k_2), 0, k'_3, k'_4) >$.
Situation 2: (If $2(k_1, k_2) = (k_1, 2k_2)$)
$< (k_1, k_2, k_3, k_4) > \cong < (0, (k_1, k_2), k'_3, k'_4) >$.
One also proves $< (2k, 0, k3, k4) > \ncong < (0, k, k'_3, k'_4) >$ by presenting a general form for a possible isomorphism (7.1) and see things do not work out.
For $k = 2l+1$, we take the following set of new generators in $< (k, 0, 0, 0) >$

$$a' = ad^{-l}, \ b' = b, \ c' = c, \ d' = d, \ \alpha' = b^{-1}\alpha$$

to see that $< (k, 0, 0, 0) > \cong < (k, 0, 1, 0) >$. When $k = 2l + 1$ it is seen that this is not the case. Also $< (0, k, 0, 0) > \ncong < (0, k, 1, 0) >$ is checked by trying a general form for the isomorphism.
13. $\alpha^2 = \beta^2 = 1$ in Q. Therefore, we only have to look at the trivial action.
16. The rank of the torsion free part of $H^2(Q, \mathbb{Z})$ is 0.

7.3 4-dimensional AB–groups, with 3–step nilpotent Fitting subgroup

For this class, it seemed adequate to present two tables: one containing the presentations of the AB–groups, the other containing the affine representations. This is done because, in the representation, we cannot longer use the same matrices for the generators a, b and c in all cases.

First, we list the presentations of all AB–groups E of this class. The order in which the groups appear is determined by the order of the underlying AC–group in the table of section 7.1. Again, the group generated by a, b, c and d corresponds to the Fitting subgroup of E. Under the table, we added some comments concerning the establishment of this table.

The general set up of one table entry is as follows:

> Number of Q as found in section 7.1   Cohomology class corresponding to Q
>
> $\qquad$ Presentation for E depending on k_1, k_2, k_3, k_4
> $\qquad\qquad$ $H^2(Q, \mathbb{Z})$ in terms of k_1, k_2, k_3, k_4
>
> AB-groups:
> The cohomology classes corresponding to AB-groups
> and isomorphism type information for these.

1. $Q =< (k) >$

$$E : < a, b, c, d \,\|\, \begin{array}{ll} [b,a] = c^k d^{k_1} & [d,a] = 1 \\ [c,a] = d^{k_2} & [d,b] = 1 \\ [c,b] = d^{k_3} & [d,c] = 1 \end{array} >$$

$$H^2(Q, \mathbb{Z}) = \mathbb{Z}^2 \oplus \mathbb{Z}_k = \mathbb{Z}^3/A, \ A = \{(k_1, k_2, k_3) \,\|\, k_2 = k_3 = 0, k_1 \in k\mathbb{Z}\}$$

AB-groups:
$< (k_1, k_2, k_3) >$.
Remark: see classification of rank 4, 3–step nilpotent groups.

---

2. $Q =< (k, 0, 0, 1) >$

$$E : < a, b, c, d, \alpha \,\|\, \begin{array}{ll} [b,a] = c^{2k} d^{2k(k_1 + k_2 + k_3)} & [d,a] = 1 \\ [c,a] = d^{2k_1} & [d,b] = 1 \\ [c,b] = d^{2k_2} & [d,c] = 1 \\ \alpha a = a^{-1} \alpha d^{k_1} & \alpha^2 = c d^{k_3} \\ \alpha b = b^{-1} \alpha d^{k_2} & \alpha d = d^{-1} \alpha \\ \alpha c = c \alpha d^{-2k_3} & \end{array} >$$

$$H^2(Q, \mathbb{Z}) = \mathbb{Z}^2 = \mathbb{Z}^3/A, \ A = \{(k_1, k_2, k_3) \,\|\, k_1 = k_2 = 0, k_3 \in \mathbb{Z}\}$$

AB-groups:

$\forall m > 0, \; < (m, 0, 0) >$.

Remark: $< (k_1, k_2, k_3) > \cong < (gcd(k_1, k_2), 0, 0) >$

3.

$\underline{Q = < (2l + 1, 1) >}$:

$$E : \; < a, b, c, d, \alpha \; \| \quad \begin{array}{ll} [b, a] = c^{2l+1} d^{lk_1 + (2l+1)k_2} & [d, a] = 1 \quad > \\ [c, a] = 1 & [d, b] = 1 \\ [c, b] = d^{k_1} & [d, c] = 1 \\ \alpha a = a\alpha c d^{k_2} & \alpha^2 = d^{k_4} \\ \alpha b = b^{-1} \alpha d^{k_3} & \alpha d = d\alpha \\ \alpha c = c^{-1} \alpha d^{-2k_2} \end{array}$$

$$H^2(Q, \mathbb{Z}) = \mathbb{Z} \oplus (\mathbb{Z}_2)^2 = \mathbb{Z}^4 / A,$$

$$A = \{ (k_1, k_2, k_3, k_4) \; \| \; k_1 = 0, \; k_3, k_4 \in 2\mathbb{Z}, \; k_2 \in \mathbb{Z} \}$$

AB-groups:

$\forall m > 0, \; m \equiv 0 \bmod 2, \; < (m, 0, 0, 1) >$.

$\underline{Q = < (2l, 1) >}$:

$$E : \; < a, b, c, d, \alpha \; \| \quad \begin{array}{ll} [b, a] = c^{2l} d^{(2l-1)k_1 + 2lk_2} & [d, a] = 1 \quad > \\ [c, a] = 1 & [d, b] = 1 \\ [c, b] = d^{2k_1} & [d, c] = 1 \\ \alpha a = a\alpha c d^{k_2} & \alpha^2 = d^{k_4} \\ \alpha b = b^{-1} \alpha d^{k_3} & \alpha d = d\alpha \\ \alpha c = c^{-1} \alpha d^{-2k_2} \end{array}$$

$$H^2(Q, \mathbb{Z}) = \mathbb{Z} \oplus (\mathbb{Z}_2)^2 = \mathbb{Z}^4 / A,$$

$$A = \{ (k_1, k_2, k_3, k_4) \; \| \; k_1 = 0, \; k_3, k_4 \in 2\mathbb{Z}, \; k_2 \in \mathbb{Z} \}$$

AB-groups:

$\forall m > 0, \; < (m, 0, 0, 1) >$.

$\underline{Q = < (2l, 0) >}$:

$$E : \; < a, b, c, d, \alpha \; \| \quad \begin{array}{ll} [b, a] = c^{2l} d^{l(k_1 - k_3)} & [d, a] = 1 \quad > \\ [c, a] = 1 & [d, b] = 1 \\ [c, b] = d^{k_1} & [d, c] = 1 \\ \alpha a = a\alpha & \alpha^2 = d^{k_4} \\ \alpha b = b^{-1} \alpha d^{k_2} & \alpha d = d\alpha \\ \alpha c = c^{-1} \alpha d^{k_3} \end{array}$$

$$H^2(Q, \mathbb{Z}) = \mathbb{Z} \oplus (\mathbb{Z}_2)^3 = \mathbb{Z}^4 / A, \; A = \{ (k_1, k_2, k_3, k_4) \; \| \; k_1 = 0, \; k_2, k_3, k_4 \in 2\mathbb{Z} \}$$

AB-groups:

$\forall m > 0, \; m \equiv 0 \bmod 2, \; < (m, 0, 0, 1) >$.

4. $Q = <(k, 0)>$
Trivial action:

$$E : < a, b, c, d, \alpha \parallel \begin{array}{ll} [b, a] = c^{2k} d^{k(k_1 - k_3)} & [d, a] = 1 \\ [c, a] = 1 & [d, b] = 1 \\ [c, b] = d^{k_1} & [d, c] = 1 \\ \alpha a = a\alpha & \alpha^2 = ad^{k_4} \\ \alpha b = b^{-1} \alpha c^{-k} d^{k_2} & \alpha d = d\alpha \\ \alpha c = c^{-1} \alpha d^{k_3} \end{array} >$$

If $k \equiv 0 \bmod 2$, then

$$H^2(Q, \mathbb{Z}) = \mathbb{Z} \oplus (\mathbb{Z}_2)^2 = \mathbb{Z}^4 / A,$$

$$A = \{(k_1, k_2, k_3, k_4) \parallel k_1 = 0, \ k_2, k_3 \in 2\mathbb{Z}, \ k_4 \in \mathbb{Z}\}$$

AB-groups:
$\forall m > 0, \ < (m, 0, 0, 0) >.$
$\forall m > 0, \ m \equiv 0 \bmod 2, \ < (m, 1, 0, 0) >.$
$\forall m > 0, \ m \equiv 0 \bmod 2, \ < (m, 0, 1, 0) >.$
Remark:
If $k_1 \not\equiv 0 \bmod 2, \ < (k_1, k_2, k_3, k_4) > \cong < (k_1, 0, 0, 0) >, \ \forall k_2, k_3, k_4 \in \mathbb{Z}$
If $m \equiv 0 \bmod 2, \ < (m, 0, 1, 0) > \cong < (m, 1, 1, 0) >$

If $k \not\equiv 0 \bmod 2$, then

$$H^2(Q, \mathbb{Z}) = \mathbb{Z} \oplus \mathbb{Z}_4 = \mathbb{Z}^4 / A,$$

$$A = \{(k_1, k_2, k_3, k_4) \parallel k_1 = 0, \ (2k_2 - k_3) \in 4\mathbb{Z}, \ k_4 \in \mathbb{Z}\}$$

AB-groups:
$\forall m > 0, \ < (m, 0, 0, 0) >.$
$\forall m > 0, \ m \equiv 0 \bmod 4, \ < (m, 1, 0, 0) >.$
$\forall m > 0, \ m \equiv 0 \bmod 2, \ < (m, 0, 1, 0) >.$
Remark:
If $k_1 \not\equiv 0 \bmod 2, \ < (k_1, k_2, k_3, k_4) > \cong < (k_1, 0, 0, 0) >, \ \forall k_2, k_3, k_4 \in \mathbb{Z}$
If $m = 4n + 2, \ n \in \mathbb{Z}, \ < (m, 0, 0, 0) > \cong < (m, 1, 0, 0) >$
If $m \equiv 0 \bmod 2, \ < (m, 0, 1, 0) > \cong < (m, 1, 1, 0) >$

Non-trivial action:

$$E : < a, b, c, d, \alpha \parallel \begin{array}{ll} [b, a] = c^{2k} d^{-2k_2 - kk_1} & [d, a] = 1 \\ [c, a] = d^{-2k_1} & [d, b] = 1 \\ [c, b] = 1 & [d, c] = 1 \\ \alpha a = a\alpha d^{-2k_3} & \alpha^2 = ad^{k_3} \\ \alpha b = b^{-1} \alpha c^{-k} d^{k_2} & \alpha d = d^{-1} \alpha \\ \alpha c = c^{-1} \alpha d^{k_1} \end{array} >$$

$$H^2(Q, \mathbb{Z}) = \mathbb{Z} \oplus \mathbb{Z}_k = \mathbb{Z}^3/A, \ A = \{(k_1, k_2, k_3) \ \| \ k_1 = 0, \ k_2 \in k\mathbb{Z}, \ k_3 \in \mathbb{Z}\}$$

AB-groups:
All groups with the remark that groups, containing isomorphic maximal normal nilpotent groups, are isomorphic themselves.

5.

$Q =< (2l, 0) >$

$$E : < a, b, c, d, \alpha \ \| \quad \begin{array}{ll} [b, a] = c^{2l}d^{-lk_3} & [d, a] = 1 \\ [c, a] = d^{k_1} & [d, b] = 1 \\ [c, b] = d^{-k_1} & [d, c] = 1 \\ \alpha a = b\alpha d^{k_2} & \alpha^2 = d^{k_4} \\ \alpha b = a\alpha d^{-k_2} & \alpha d = d\alpha \\ \alpha c = c^{-1}\alpha d^{k_3} & \end{array} \quad >$$

$$H^2(Q, \mathbb{Z}) = \mathbb{Z} \oplus (\mathbb{Z}_2)^2 = \mathbb{Z}^4/A,$$

$$A = \{(k_1, k_2, k_3, k_4) \ \| \ k_1 = 0, \ k_3, k_4 \in 2\mathbb{Z}, \ k_2 \in \mathbb{Z}\}$$

AB-groups:

$\forall m > 0, \ < (m, 0, 0, 1) >$

$Q =< (2l + 1, 0) >$

$$E : < a, b, c, d, \alpha \ \| \quad \begin{array}{ll} [b, a] = c^{2l+1}d^{-(2l+1)k_3} & [d, a] = 1 \\ [c, a] = d^{k_1} & [d, b] = 1 \\ [c, b] = d^{-k_1} & [d, c] = 1 \\ \alpha a = b\alpha d^{k_2} & \alpha^2 = d^{k_4} \\ \alpha b = a\alpha d^{-k_2} & \alpha d = d\alpha \\ \alpha c = c^{-1}\alpha d^{2k_3} & \end{array} \quad >$$

$$H^2(Q, \mathbb{Z}) = \mathbb{Z} \oplus \mathbb{Z}_2 = \mathbb{Z}^4/A,$$

$$A = \{(k_1, k_2, k_3, k_4) \ \| \ k_1 = 0, \ k_4 \in 2\mathbb{Z}, \ k_2, k_3 \in \mathbb{Z}\}$$

AB-groups:

$\forall m > 0, \ < (m, 0, 0, 1) >$

7.

$Q =< (2l, 0, 1, 0) >$

$$E : < a, b, c, d, \alpha, \beta \ \| \quad \begin{array}{ll} [b, a] = c^{4l}d^{4l(k_1+k_2)} & [d, a] = 1 \\ [c, a] = 1 & [d, b] = 1 \\ [c, b] = d^{2k_1} & [d, c] = 1 \\ \alpha a = a^{-1}\alpha & \alpha^2 = cd^{k_2} \\ \alpha b = b^{-1}\alpha d^{k_1} & \alpha d = d^{-1}\alpha \\ \alpha c = c\alpha d^{-2k_2} & \\ \beta a = a\beta c^{-2l}d^{l(k_1-2k_2)} & \beta^2 = d^{k_3} \\ \beta b = b^{-1}\beta d^{k_1-2k_2+2k_3-2k_4} & \beta d = d\beta \\ \beta c = c^{-1}\beta d^{k_1-2k_2} & \alpha\beta = b^{-1}\beta\alpha c^{-1}d^{k_4} \end{array} \quad >$$

$$H^2(Q, \mathbb{Z}) = \mathbb{Z} \oplus \mathbb{Z}_2 = \mathbb{Z}^4/A,$$

$$A = \{(k_1, k_2, k_3, k_4) \,\|\, k_1 = 0, \ k_3 \in 2\mathbb{Z}, \ k_2, k_4 \in \mathbb{Z}\}$$

AB-groups:
$\forall m > 0, \ m \cong 0 \bmod 2, \ < (m, 0, 1, 0) >$
$Q = < (2l + 1, 0, 1, 0) >$

$E : < a, b, c, d, \alpha, \beta \,\|\,$

$[b, a] = c^{4l+2} d^{(4l+2)(2k_1 + k_2)}$ $[d, a] = 1$ $>$
$[c, a] = 1$ $[d, b] = 1$
$[c, b] = d^{4k_1}$ $[d, c] = 1$
$\alpha a = a^{-1} \alpha$ $\alpha^2 = cd^{k_2}$
$\alpha b = b^{-1} \alpha d^{2k_1}$ $\alpha d = d^{-1} \alpha$
$\alpha c = c \alpha d^{-2k_2}$
$\beta a = a \beta c^{-2l-1} d^{(2l+1)(k_1 - k_2)}$ $\beta^2 = d^{k_3}$
$\beta b = b^{-1} \beta d^{2(k_1 - k_2 + k_3 - k_4)}$ $\beta d = d \beta$
$\beta c = c^{-1} \beta d^{2(k_1 - k_2)}$ $\alpha \beta = b^{-1} \beta \alpha c^{-1} d^{k_4}$

$$H^2(Q, \mathbb{Z}) = \mathbb{Z} \oplus \mathbb{Z}_2 = \mathbb{Z}^4/A,$$

$$A = \{(k_1, k_2, k_3, k_4) \,\|\, k_1 = 0, \ k_3 \in 2\mathbb{Z}, \ k_2, k_4 \in \mathbb{Z}\}$$

AB-groups:
$\forall m > 0, \ < (m, 0, 1, 0) >$

---

8. $Q = < (k, 0, 0, 1) >$
$^\alpha d = d^{-1}$ and $^\beta d = d$

$E : < a, b, c, d, \alpha, \beta \,\|\,$

$[b, a] = c^{4k} d^{2k(k_1 + 2k_2)}$ $[d, a] = 1$ $>$
$[c, a] = 1$ $[d, b] = 1$
$[c, b] = d^{2k_1}$ $[d, c] = 1$
$\alpha a = a^{-1} \alpha c^{2k} d^{2k k_2}$ $\alpha^2 = cd^{k_2}$
$\alpha b = b^{-1} \alpha c^{-2k} d^{(1 - 2k)k_1 - 2k k_2}$ $\alpha d = d^{-1} \alpha$
$\alpha c = c \alpha d^{-2k_2}$
$\beta a = a \beta$ $\beta^2 = a d^{k_3}$
$\beta b = b^{-1} \beta c^{-2k} d^{k_1(1 + 2k) - 2k_2(1 + 3k) + 2(k_3 - k_4)}$ $\beta d = d \beta$
$\alpha \beta = a^{-1} b^{-1} \beta \alpha c^{-(1 + 2k)} d^{k_4}$
$\beta c = c^{-1} \beta d^{k_1 - 2k_2}$

$$H^2(Q, \mathbb{Z}) = \mathbb{Z} = \mathbb{Z}^4/A, \ A = \{(k_1, k_2, k_3, k_4) \,\|\, k_1 = 0, \ k_2, k_3, k_4 \in \mathbb{Z}\}$$

AB-groups:
$\forall m > 0, \ < (m, 0, 0, 0) >$

$^{\alpha}d = d^{-1}$ and $^{\beta}d = d^{-1}$

$$E : < a, b, c, d, \alpha, \beta \parallel$$

$[b, a] = c^{4k}d^{-2kk_1 + 4kk_2}$	$[d, a] = 1$	$>$
$[c, a] = d^{-2k_1}$	$[d, b] = 1$	
$[c, b] = 1$	$[d, c] = 1$	
$\alpha a = a^{-1}\alpha c^{2k}d^{-(1+2k)k_1 + 2kk_2}$	$\alpha^2 = cd^{k_2}$	
$\alpha b = b^{-1}\alpha c^{-2k}d^{-2kk_2}$	$\alpha d = d^{-1}\alpha$	
$\alpha c = c\alpha d^{-2k_2}$		
$\beta a = a\beta d^{-2k_3}$	$\beta^2 = ad^{k_3}$	
$\beta b = b^{-1}\beta c^{-2k}d^{-2kk_2}$	$\beta d = d^{-1}\beta$	
$\beta c = c^{-1}\beta d^{k_1}$		
$\alpha\beta = a^{-1}b^{-1}\beta\alpha c^{-(1+2k)}d^{k_4}$		

$$H^2(Q, \mathbb{Z}) = \mathbb{Z} = \mathbb{Z}^4 / A, \quad A = \{(k_1, k_2, k_3, k_4) \parallel k_1 = 0, \ k_2, k_3, k_4 \in \mathbb{Z}\}$$

AB-groups:

All groups are isomorphic to one of the previous case.

Some comments

1. This group leads to all rank 4, 3–step nilpotent groups.

2. Using the results of the previous section we have to require that α acts non-trivial on $\mathbb{Z}$. This means that we only have to consider those 3-dimensional AC–groups of category 2, which are torsion free. In fact, there's only one such a group, namely $Q = < (k, 0, 0, 1) >$, for $k \equiv 0 \bmod 2$.

3. For category 3, both actions on $\mathbb{Z}$ are allowed. However, since any AC–group Q of this category has torsion, we only have to investigate the trivial action. We notice that for $k_1 \not\equiv 0 \bmod 2 < (k, 0) > \cong < (k, 1) >$, but for $k \equiv 0 \bmod 2 < (k, 0) > \not\cong < (k, 1) >$

4. For this category, both the trivial and the non-trivial action are to be considered.

5. Since all groups Q have torsion, we only investigate the trivial action.

6. In Q we can compute $(\alpha\beta)^2 = 1$. This forces the fact that for the only allowable action of Q on $\mathbb{Z}$, there is a torsion element which acts non-trivially.

7. Because $\beta^2 = 1$ we have to require that β acts trivially. The only allowable action for α is a non-trivial one, so we have to investigate those groups without torsion elements "involving" α. Therefore we must consider $k_3 = 1$ (else $\alpha^2 = 1$) and $k_2 = 0$ (else $(ac^{-1}\alpha)^2 = 1$). Conclusion: the only group to be considered is $Q = < (k, 0, 1, 0) >$.

8. α has to act non-trivially (So consider only the case $k_4 = 1$), but for β there are two choices.

9. The action of α is a non-trivial one. Since $(\beta^2) = 1$ we are only inter-

ested in a trivial action for β. But in this case $(\alpha\beta)^2 = 1$, which implies that a torsion element does act non-trivially. Conclusion: this category does not lead to AB-groups.

Now, we also indicate an affine representation for any of these groups. We again write down this representation $\lambda : E \to \mathrm{Aff}(\mathbb{R}^4)$ by giving the images of the generators a, b, c, d, α and if necessary β. In each case we use

$$\lambda(d) = \begin{pmatrix} 1 & 0 & 0 & 0 & 1 \\ 0 & 1 & 0 & 0 & 0 \\ 0 & 0 & 1 & 0 & 0 \\ 0 & 0 & 0 & 1 & 0 \\ 0 & 0 & 0 & 0 & 1 \end{pmatrix}.$$

The numbers of each representation described below, correspond to the numbers of groups in the table above:

1.

$$\lambda(a) = \begin{pmatrix} 1 & \frac{-2k_2}{3} & 0 & \frac{-k_1}{2}+\frac{2kk_2}{3} & 0 \\ 0 & 1 & 0 & -\frac{k}{2} & 0 \\ 0 & 0 & 1 & 0 & 1 \\ 0 & 0 & 0 & 1 & 0 \\ 0 & 0 & 0 & 0 & 1 \end{pmatrix} \quad \lambda(b) = \begin{pmatrix} 1 & \frac{-2k_3}{3} & \frac{k_1}{2}-\frac{2kk_3}{3} & 0 & 0 \\ 0 & 1 & \frac{k}{2} & 0 & 0 \\ 0 & 0 & 1 & 0 & 0 \\ 0 & 0 & 0 & 1 & 1 \\ 0 & 0 & 0 & 0 & 1 \end{pmatrix}$$

$$\lambda(c) = \begin{pmatrix} 1 & 0 & \frac{k_2}{3} & \frac{k_3}{3} & 0 \\ 0 & 1 & 0 & 0 & 1 \\ 0 & 0 & 1 & 0 & 0 \\ 0 & 0 & 0 & 1 & 0 \\ 0 & 0 & 0 & 0 & 1 \end{pmatrix}$$

2.

$$\lambda(a) = \begin{pmatrix} 1 & \frac{-4k_1}{3} & 0 & \frac{2kk_1}{3}-kk_3 & 0 \\ 0 & 1 & 0 & -k & 0 \\ 0 & 0 & 1 & 0 & 1 \\ 0 & 0 & 0 & 1 & 0 \\ 0 & 0 & 0 & 0 & 1 \end{pmatrix} \quad \lambda(b) = \begin{pmatrix} 1 & \frac{-4k_2}{3} & \frac{-2kk_2}{3}+kk_3 & 0 & 0 \\ 0 & 1 & k & 0 & 0 \\ 0 & 0 & 1 & 0 & 0 \\ 0 & 0 & 0 & 1 & 1 \\ 0 & 0 & 0 & 0 & 1 \end{pmatrix}$$

$$\lambda(c) = \begin{pmatrix} 1 & 0 & \frac{2k_1}{3} & \frac{2k_2}{3} & 0 \\ 0 & 1 & 0 & 0 & 1 \\ 0 & 0 & 1 & 0 & 0 \\ 0 & 0 & 0 & 1 & 0 \\ 0 & 0 & 0 & 0 & 1 \end{pmatrix} \quad \lambda(\alpha) = \begin{pmatrix} -1 & 2k_3 & \frac{-k_1}{3} & \frac{-k_2}{3} & 0 \\ 0 & 1 & 0 & 0 & \frac{1}{2} \\ 0 & 0 & -1 & 0 & 0 \\ 0 & 0 & 0 & -1 & 0 \\ 0 & 0 & 0 & 0 & 1 \end{pmatrix}$$

3.

$Q = <(2l+1, 1)>$

$$\lambda(a) = \begin{pmatrix} 1 & 0 & 0 & \frac{k_1}{6} & 0 \\ 0 & 1 & 0 & -\frac{1}{2} - l & 0 \\ 0 & 0 & 1 & 0 & 1 \\ 0 & 0 & 0 & 1 & 0 \\ 0 & 0 & 0 & 0 & 1 \end{pmatrix} \quad \lambda(b) = \begin{pmatrix} 1 & \frac{-2k_1}{3} & \frac{-k_1}{2} - \frac{k_1 l}{3} & -k_3 & 0 \\ 0 & 1 & \frac{1}{2} + l & 0 & 0 \\ 0 & 0 & 1 & 0 & 0 \\ 0 & 0 & 0 & 1 & 1 \\ 0 & 0 & 0 & 0 & 1 \end{pmatrix}$$

$$\lambda(c) = \begin{pmatrix} 1 & 0 & 0 & \frac{k_1}{3} & -k_2 \\ 0 & 1 & 0 & 0 & 1 \\ 0 & 0 & 1 & 0 & 0 \\ 0 & 0 & 0 & 1 & 0 \\ 0 & 0 & 0 & 0 & 1 \end{pmatrix} \quad \lambda(\alpha) = \begin{pmatrix} 1 & 0 & 0 & 0 & \frac{k_4}{2} \\ 0 & -1 & -1 & 0 & 0 \\ 0 & 0 & 1 & 0 & 0 \\ 0 & 0 & 0 & -1 & 0 \\ 0 & 0 & 0 & 0 & 1 \end{pmatrix}$$

$Q = <(2l, 1)>$

$$\lambda(a) = \begin{pmatrix} 1 & 0 & 0 & \frac{k_1}{3} - k_2 l & 0 \\ 0 & 1 & 0 & -l & 0 \\ 0 & 0 & 1 & 0 & 1 \\ 0 & 0 & 0 & 1 & 0 \\ 0 & 0 & 0 & 0 & 1 \end{pmatrix}$$

$$\lambda(b) = \begin{pmatrix} 1 & \frac{-4k_1}{3} & k_2 l - \frac{2k_1}{3} - \frac{2k_1 l}{3} & -k_3 & 0 \\ 0 & 1 & l & 0 & 0 \\ 0 & 0 & 1 & 0 & 0 \\ 0 & 0 & 0 & 1 & 1 \\ 0 & 0 & 0 & 0 & 1 \end{pmatrix}$$

$$\lambda(c) = \begin{pmatrix} 1 & 0 & 0 & \frac{2k_1}{3} & 0 \\ 0 & 1 & 0 & 0 & 1 \\ 0 & 0 & 1 & 0 & 0 \\ 0 & 0 & 0 & 1 & 0 \\ 0 & 0 & 0 & 0 & 1 \end{pmatrix} \quad \lambda(\alpha) = \begin{pmatrix} 1 & -2k_2 & -k_2 & 0 & \frac{k_4}{2} \\ 0 & -1 & -1 & 0 & 0 \\ 0 & 0 & 1 & 0 & 0 \\ 0 & 0 & 0 & -1 & 0 \\ 0 & 0 & 0 & 0 & 1 \end{pmatrix}$$

$Q = <(2l, 0)>$

$$\lambda(a) = \begin{pmatrix} 1 & 0 & 0 & \frac{k_3 l}{2} & 0 \\ 0 & 1 & 0 & -l & 0 \\ 0 & 0 & 1 & 0 & 1 \\ 0 & 0 & 0 & 1 & 0 \\ 0 & 0 & 0 & 0 & 1 \end{pmatrix} \quad \lambda(b) = \begin{pmatrix} 1 & \frac{-2k_1}{3} & -\frac{k_1 l}{3} - \frac{k_3 l}{2} & -k_2 & 0 \\ 0 & 1 & l & 0 & 0 \\ 0 & 0 & 1 & 0 & 0 \\ 0 & 0 & 0 & 1 & 1 \\ 0 & 0 & 0 & 0 & 1 \end{pmatrix}$$

$$\lambda(c) = \begin{pmatrix} 1 & 0 & 0 & \frac{k_1}{3} & 0 \\ 0 & 1 & 0 & 0 & 1 \\ 0 & 0 & 1 & 0 & 0 \\ 0 & 0 & 0 & 1 & 0 \\ 0 & 0 & 0 & 0 & 1 \end{pmatrix} \qquad \lambda(\alpha) = \begin{pmatrix} 1 & k_3 & 0 & 0 & \frac{k_4}{2} \\ 0 & -1 & 0 & 0 & 0 \\ 0 & 0 & 1 & 0 & 0 \\ 0 & 0 & 0 & -1 & 0 \\ 0 & 0 & 0 & 0 & 1 \end{pmatrix}$$

4.

Trivial action

$$\lambda(a) = \begin{pmatrix} 1 & 0 & 0 & \frac{kk_3}{2} & -k_4 \\ 0 & 1 & 0 & -k & 0 \\ 0 & 0 & 1 & 0 & 1 \\ 0 & 0 & 0 & 1 & 0 \\ 0 & 0 & 0 & 0 & 1 \end{pmatrix} \qquad \lambda(\alpha) = \begin{pmatrix} 1 & k_3 & 0 & 0 & 0 \\ 0 & -1 & 0 & \frac{k}{2} & 0 \\ 0 & 0 & 1 & 0 & \frac{1}{2} \\ 0 & 0 & 0 & -1 & 0 \\ 0 & 0 & 0 & 0 & 1 \end{pmatrix}$$

$$\lambda(c) = \begin{pmatrix} 1 & 0 & 0 & \frac{k_1}{3} & 0 \\ 0 & 1 & 0 & 0 & 1 \\ 0 & 0 & 1 & 0 & 0 \\ 0 & 0 & 0 & 1 & 0 \\ 0 & 0 & 0 & 0 & 1 \end{pmatrix} \qquad \lambda(b) = \begin{pmatrix} 1 & \frac{-2k_1}{3} & -\frac{k_1 k}{3} & -\frac{k_3 k}{2} & \frac{3kk_3}{4} & -\frac{kk_1}{2} & -k_2 & 0 \\ 0 & 1 & k & 0 & 0 \\ 0 & 0 & 1 & 0 & 0 \\ 0 & 0 & 0 & 1 & 1 \\ 0 & 0 & 0 & 0 & 1 \end{pmatrix}$$

Non-trivial action

$$\lambda(a) = \begin{pmatrix} 1 & \frac{4k_1}{3} & 0 & -\frac{kk_1}{3} & -k_3 \\ 0 & 1 & 0 & -k & 0 \\ 0 & 0 & 1 & 0 & 1 \\ 0 & 0 & 0 & 1 & 0 \\ 0 & 0 & 0 & 0 & 1 \end{pmatrix} \qquad \lambda(b) = \begin{pmatrix} 1 & 0 & \frac{4k_1 k}{3} - 2k_2 & 0 & 0 \\ 0 & 1 & k & 0 & 0 \\ 0 & 0 & 1 & 0 & 0 \\ 0 & 0 & 0 & 1 & 1 \\ 0 & 0 & 0 & 0 & 1 \end{pmatrix}$$

$$\lambda(c) = \begin{pmatrix} 1 & 0 & -\frac{2k_1}{3} & 0 & 0 \\ 0 & 1 & 0 & 0 & 1 \\ 0 & 0 & 1 & 0 & 0 \\ 0 & 0 & 0 & 1 & 0 \\ 0 & 0 & 0 & 0 & 1 \end{pmatrix} \qquad \lambda(\alpha) = \begin{pmatrix} -1 & \frac{-2k_1}{3} & 0 & 0 & 0 \\ 0 & -1 & 0 & \frac{k}{2} & 0 \\ 0 & 0 & 1 & 0 & \frac{1}{2} \\ 0 & 0 & 0 & -1 & 0 \\ 0 & 0 & 0 & 0 & 1 \end{pmatrix}$$

5.

$$Q = < (2l, 0) >$$

$$\lambda(a) = \begin{pmatrix} 1 & \frac{-2k_1}{3} & 0 & k_3 l & k_2 \\ 0 & 1 & 0 & -l & 0 \\ 0 & 0 & 1 & 0 & 1 \\ 0 & 0 & 0 & 1 & 0 \\ 0 & 0 & 0 & 0 & 1 \end{pmatrix} \qquad \lambda(b) = \begin{pmatrix} 1 & \frac{2k_1}{3} & 0 & 0 & 0 \\ 0 & 1 & l & 0 & 0 \\ 0 & 0 & 1 & 0 & 0 \\ 0 & 0 & 0 & 1 & 1 \\ 0 & 0 & 0 & 0 & 1 \end{pmatrix}$$

$$\lambda(c) = \begin{pmatrix} 1 & 0 & \frac{k_1}{3} & -\frac{k_1}{3} & 0 \\ 0 & 1 & 0 & 0 & 1 \\ 0 & 0 & 1 & 0 & 0 \\ 0 & 0 & 0 & 1 & 0 \\ 0 & 0 & 0 & 0 & 1 \end{pmatrix} \quad \lambda(\alpha) = \begin{pmatrix} 1 & k_3 & 0 & 0 & \frac{k_4}{2} \\ 0 & -1 & 0 & 0 & 0 \\ 0 & 0 & 0 & 1 & 0 \\ 0 & 0 & 1 & 0 & 0 \\ 0 & 0 & 0 & 0 & 1 \end{pmatrix}$$

$Q = < (2l + 1, 0) >$

$$\lambda(a) = \begin{pmatrix} 1 & \frac{-2k_1}{3} & 0 & k_3 + 2k_3l & k_2 \\ 0 & 1 & 0 & -\frac{1}{2} - l & 0 \\ 0 & 0 & 1 & 0 & 1 \\ 0 & 0 & 0 & 1 & 0 \\ 0 & 0 & 0 & 0 & 1 \end{pmatrix} \quad \lambda(b) = \begin{pmatrix} 1 & \frac{2k_1}{3} & 0 & 0 & 0 \\ 0 & 1 & \frac{1}{2} + l & 0 & 0 \\ 0 & 0 & 1 & 0 & 0 \\ 0 & 0 & 0 & 1 & 1 \\ 0 & 0 & 0 & 0 & 1 \end{pmatrix}$$

$$\lambda(c) = \begin{pmatrix} 1 & 0 & \frac{k_1}{3} & -\frac{k_1}{3} & 0 \\ 0 & 1 & 0 & 0 & 1 \\ 0 & 0 & 1 & 0 & 0 \\ 0 & 0 & 0 & 1 & 0 \\ 0 & 0 & 0 & 0 & 1 \end{pmatrix} \quad \lambda(\alpha) = \begin{pmatrix} 1 & 2k_3 & 0 & 0 & \frac{k_4}{2} \\ 0 & -1 & 0 & 0 & 0 \\ 0 & 0 & 0 & 1 & 0 \\ 0 & 0 & 1 & 0 & 0 \\ 0 & 0 & 0 & 0 & 1 \end{pmatrix}$$

7.

$Q = < (2l, 0, 1, 0) >$

$$\lambda(a) = \begin{pmatrix} 1 & 0 & 0 & -2k_2l & 0 \\ 0 & 1 & 0 & -2l & 0 \\ 0 & 0 & 1 & 0 & 1 \\ 0 & 0 & 0 & 1 & 0 \\ 0 & 0 & 0 & 0 & 1 \end{pmatrix}$$

$$\lambda(b) = \begin{pmatrix} 1 & \frac{-4k_1}{3} & \frac{-4k_1l}{3} + 2k_2l & 0 & \frac{k_1}{2} - k_2 - k_4 \\ 0 & 1 & 2l & 0 & 0 \\ 0 & 0 & 1 & 0 & 0 \\ 0 & 0 & 0 & 1 & 1 \\ 0 & 0 & 0 & 0 & 1 \end{pmatrix}$$

$$\lambda(c) = \begin{pmatrix} 1 & 0 & 0 & \frac{2k_1}{3} & 0 \\ 0 & 1 & 0 & 0 & 1 \\ 0 & 0 & 1 & 0 & 0 \\ 0 & 0 & 0 & 1 & 0 \\ 0 & 0 & 0 & 0 & 1 \end{pmatrix} \quad \lambda(\alpha) = \begin{pmatrix} -1 & 2k_2 & 0 & \frac{-k_1}{3} & 0 \\ 0 & 1 & 0 & 0 & \frac{1}{2} \\ 0 & 0 & -1 & 0 & 0 \\ 0 & 0 & 0 & -1 & 0 \\ 0 & 0 & 0 & 0 & 1 \end{pmatrix}$$

$$\lambda(\beta) = \begin{pmatrix} 1 & \frac{2k_1}{3} - 2k_2 & \frac{-k_1 l}{3} + k_2 l & 2k_3 & 0 \\ 0 & -1 & l & 0 & 0 \\ 0 & 0 & 1 & 0 & 0 \\ 0 & 0 & 0 & -1 & \frac{1}{2} \\ 0 & 0 & 0 & 0 & 1 \end{pmatrix}$$

$Q = < (2l+1, 0, 1, 0) >$

$$\lambda(a) = \begin{pmatrix} 1 & 0 & 0 & -k_2 - 2k_2 l & 0 \\ 0 & 1 & 0 & -1 - 2l & 0 \\ 0 & 0 & 1 & 0 & 1 \\ 0 & 0 & 0 & 1 & 0 \\ 0 & 0 & 0 & 0 & 1 \end{pmatrix} \quad \lambda(\alpha) = \begin{pmatrix} -1 & 2k_2 & 0 & \frac{-2k_1}{3} & 0 \\ 0 & 1 & 0 & 0 & \frac{1}{2} \\ 0 & 0 & -1 & 0 & 0 \\ 0 & 0 & 0 & -1 & 0 \\ 0 & 0 & 0 & 0 & 1 \end{pmatrix}$$

$$\lambda(b) = \begin{pmatrix} 1 & \frac{-8k_1}{3} & \frac{-4k_1}{3} + k_2 - \frac{8k_1 l}{3} + 2k_2 l & 0 & k_1 - k_2 - k_4 \\ 0 & 1 & 1 + 2l & 0 & 0 \\ 0 & 0 & 1 & 0 & 0 \\ 0 & 0 & 0 & 1 & 1 \\ 0 & 0 & 0 & 0 & 1 \end{pmatrix}$$

$$\lambda(c) = \begin{pmatrix} 1 & 0 & 0 & \frac{4k_1}{3} & 0 \\ 0 & 1 & 0 & 0 & 1 \\ 0 & 0 & 1 & 0 & 0 \\ 0 & 0 & 0 & 1 & 0 \\ 0 & 0 & 0 & 0 & 1 \end{pmatrix}$$

$$\lambda(\beta) = \begin{pmatrix} 1 & \frac{4k_1}{3} - 2k_2 & \frac{-k_1}{3} + \frac{k_2}{2} - \frac{2k_1 l}{3} + k_2 l & 2k_3 & 0 \\ 0 & -1 & \frac{1}{2} + l & 0 & 0 \\ 0 & 0 & 1 & 0 & 0 \\ 0 & 0 & 0 & -1 & \frac{1}{2} \\ 0 & 0 & 0 & 0 & 1 \end{pmatrix}$$

8.
$^\alpha d = d^{-1}, \; ^\beta d = d$

$$\lambda(a) = \begin{pmatrix} 1 & 0 & 0 & \frac{2kk_1}{3} - 2kk_2 & \frac{k_1}{2} + 2kk_1 - k_2 - 2kk_2 - k_4 \\ 0 & 1 & 0 & -2k & 0 \\ 0 & 0 & 1 & 0 & 1 \\ 0 & 0 & 0 & 1 & 0 \\ 0 & 0 & 0 & 0 & 1 \end{pmatrix}$$

$$\lambda(b) = \begin{pmatrix} 1 & \frac{-4k_1}{3} & \frac{-8kk_1}{3} + 2kk_2 & \frac{2kk_1}{3} & 0 \\ 0 & 1 & 2k & 0 & 0 \\ 0 & 0 & 1 & 0 & 0 \\ 0 & 0 & 0 & 1 & 1 \\ 0 & 0 & 0 & 0 & 1 \end{pmatrix}$$

$$\lambda(c) = \begin{pmatrix} 1 & 0 & 0 & \frac{2k_1}{3} & 0 \\ 0 & 1 & 0 & 0 & 1 \\ 0 & 0 & 1 & 0 & 0 \\ 0 & 0 & 0 & 1 & 0 \\ 0 & 0 & 0 & 0 & 1 \end{pmatrix} \quad \lambda(\alpha) = \begin{pmatrix} -1 & 2k_2 & 2kk_2 & \frac{-k_1}{3} - 2kk_2 & 0 \\ 0 & 1 & 2k & -2k & \frac{1}{2} \\ 0 & 0 & -1 & 0 & 0 \\ 0 & 0 & 0 & -1 & 0 \\ 0 & 0 & 0 & 0 & 1 \end{pmatrix}$$

$$\lambda(\beta) = \begin{pmatrix} 1 & \frac{2k_1}{3} - 2k_2 & \frac{kk_1}{3} - kk_2 & k_1 + \frac{11kk_1}{3} - 2k_2 - 3kk_2 + 2k_3 - 2k_4 & 0 \\ 0 & -1 & -k & k & 0 \\ 0 & 0 & 1 & 0 & \frac{1}{2} \\ 0 & 0 & 0 & -1 & \frac{1}{2} \\ 0 & 0 & 0 & 0 & 1 \end{pmatrix}$$

$$^{\alpha}d = d^{-1}, \ ^{\beta}d = d^{-1}$$

$$\lambda(a) = \begin{pmatrix} 1 & \frac{4k_1}{3} & \frac{2kk_1}{3} & \frac{-8kk_1}{3} & \frac{kk_1}{2} - k_3 \\ 0 & 1 & 0 & -2k & 0 \\ 0 & 0 & 1 & 0 & 1 \\ 0 & 0 & 0 & 1 & 0 \\ 0 & 0 & 0 & 0 & 1 \end{pmatrix} \quad \lambda(c) = \begin{pmatrix} 1 & 0 & \frac{-2k_1}{3} & 0 & -k_2 \\ 0 & 1 & 0 & 0 & 1 \\ 0 & 0 & 1 & 0 & 0 \\ 0 & 0 & 0 & 1 & 0 \\ 0 & 0 & 0 & 0 & 1 \end{pmatrix}$$

$$\lambda(b) = \begin{pmatrix} 1 & 0 & \frac{2kk_1}{3} & 0 & -\frac{k_1}{2} - \frac{kk_1}{2} + k_2 + 2kk_2 + k_3 + k_4 \\ 0 & 1 & 2k & 0 & 0 \\ 0 & 0 & 1 & 0 & 0 \\ 0 & 0 & 0 & 1 & 1 \\ 0 & 0 & 0 & 0 & 1 \end{pmatrix}$$

$$\lambda(\alpha) = \begin{pmatrix} -1 & 0 & \frac{k_1}{3} & 0 & 0 \\ 0 & 1 & 2k & -2k & \frac{1}{2} \\ 0 & 0 & -1 & 0 & 0 \\ 0 & 0 & 0 & -1 & 0 \\ 0 & 0 & 0 & 0 & 1 \end{pmatrix} \quad \lambda(\beta) = \begin{pmatrix} -1 & \frac{-2k_1}{3} & 0 & kk_1 & 0 \\ 0 & -1 & -k & k & 0 \\ 0 & 0 & 1 & 0 & \frac{1}{2} \\ 0 & 0 & 0 & -1 & \frac{1}{2} \\ 0 & 0 & 0 & 0 & 1 \end{pmatrix}$$

Appendix

The use of Mathematica®

We are going to illustrate the use of Mathematica® with a concrete example of the construction of a 4–dimensional AB–group. We will execute all the computations needed in case number 3 in the table of section 7.2.

A.1 Choose a crystallographic group Q

The group E to be constructed fits in a short exact sequence

$$1 \to \mathbb{Z} \to E \to Q \to 1$$

where Q is the 3–dimensional crystallographic group listed in [10] on page 62 as follows (the numbering of the lines is added):

```
1    FAMILY II: ....

2       CRYSTAL SYSTEM 2: ...

3          Q-CLASS 2/1: ORDER 2; ISOM TYPE 2.1; 2 Z-CLASSES ...

4                REL: A2=I

5          Z-CLASS 2/1/1: Z(P2); ...

6             GEN:        A -1  0  0
7                            0  1  0
8                            0  0 -1

9             SPGR: 01 A [0,0,0]      IT 3; OBT 1
10            FF    02 A [0,1,0]/2    IT 4; OBT 1
```

---

11   Z-CLASS ...

We extract the information we need in the following way:

- On line 3 we see that the holonomy group F of Q is of order 2 and so F is isomorphic to $\mathbb{Z}_2$ (= Isomorphism Type 2.1 in [10, Table 6B]).

- Line 4 describes the holonomy group F. There is one generator A and one relation $A^2 = 1$. We will use α (or **alfa**) to denote this generator in the sequel.

- The group Q has an affine representation (seen in $Gl(4, \mathbb{R})$) as follows:
  First there are the three translations, which are always the same and which we denote by a, b, c:

$$a = \begin{pmatrix} 1 & 0 & 0 & 1 \\ 0 & 1 & 0 & 0 \\ 0 & 0 & 1 & 0 \\ 0 & 0 & 0 & 1 \end{pmatrix} \quad b = \begin{pmatrix} 1 & 0 & 0 & 0 \\ 0 & 1 & 0 & 1 \\ 0 & 0 & 1 & 0 \\ 0 & 0 & 0 & 1 \end{pmatrix} \quad c = \begin{pmatrix} 1 & 0 & 0 & 0 \\ 0 & 1 & 0 & 0 \\ 0 & 0 & 1 & 1 \\ 0 & 0 & 0 & 1 \end{pmatrix}.$$

  Further, the affine transformation corresponding to α is indicated in lines 6,7,8 (the rotational part) and in line 9 (the translational part). So

$$\alpha = \begin{pmatrix} -1 & 0 & 0 & 0 \\ 0 & 1 & 0 & 0 \\ 0 & 0 & -1 & 0 \\ 0 & 0 & 0 & 1 \end{pmatrix}.$$

- A presentation for Q can be written down now very easily following section 5.2. One should keep in mind that the action of F on $\mathbb{Z}^3$ is given by the rotational part of α. For this group, we can see, since the translational part of α equals zero, that $Q = \mathbb{Z}^3 \rtimes \mathbb{Z}_2$. For other groups, one will have to use the matrix representation of Q to complete its presentation. Conclusion:

$$Q = <a, b, c, \alpha \parallel \begin{array}{ll} [b, a] = 1 & [c, a] = 1 \\ [c, b] = 1 & \alpha^2 = 1 \\ \alpha a = a^{-1}\alpha & \alpha b = b\alpha \\ \alpha c = c^{-1}\alpha \end{array} >.$$

- Line 10 denotes the next crystallographic group (number 4) and the symbol FF in front of it, indicates that this group is torsion free. (Fixedpoint Free space group).

Continuing as in section 5.2 we now describe any extension of Q by $\mathbb{Z}$ compatible with the trivial action of Q on $\mathbb{Z}$ (which factors through $\mathbb{Z}_2$). (Since $\alpha^2 = 1$, we cannot hope that extensions compatible with the non trivial action, will be torsion free). Such a group E has a presentation as follows:

$$E : < a, b, c, d, \alpha \parallel \begin{array}{ll} [b,a] = d^{l_1} & [d,a] = 1 \\ [c,a] = d^{l_2} & [d,b] = 1 \\ [c,b] = d^{l_3} & [d,c] = 1 \\ \alpha a = a^{-1}\alpha d^{l_4} & \alpha^2 = d^{l_7} \\ \alpha b = b\alpha d^{l_5} & \alpha d = d\alpha \\ \alpha c = c^{-1}\alpha d^{l_6} \end{array} > .$$

for some integers $l_1, l_2, \ldots, l_7$. We use Mathematica® to find the computational consistent ones, which is described in the following section.

A.2   Determination of computational consistent presentations

The following little program contains a function "matrixmacht" which computes formal powers of a unitriangular matrix, based on lemma 4.4.5. This program is runned automatically each time we start a Mathematica® session.

```
(* Personal Commands Used in the Infra-nil computations *)
(* ----------------------------------------------------- *)

(* Remark: All variables are written with 3 identical symbols to *)
(*         prevent confusion with variables used within a        *)
(*         Mathematica session.                                  *)


(* Definition of the binomial coefficients *)
(* xxx : is a formal parameter *)
(* nnn : is an integer *)

bin[xxx_,nnn_ ]:=(xxx-nnn+1)/nnn bin[xxx,nnn-1]
bin[xxx_,0 ]:=1
```

```
(* matrixmacht : Power of an uppertringular matrix "aaa" with *)
(*              formal power "xxx".                              *)
(*              Based on the formula:                           *)
(*              $A^x=((A-I)+I)^x=\sum_x \bn{x}{1} (A-I)^x$      *)

matrixmacht[aaa_ , xxx_ ]:=Sum[ (bin[xxx , 111]*
                   MatrixPower[aaa-IdentityMatrix[ Length[aaa]],111]),
                   {111,0,Length[aaa]}]


(* Some convenient abbreviations *)

mat[aaa_]:=TableForm[Expand[aaa]]
com[aaa_,bbb_]:=Expand[Inverse[aaa].Inverse[bbb].aaa.bbb]
```

Now we are ready to start the program used do determine the computational consistent presentations for E. The program is stored in a file with the name **groep.3** and looks like

```
(* Determining the computational consistent presentations and affine *)
(* representations for this class of AC-groups.                       *)

(* A general canonical type representation built up from the data    *)
(* in the book of Neubueser e.a.                                      *)

a={{1,A1,A2,A3,A4},
   {0,1,0,0,1},
   {0,0,1,0,0},
   {0,0,0,1,0},
   {0,0,0,0,1}}

b={{1,B1,B2,B3,B4},
   {0,1,0,0,0},
   {0,0,1,0,1},
   {0,0,0,1,0},
   {0,0,0,0,1}}

c={{1,C1,C2,C3,C4},
   {0,1,0,0,0},
   {0,0,1,0,0},
   {0,0,0,1,1},
   {0,0,0,0,1}}

d={{1,0,0,0,1},
   {0,1,0,0,0},
   {0,0,1,0,0},
```

```
    {0,0,0,1,0},
    {0,0,0,0,1}}

alfa={{1,alf1,alf2,alf3,alf4},(* The "1" on this line indicates that we*)
      {0,-1, 0, 0, 0},       (* are dealing with a trivial action of  *)
      {0, 0, 1, 0, 0},       (* alfa on d                             *)
      {0, 0, 0,-1, 0},
      {0, 0, 0, 0, 1}}

(*The conditions, which should be satisfied by the unknowns A1,...,alf4*)
(*And the conditions on 11,12,...,17 for computational consistency.   *)
(*All matrices printed should be zero.                                *)

Print[ Expand[ com[b,a]-matrixmacht[d,11] ]]
Print[ Expand[ com[c,a]-matrixmacht[d,12] ]]
Print[ Expand[ com[c,b]-matrixmacht[d,13] ]]
Print[ Expand[ alfa.a-Inverse[a].alfa.matrixmacht[d,14] ]]
Print[ Expand[ alfa.b-b.alfa.matrixmacht[d,15] ]]
Print[ Expand[ alfa.c-Inverse[c].alfa.matrixmacht[d,16] ]]
Print[ Expand[ alfa.alfa-matrixmacht[d,17] ]]
```

We now list a recorded version of a Mathematica® session to show how we use this program:

```
euler% math
Mathematica 2.1 for SPARC
Copyright 1988-92 Wolfram Research, Inc.

In[1]:= <<groep3.1
{{0, 0, 0, 0, -A2 + B1 - 11}, {0, 0, 0, 0, 0}, {0, 0, 0, 0, 0},

>   {0, 0, 0, 0, 0}, {0, 0, 0, 0, 0}}
{{0, 0, 0, 0, -A3 + C1 - 12}, {0, 0, 0, 0, 0}, {0, 0, 0, 0, 0},

>   {0, 0, 0, 0, 0}, {0, 0, 0, 0, 0}}
{{0, 0, 0, 0, -B3 + C2 - 13}, {0, 0, 0, 0, 0}, {0, 0, 0, 0, 0},

>   {0, 0, 0, 0, 0}, {0, 0, 0, 0, 0}}
{{0, 0, 2 A2, 0,alf1 - A1 + 2 A4 - 14}, {0, 0, 0, 0, 0}, {0, 0, 0, 0, 0},

>   {0, 0, 0, 0, 0}, {0, 0, 0, 0, 0}}
{{0, 2 B1, 0, 2 B3, alf2 - 15}, {0, 0, 0, 0, 0}, {0, 0, 0, 0, 0},

>   {0, 0, 0, 0, 0}, {0, 0, 0, 0, 0}}
{{0, 0, 2 C2, 0,alf3 - C3 + 2 C4 - 16}, {0, 0, 0, 0, 0}, {0, 0, 0, 0, 0},

>   {0, 0, 0, 0, 0}, {0, 0, 0, 0, 0}}
```

```
{{0, 0, 2 alf2, 0, 2 alf4 - 17}, {0, 0, 0, 0, 0}, {0, 0, 0, 0, 0},

>    {0, 0, 0, 0, 0}, {0, 0, 0, 0, 0}}

In[2]:= A2=B1-11;A3=C1-12;B3=C2-13;alf2=15;alf4=17/2;

In[3]:= <<groep3.1
{{0, 0, 0, 0, 0}, {0, 0, 0, 0, 0}, {0, 0, 0, 0, 0}, {0, 0, 0, 0, 0},

>    {0, 0, 0, 0, 0}}
{{0, 0, 0, 0, 0}, {0, 0, 0, 0, 0}, {0, 0, 0, 0, 0}, {0, 0, 0, 0, 0},

>    {0, 0, 0, 0, 0}}
{{0, 0, 0, 0, 0}, {0, 0, 0, 0, 0}, {0, 0, 0, 0, 0}, {0, 0, 0, 0, 0},

>    {0, 0, 0, 0, 0}}
{{0, 0, 2 B1 - 2 11, 0, alf1 - A1 + 2 A4 - 14}, {0, 0, 0, 0, 0},

>    {0, 0, 0, 0, 0}, {0, 0, 0, 0, 0}, {0, 0, 0, 0, 0}}
{{0, 2 B1, 0, 2 C2 - 2 13, 0}, {0, 0, 0, 0, 0}, {0, 0, 0, 0, 0},

>    {0, 0, 0, 0, 0}, {0, 0, 0, 0, 0}}
{{0, 0, 2 C2, 0,alf3 - C3 + 2 C4 - 16}, {0, 0, 0, 0, 0}, {0, 0, 0, 0, 0},

>    {0, 0, 0, 0, 0}, {0, 0, 0, 0, 0}}
{{0, 0, 2 15, 0, 0}, {0, 0, 0, 0, 0}, {0, 0, 0, 0, 0}, {0, 0, 0, 0, 0},

>    {0, 0, 0, 0, 0}}

In[4]:= B1=0;11=0;C2=0;13=0;15=0;alf1=A1-2 A4 + 14;

In[5]:= <<groep3.1
{{0, 0, 0, 0, 0}, {0, 0, 0, 0, 0}, {0, 0, 0, 0, 0}, {0, 0, 0, 0, 0},

>    {0, 0, 0, 0, 0}}
{{0, 0, 0, 0, 0}, {0, 0, 0, 0, 0}, {0, 0, 0, 0, 0}, {0, 0, 0, 0, 0},

>    {0, 0, 0, 0, 0}}
{{0, 0, 0, 0, 0}, {0, 0, 0, 0, 0}, {0, 0, 0, 0, 0}, {0, 0, 0, 0, 0},

>    {0, 0, 0, 0, 0}}
{{0, 0, 0, 0, 0}, {0, 0, 0, 0, 0}, {0, 0, 0, 0, 0}, {0, 0, 0, 0, 0},

>    {0, 0, 0, 0, 0}}
{{0, 0, 0, 0, 0}, {0, 0, 0, 0, 0}, {0, 0, 0, 0, 0}, {0, 0, 0, 0, 0},

>    {0, 0, 0, 0, 0}}
{{0, 0, 0, 0,alf3 - C3 + 2 C4 - 16}, {0, 0, 0, 0, 0}, {0, 0, 0, 0, 0},
```

```
>   {0, 0, 0, 0, 0}, {0, 0, 0, 0, 0}}
{{0, 0, 0, 0, 0}, {0, 0, 0, 0, 0}, {0, 0, 0, 0, 0}, {0, 0, 0, 0, 0},

>   {0, 0, 0, 0, 0}}

In[6]:= alf3=C3 -2 C4 + 16;

In[7]:= <<groep3.1
{{0, 0, 0, 0, 0}, {0, 0, 0, 0, 0}, {0, 0, 0, 0, 0}, {0, 0, 0, 0, 0},

>   {0, 0, 0, 0, 0}}
{{0, 0, 0, 0, 0}, {0, 0, 0, 0, 0}, {0, 0, 0, 0, 0}, {0, 0, 0, 0, 0},

>   {0, 0, 0, 0, 0}}
{{0, 0, 0, 0, 0}, {0, 0, 0, 0, 0}, {0, 0, 0, 0, 0}, {0, 0, 0, 0, 0},

>   {0, 0, 0, 0, 0}}
{{0, 0, 0, 0, 0}, {0, 0, 0, 0, 0}, {0, 0, 0, 0, 0}, {0, 0, 0, 0, 0},

>   {0, 0, 0, 0, 0}}
{{0, 0, 0, 0, 0}, {0, 0, 0, 0, 0}, {0, 0, 0, 0, 0}, {0, 0, 0, 0, 0},

>   {0, 0, 0, 0, 0}}
{{0, 0, 0, 0, 0}, {0, 0, 0, 0, 0}, {0, 0, 0, 0, 0}, {0, 0, 0, 0, 0},

>   {0, 0, 0, 0, 0}}
{{0, 0, 0, 0, 0}, {0, 0, 0, 0, 0}, {0, 0, 0, 0, 0}, {0, 0, 0, 0, 0},

>   {0, 0, 0, 0, 0}}

In[8]:= Print[l1," ",l2," ",l3," ",l4," ",l5," ",l6," ",l7]
0 12 0 14 0 16 17

In[9]:= 12=k1;14=k2;16=k3;17=k4;

In[10]:= mat[a]

Out[10]//TableForm= 1    A1    0    C1 - k1    A4

                    0    1     0    0          1

                    0    0     1    0          0

                    0    0     0    1          0

                    0    0     0    0          1

In[11]:= A1=0;C1=k1/2;A4=0;
```

In[12] := mat[b]

Out[12]//TableForm=

1	0	B2	0	B4
0	1	0	0	0
0	0	1	0	1
0	0	0	1	0
0	0	0	0	1

In[13] := B2=0;B4=0;

In[14] := mat[c]

Out[14]//TableForm=

$$\frac{k1}{--}$$

1	2	0	C3	C4
0	1	0	0	0
0	0	1	0	0
0	0	0	1	1
0	0	0	0	1

In[15] := C3=0;C4=0;

In[16] := mat[alfa]

Out[16]//TableForm=

$$\frac{k4}{--}$$

1	k2	0	k3	2
0	-1	0	0	0
0	0	1	0	0
0	0	0	-1	0
0	0	0	0	1

In[17] := mat[a]

Out[17]//TableForm=

$$\frac{-k1}{---}$$

1	0	0	2	0

$$
\begin{array}{ccccc}
0 & 1 & 0 & 0 & 1 \\
0 & 0 & 1 & 0 & 0 \\
0 & 0 & 0 & 1 & 0 \\
0 & 0 & 0 & 0 & 1
\end{array}
$$

In[18] := mat[b]

Out[18]//TableForm=

1	0	0	0	0
0	1	0	0	0
0	0	1	0	1
0	0	0	1	0
0	0	0	0	1

In[19] := mat[c]

Out[19]//TableForm=

k1				
--				
1	2	0	0	0
0	1	0	0	0
0	0	1	0	0
0	0	0	1	1
0	0	0	0	1

In[20] := Save["representatie3.1",a,b,c,d,alfa]

In[21] := Quit
euler%

The program specifies the computational consistent groups E, which can all be presented by means of the four parameters k_1, k_2, k_3 and k_4:

$$E : \; < a, b, c, d, \alpha \; \| \quad \begin{array}{ll} [b, a] = 1 & [d, a] = 1 \\ [c, a] = d^{k_1} & [d, b] = 1 \\ [c, b] = 1 & [d, c] = 1 \\ \alpha a = a^{-1}\alpha d^{k_2} & \alpha^2 = d^{k_4} \\ \alpha b = b\alpha & \alpha d = d\alpha \\ \alpha c = c^{-1}\alpha d^{k_3} & \end{array} \quad > .$$

A.3 Computation of $H^2(Q, Z)$

The previous section also shows that the set of standard cocycles

$$SZ^2_\varphi(Q, \mathbb{Z}) \cong \mathbb{Z}^4$$

and that a canonical representation for each group E is saved in a file with the name representatie 3.1.

The following step in the process is to determine which standard cocycles are cohomologous to zero. This is done with a program having the name cocyk3.1:

```
(* Which standard cocycles are cohomologous to zero?   *)
(* By theorem 5.2.2 we may suppose that                *)
k1=0
<<representatie3.1

(* The action of alfa on d can be read of as the first entry of the *)
(* matrix which represents alfa                        *)
actalfa=alfa[[1,1]]


For[kk=0,kk<=1,kk++,
  For[mm=0,mm<=1,mm++,
    {product=matrixmacht[a,x1].matrixmacht[b,x2].
            matrixmacht[c,x3].MatrixPower[alfa,kk].
            matrixmacht[a,y1].matrixmacht[b,y2].
            matrixmacht[c,y3].MatrixPower[alfa,mm],
            (* Product is a general product of two elements.  *)
            (* Product can be written in the form:            *)
            (* a^acoef b^bcoef c^ccoef d^f(x,y) alfa^alfacoef *)
            (* The following commands determine all these     *)
            (* coefficients.                                   *)
        alfacoef=Mod[kk+mm,2], (* Computation in Z-modulo 2 *)
        product=product.MatrixPower[alfa,-alfacoef],
        acoef=product[[2,5]], (* This is a direct consequence *)
```

```
        bcoef=product[[3,5]], (* of the canonical type        *)
        ccoef=product[[4,5]], (* representation               *)
product=matrixmacht[c,-ccoef].matrixmacht[b,-bcoef].
            matrixmacht[a,-acoef].product,
        (* The power of d equals the cocycle *)
        f=Simplify[product[[1,5]]],
        (* A check for exactness of the program: *)
product[[1,5]]=0,
If[product==IdentityMatrix[5],Print[],Print["Something's wrong!"]],
      Print["Cocykel:"],
Print[f],
term=(actalfa)^(kk), (* term= the action of x on d *)
        (* The general form of delta gamma for x and y *)
deltagamma=( (term y1 + x1 -acoef) gammaa +
            (term y2 + x2 -bcoef) gammab +
            (term y3 + x3 -ccoef) gammac +
            (term mm + kk -alfacoef )gammaalfa ),
difference=Expand[f-deltagamma],
Print[],
Print["f(x,y)-deltagamma(x,y):"],
Print[difference],
Print["_____",
        "_____"],
} ]]
Print["=====================================",
    "====================================="]
```

The recorded Mathematica® session is as follows:

```
euler% math
Mathematica 2.1 for SPARC
Copyright 1988-92 Wolfram Research, Inc.

In[1]:= <<cocyk3.1
Cocykel:
0
f(x,y)-deltagamma(x,y):
0

------------------------------------------------------------------------
Cocykel:
0
f(x,y)-deltagamma(x,y):
0

------------------------------------------------------------------------
Cocykel:
k2 y1 + k3 y3
f(x,y)-deltagamma(x,y):
```

```
-2 gammaa y1 + k2 y1 - 2 gammac y3 + k3 y3
-----------------------------------------------------------------
Cocykel:
k4 + k2 y1 + k3 y3
f(x,y)-deltagamma(x,y):
-2 gammaalfa + k4 - 2 gammaa y1 + k2 y1 - 2 gammac y3 + k3 y3

-----------------------------------------------------------------
=================================================================


In[2]:= gammaa=k2/2;gammac=k3/2;gammaalfa=k4/2;

In[3]:= <<cocyk3.1
Cocykel:
0
f(x,y)-deltagamma(x,y):
0

-----------------------------------------------------------------
Cocykel:
0
f(x,y)-deltagamma(x,y):
0

-----------------------------------------------------------------
Cocykel:
k2 y1 + k3 y3
f(x,y)-deltagamma(x,y):
0

-----------------------------------------------------------------
Cocykel:
k4 + k2 y1 + k3 y3
f(x,y)-deltagamma(x,y):
0

-----------------------------------------------------------------
=================================================================


In[4]:= Quit
euler%
```

This shows that a cocycle f depending on the four variables k_1, k_2, k_3 and k_4 is cohomologous to 0 if and only if

$$k_1 = 0, \ k_2, k_3, k_4 \in 2\mathbb{Z}$$

This allows us to conclude that

$$H_\varphi^2(Q, \mathbb{Z}) \cong \mathbb{Z} \oplus (\mathbb{Z}_2)^3$$

A.4   Investigation of the torsion

The following problem is to discover in which groups E there is torsion. To achieve this, we first look for all torsion elements in the crystallographic group Q. Since the order of the holonomy group is 2, the order of a torsion element will also be equal to two. The program used to find torsion in Q is called **torsion** and is the following:

```
(* Searching the torsion elements of the crystallographic group Q *)

(* Loading the representation for E and so for Q *)
<<representatie3.1

(* order= The order of a possible torsion element *)
order=2

Print[" "]
(* A general element having a possible finite order *)
test=matrixmacht[a,m1].matrixmacht[b,m2].matrixmacht[c,m3].alfa

(* testelement to the power=order *)
testpower=MatrixPower[test,order]

(* Looking at the element in Q rather than in E *)
testcutoff=IdentityMatrix[4]
For[i=1,i<=4,i++,
 For[j=1,j<=4,j++,
   testcutoff[[i,j]]=testpower[[i+1,j+1]] ]]

(* What are the possible torsions? *)
Print[mat[testcutoff]]
Print[" "]
Print[Solve[testcutoff==IdentityMatrix[4],{m1,m2,m3}] ]
Print[" "]
```

The result of this program is:

```
euler% math
Mathematica 2.1 for SPARC
Copyright 1988-92 Wolfram Research, Inc.

In[1]:= <<torsion

1    0    0    0

0    1    0    2 m2
```

```
0    0    1    0

0    0    0    1

{{m2 -> 0}}


In[2]:= Quit
euler%
```

Before we continue, we have to notice that the file **representatie3.1**
is in a dangerous form, since it contains a lot of redundancy, which may
cause errors. Therefore we do the following:

```
euler% cat representatie3.1
a = {{1, A1, 0, C1 - 12, A4}, {0, 1, 0, 0, 1}, {0, 0, 1, 0, 0},
     {0, 0, 0, 1, 0}, {0, 0, 0, 0, 1}}

A1 = 0

C1 = k1/2

12 = k1

A4 = 0

b = {{1, 0, B2, 0,B4}, {0, 1, 0, 0, 0}, {0, 0, 1, 0, 1}, {0, 0, 0, 1, 0},
     {0, 0, 0, 0, 1}}

B2 = 0

B4 = 0

c = {{1,C1, 0,C3, C4}, {0, 1, 0, 0, 0}, {0, 0, 1, 0, 0}, {0, 0, 0, 1, 1},
     {0, 0, 0, 0, 1}}

C3 = 0

C4 = 0

d = {{1, 0, 0, 0, 1}, {0, 1, 0, 0, 0}, {0, 0, 1, 0, 0}, {0, 0, 0, 1, 0},
     {0, 0, 0, 0, 1}}

alfa = {{1, A1 - 2*A4 + 14, 0, C3 - 2*C4 + 16, 17/2}, {0, -1, 0, 0, 0},
        {0, 0, 1, 0, 0}, {0, 0, 0, -1, 0}, {0, 0, 0, 0, 1}}

14 = k2
```

```
16 = k3

17 = k4
euler% math
Mathematica 2.1 for SPARC
Copyright 1988-92 Wolfram Research, Inc.

In[1]:= <<representatie3.1

Out[1]= k4

In[2]:= DeleteFile["representatie3.1"]

In[3]:= a>>representatie3.1

In[4]:= b>>>representatie3.1

In[5]:= c>>>representatie3.1

In[6]:= d>>>representatie3.1

In[7]:= alfa>>>representatie3.1

In[8]:= Quit
euler% cat representatie3.1
{{1, 0, 0,-k1/2, 0}, {0, 1, 0, 0, 1}, {0, 0, 1, 0, 0}, {0, 0, 0, 1, 0},
   {0, 0, 0, 0, 1}}
{{1, 0, 0, 0, 0}, {0, 1, 0, 0, 0}, {0, 0, 1, 0, 1}, {0, 0, 0, 1, 0},
   {0, 0, 0, 0, 1}}
{{1, k1/2, 0, 0, 0}, {0, 1, 0, 0, 0}, {0, 0, 1, 0, 0}, {0, 0, 0, 1, 1},
   {0, 0, 0, 0, 1}}
{{1, 0, 0, 0, 1}, {0, 1, 0, 0, 0}, {0, 0, 1, 0, 0}, {0, 0, 0, 1, 0},
   {0, 0, 0, 0, 1}}
{{1,k2, 0,k3, k4/2}, {0, -1, 0, 0, 0}, {0, 0, 1, 0, 0}, {0, 0, 0, -1, 0},
   {0, 0, 0, 0, 1}}
euler% vi representatie3.1
       (--> a little bit of file editing <--)
euler% cat representatie3.1
a={{1, 0, 0,-k1/2, 0}, {0, 1, 0, 0, 1}, {0, 0, 1, 0, 0}, {0, 0, 0, 1, 0},
   {0, 0, 0, 0, 1}}
b={{1, 0, 0, 0, 0}, {0, 1, 0, 0, 0}, {0, 0, 1, 0, 1}, {0, 0, 0, 1, 0},
   {0, 0, 0, 0, 1}}
c={{1, k1/2, 0, 0, 0}, {0, 1, 0, 0, 0}, {0, 0, 1, 0, 0}, {0, 0, 0, 1, 1},
   {0, 0, 0, 0, 1}}
d={{1, 0, 0, 0, 1}, {0, 1, 0, 0, 0}, {0, 0, 1, 0, 0}, {0, 0, 0, 1, 0},
   {0, 0, 0, 0, 1}}
alfa={{1, k2, 0, k3, k4/2}, {0, -1, 0, 0, 0}, {0, 0, 1, 0, 0},
      {0, 0, 0, -1, 0} , {0, 0, 0, 0, 1}}
euler%
```

Now we are really ready to start the computations concerning torsion. We know from corollary 6.1.2 that we only have to deal with $k_1, k_2, k_3, k_4 \in \{0,1\}$. Let $q = a^{m_1} b^{m_2} c^{m_3} \alpha$ be a torsion element in Q (of order 2), then for the lift $\tilde{q} = a^{m_1} b^{m_2} c^{m_3} \alpha$ of q in E we have that

$$\tilde{q}^2 = (a^{m_1} b^{m_2} c^{m_3} \alpha)^2 = d^{P(m_1, m_2, m_3)}$$

where $P(m_1, m_2, m_3)$ is some polynomial function in the variables m_1, m_2 and m_3. If q acts non trivially on d, we know that $\tilde{q}$ is indeed a torsion element. So suppose that the action of q on d is trivial. Now it's easy to see that there exists a lift $\tilde{q}$ of q which is a torsion element of order 2 if and only if

$$P(m_1, m_2, m_3) \equiv 0 \bmod 2.$$

To investigate this, the following lemma is very helpful:

Lemma A.4.1 *Let $P(x)$ be an integer valued polynomial function (i.e. $P(z) \in \mathbb{Z}, \ \forall z \in \mathbb{Z}$) of total degree $\leq M$. Then*

$$\forall z, k, n \in \mathbb{Z} : \ P(z) \equiv P(z + kn(M!)) \bmod n.$$

<u>Proof:</u> Every integer valued polynomial in the variable x of degree $\leq M$, is a $\mathbb{Z}$–linear combination of polynomials of the form

$$\binom{x}{i}, \ (0 \leq i \leq M).$$

We remark that $\binom{x}{0}$ denotes the constant 1. For $z, k, n \in \mathbb{Z}$, $1 \leq i \leq M$:

$$\binom{z + kn(M!)}{i} = \frac{(z + M!kn)((z-1) + M!kn)\ldots((z-i+1) + M!kn)}{1.2.3\ldots i}$$

$$\equiv \frac{z(z-1)(z-2)\ldots(z-i+1)}{1.2.3\ldots i} \bmod n.$$

This finishes the proof.   ∎

The lemma tells us that in checking if $P(m_1, m_2, m_3) \equiv 0 \bmod 2$, we may restrict ourselves to $m_i \in \{0, 1, \ldots, 2(\text{degree of } P \text{ in } m_i)! - 1\}$, for $i = 1, 2, 3$. We used all of this in writing the following program (torsion3.1) to find the AB–groups:

```
(* maxki= the maximum value of ki < the order of the torsion we are *)
(*        looking for.                                              *)
maxk1=1
maxk2=1
maxk3=1
maxk4=1
order=2
   For[k1=0,k1<=maxk1,k1++,
    For[k2=0,k2<=maxk2,k2++,
     For[k3=0,k3<=maxk3,k3++,
      For[k4=0,k4<=maxk4,k4++,
      {<<representatie3.1,
      (* A general torsionelement in Q, lifted to E *)
      torsionelement[m1_,m2_,m3_]:=matrixmacht[a,m1]. (* Fill in the  *)
                                  matrixmacht[b, 0]. (* solutions     *)
                                  matrixmacht[c,m3]. (* found with    *)
                                  alfa,              (* program torsion *)
      (* If a torsion element acts non trvially on Z, *)
      (* then any lift to E has torsion               *)
      If[(torsionelement[m1,m2,m3])[[1,1]]==-1,
          Print["Ignore the rest, group has torsion!"] ],
      Print["group: (k1=",k1,",k2=",k2,",k3=",k3,",k4=",k4,")"],
      (* We compute the power of d in torsionelement^order *)
      f[m1_,m2_,m3_]:=(
          Expand[(MatrixPower[torsionelement[m1,m2,m3],order])[[1,5]] ]),
      test=f[x,y,z], (* test is a polynomial function in x,y,z *)
      maxm1=(Exponent[test,x]), (* The degree of f in m1 *)
      maxm2=(Exponent[test,y]), (* The degree of f in m2 *)
      maxm3=(Exponent[test,z]), (* The degree of f in m3 *)
      If[maxm1<=0,maxm1=1,maxm1=order (maxm1!)],
      If[maxm2<=0,maxm2=1,maxm2=order (maxm2!)],
      If[maxm3<=0,maxm3=1,maxm3=order (maxm3!)],
      torsion=0,
      For[111=0,111<maxm1,111++,
        For[112=0,112<maxm2,112++,
          For[113=0,113<maxm3,113++,{
             If[Mod[f[111,112,113],order]==0,torsion=torsion+1]} ]]],
      If[torsion>0,Print["torsion"],
          Print["These kind of elements have infinite order!"] ],
      Print[" "],
      m1=., (* Cleaning up *)
      m2=.,
      m3=.,
      f=.,
      a=.,
      b=.,
      c=.,
      d=.,
      alfa=.}
```

]]]]

The output with Mathematica® looks like:

```
euler% math
Mathematica 2.1 for SPARC
Copyright 1988-92 Wolfram Research, Inc.

In[1]:= <<torsion3.1
group: (k1=0,k2=0,k3=0,k4=0)
torsion

group: (k1=0,k2=0,k3=0,k4=1)
These kind of elements have infinite order!

group: (k1=0,k2=0,k3=1,k4=0)
torsion

group: (k1=0,k2=0,k3=1,k4=1)
torsion

group: (k1=0,k2=1,k3=0,k4=0)
torsion

group: (k1=0,k2=1,k3=0,k4=1)
torsion

group: (k1=0,k2=1,k3=1,k4=0)
torsion

group: (k1=0,k2=1,k3=1,k4=1)
torsion

group: (k1=1,k2=0,k3=0,k4=0)
torsion

group: (k1=1,k2=0,k3=0,k4=1)
torsion

group: (k1=1,k2=0,k3=1,k4=0)
torsion

group: (k1=1,k2=0,k3=1,k4=1)
torsion

group: (k1=1,k2=1,k3=0,k4=0)
torsion

group: (k1=1,k2=1,k3=0,k4=1)
torsion
```

```
group: (k1=1,k2=1,k3=1,k4=0)
torsion

group: (k1=1,k2=1,k3=1,k4=1)
torsion


In[2]:= Quit
euler%
```

We have to interpret this as follows. A group E determined by a 4–tuple (k_1, k_2, k_3, k_4) is an AB–group if and only if

$$\begin{cases} k_1 \equiv 0 \bmod 2 \\ k_2 \equiv 0 \bmod 2 \\ k_3 \equiv 0 \bmod 2 \\ k_4 \equiv 1 \bmod 2. \end{cases}$$

We denote (the cohomology class of) the group E (which is determined by k_1, k_2, k_3, k_4) as $< (k_1, k_2, k_3, k_4) >$. Since $< (k_1, k_2, k_3, k_4) > \cong < (-k_1, -k_2, -k_3, -k_4) >$, we may always suppose that k_1 is positive. (We exclude $k_1 = 0$, since in this case Fitt (E) would be abelian). Of course, the relations on the cohomology–level show that we may restrict ourselves to $k_2, k_3, k_4 \in \{0, 1\}$. All this allows us to conclude that for a fixed k_1 (i.e. a fixed nilmanifold), there is at most one AB–group of this kind, namely if k_1 is even, we may take $E = < (k_1, 0, 0, 1) >$.

A.5 Summary

The previous sections show that any extension of Q by Z has a presentation of the form

$$E : < a, b, c, d, \alpha \parallel \begin{array}{ll} [b, a] = 1 & [d, a] = 1 \\ [c, a] = d^{k_1} & [d, b] = 1 \\ [c, b] = 1 & [d, c] = 1 \\ \alpha a = a^{-1} \alpha d^{k_2} & \alpha^2 = d^{k_4} \\ \alpha b = b \alpha & \alpha d = d \alpha \\ \alpha c = c^{-1} \alpha d^{k_3} \end{array} > .$$

A canonical type affine representation $\lambda : E \to \text{Aff}(\mathbb{R}^4)$ for such an E is given by

$$
\lambda(a) = \begin{pmatrix} 1 & 0 & 0 & \frac{-k_1}{2} & 0 \\ 0 & 1 & 0 & 0 & 1 \\ 0 & 0 & 1 & 0 & 0 \\ 0 & 0 & 0 & 1 & 0 \\ 0 & 0 & 0 & 0 & 1 \end{pmatrix}
\lambda(b) = \begin{pmatrix} 1 & 0 & 0 & 0 & 0 \\ 0 & 1 & 0 & 0 & 0 \\ 0 & 0 & 1 & 0 & 1 \\ 0 & 0 & 0 & 1 & 0 \\ 0 & 0 & 0 & 0 & 1 \end{pmatrix}
\lambda(c) = \begin{pmatrix} 1 & \frac{k_1}{2} & 0 & 0 & 0 \\ 0 & 1 & 0 & 0 & 0 \\ 0 & 0 & 1 & 0 & 0 \\ 0 & 0 & 0 & 1 & 1 \\ 0 & 0 & 0 & 0 & 1 \end{pmatrix}
$$

and

$$
\lambda(\alpha) = \begin{pmatrix} 1 & k_2 & 0 & k_3 & \frac{k_4}{2} \\ 0 & -1 & 0 & 0 & 0 \\ 0 & 0 & 1 & 0 & 0 \\ 0 & 0 & 0 & -1 & 0 \\ 0 & 0 & 0 & 0 & 1 \end{pmatrix} .
$$

We also showed that the group $H^2(Q, \mathbb{Z}) \cong \mathbb{Z} \oplus (\mathbb{Z}_2)^3$.

Finally, we indicated that for a fixed nilpotent group, there is maximal one AB–group of this kind, containing this nilpotent group as its maximal nilpotent normal subgroup. More precise, for $k_1 > 0$ even, there is one AB-group, determined by the parameters $(k_1, k_2, k_3, k_4) = (k_1, 0, 0, 1)$, containing

$$ N :< a, b, c, d \parallel [c, a] = d^{k_1} > $$

as its maximal nilpotent normal subgroup.

Bibliography

[1] Auslander, L. *Bieberbach's Theorem on Space Groups and Discrete Uniform Subgroups of Lie Groups.* Ann. of Math. (2), 1960, 71 (3), pp. 579–590.

[2] Auslander, L. *The structure of complete locally affine manifolds.* Topology, 1964, 3 Suppl. 1., pp. 131–139.

[3] Auslander, L. and Markus, L. *Holonomy of Flat Affinely Connected Manifolds.* Ann. of Math., 1955, 62 (1), pp. 139–151.

[4] Benoist, Y. *Une nilvariété non affine.* C. R. Acad. Sci. Paris Sér. I Math., 1992, 315 pp. 983–986.

[5] Benoist, Y. *Une nilvariété non affine.* J. Differential Geom., 1995, 41 pp. 21–52.

[6] Bieberbach, L. *Über die Bewegungsgruppen der Euklidischen Raume I.* Math. Ann., 1911, 70, pp. 297–336.

[7] Bieberbach, L. *Über die Bewegungsgruppen der Euklidischen Raume II.* Math. Ann., 1912, 72, pp. 400–412.

[8] Borel, A. and Harish-Chandra. *Arithmetic subgroups of algebraic groups.* Ann. of Math. (2), 1962, 75, pp. 485–535.

[9] Boyom, N. B. *The lifting problem for affine structures in nilpotent Lie groups.* Trans. Amer. Math. Soc., 1989, 313, pp. 347–379.

[10] Brown, H., Bülow, R., Neubüser, J., Wondratscheck, H., and Zassenhaus, H. *Crystallographic groups of four–dimensional Space.* Wiley New York, 1978.

[11] Brown, K. S. *Cohomology of groups*, volume 87 of *Grad. Texts in Math.* Springer–Verlag New York Inc., 1982.

[12] Burde, D. and Grunewald, F. *Modules for certain Lie algebras of maximal class.* J. Pure Appl. Algebra, 1995, 99 pp. 239–254.

[13] Charlap, L. S. *Bieberbach Groups and Flat Manifolds.* Universitext. Springer–Verlag, New York Inc., 1986.

[14] Conner, P. E. and Raymond, F. *Deforming Homotopy Equivalences to Homeomorphisms in Aspherical Manifolds.* Bull. A.M.S., 1977, 83 (1), pp. 36–85.

[15] Dekimpe, K. *Almost Bieberbach Groups: cohomology, construction and classification.* Doctoral Thesis, K.U.Leuven, 1993.

[16] Dekimpe, K. *Polynomial structures and the uniqueness of affinely flat infra-nilmanifolds.* 1994. preprint, to appear in Math. Zeitschrift.

[17] Dekimpe, K. *A note on the torsion elements in the centralizer of a finite index subgroup.* 1995. To appear in the Bull. of the Belgian Math. Soc. – Simon Stevin.

[18] Dekimpe, K. *The construction of affine structures on virtually nilpotent groups.* Manuscripta Math., 1995, 87 pp. 71 – 88.

[19] Dekimpe, K. and Hartl, M. *Affine Structures on 4–step Nilpotent Lie Algebras.* 1994. To appear in Journal of Pure and Applied Algebra.

[20] Dekimpe, K. and Igodt, P. *Computational aspects of affine representations for torsion free nilpotent groups via the Seifert construction.* J. Pure Applied Algebra, 1993, 84, pp. 165–190.

[21] Dekimpe, K. and Igodt, P. *The structure and topological meaning of almost–torsion free groups.* Comm. Algebra, 1994, (7), pp. 2547–2558.

[22] Dekimpe, K. and Igodt, P. *Polynomial structures on polycyclic groups.* 1995. Prepint.

[23] Dekimpe, K. and Igodt, P. *Polycyclic-by-finite groups admit a bounded-degree polynomial structure.* 1996. Prepint.

[24] Dekimpe, K. and Igodt, P. *Polynomial structures for iterated central extensions of abelian-by-nilpotent groups.* Algebraic Topology: New Trends in Localization and Periodicitiy, 1996, pages pp. 155–166. Progress in Mathematics, Birkhäuser.

[25] Dekimpe, K., Igodt, P., Kim, S., and Lee, K. B. *Affine structures for closed 3-dimensional manifolds with NIL-geometry.* Quart. J. Math. Oxford (2), 1995, 46, pp. 141–167.

[26] Dekimpe, K., Igodt, P., and Lee, K. B. *Polynomial structures for nilpotent groups.* Trans. Amer. Math. Soc., 1996, 348, pp. 77–97.

[27] Dekimpe, K., Igodt, P., and Malfait, W. *On the Fitting subgroup of almost crystallographic groups.* Tijdschrift van het Belgisch Wiskundig Genootschap, 1993, B 1, pp. 35–47.

[28] Dekimpe, K., Igodt, P., and Malfait, W. *There are only finitely many infra-nilmanifolds under each nilmanifold: a new proof.* Indagationes Math., 1994, 5 (3), pp. 259–266.

[29] Dekimpe, K. and Malfait, W. *Affine structures on a class of virtually nilpotent groups.* To appear in Topology and its Applications.

[30] Fried, D., Goldman, W., and Hirsch, M. *Affine manifolds with nilpotent holonomy.* Comment. Math. Helv., 1981, 56 pp. 487–523.

[31] Fried, D. and Goldman, W. M. *Three-Dimensional Affine Crystallographic Groups.* Adv. in Math., 1983, 47 1, pp. 1–49.

[32] Frobenius, G. *Über die unzerlegbaren diskreten Bewegungsgruppen.* Sitzungsber. Akad. Wiss. Berlin, 1911, 29 pp. 654–665.

[33] Gorbacevič, V. V. *Discrete subgroups of solvable Lie groups of type (E).* Math. USSR Sbornik, 1971, 14 N° 2, pp. 233–251.

[34] Gorbacevič, V. V. *Lattices in solvable Lie groups and deformations of homogeneous spaces.* Math. USSR Sbornik, 1973, 20 2, pp. 249–266.

[35] Gromov, M. *Almost flat manifolds.* J. Differential Geometry, 1978, 13, pp. 231–241.

[36] Grunewald, F. and Margulis, G. *Transitive and quasitransitive actions of affine groups preserving a generalised Lorentz-structure.* J. Geometry and Physics, 1988, 5 (4), pp. 493–531.

[37] Grunewald, F. and Segal, D. *On affine crystallographic groups.* J. Differential Geom., 1994, 40 (3), pp. 563–594.

[38] Igodt, P. and Lee, K. B. *Applications of group cohomology to space constructions.* Trans. A.M.S., 1987, 304 pp. 69–82.

[39] Igodt, P. and Lee, K. B. *On the uniqueness of certain affinely flat infra-nilmanifolds*. Math. Zeitschrift, 1990, 204 pp. 605–613.

[40] Igodt, P. and Malfait, W. *Extensions realising a faithful abstract kernel and their automorphisms*. Manuscripta Math., 1994, 84, pp. 135–161.

[41] Igodt, P. and Malfait, W. *Representing the automorphism group of an almost crystallographic group*. Proc. Amer. Math. Soc., 1996, 124 (2), pp. 331–340.

[42] Kamishima, Y., Lee, K. B., and Raymond, F. *The Seifert construction and its applications to infra-nilmanifolds*. Quarterly J. of Math. Oxford (2), 1983, 34, pp. 433–452.

[43] Kim, H. *Complete left-invariant affine structures on nilpotent Lie groups*. J. Differential Geom., 1986, 24, pp. 373–394.

[44] Lee, K. B. *Aspherical manifolds with virtually 3-step nilpotent fundamental group*. Amer. J. Math., 1983, 105 pp. 1435–1453.

[45] Lee, K. B. *There are only finitely many infra-nilmanifolds under each nilmanifold*. Quart. J. Math. Oxford (2), 1988, 39, pp. 61–66.

[46] Lee, K. B. and Raymond, F. *Topological, affine and isometric actions on flat Riemannian manifolds*. J. Differential geometry, 1981, 16 pp. 255–269.

[47] Lee, K. B. and Raymond, F. *Geometric realization of group extensions by the Seifert construction*. Contemporary Math. A. M. S., 1984, 33, pp. 353–411.

[48] Lee, K. B. and Raymond, F. *Rigidity of almost crystallographic groups*. Contemporary Math. A. M. S., 1985, 44, pp. 73–78.

[49] Lee, K. B. and Raymond, F. *Examples of solvmanifolds without certain affine structure*. Illinois J. Math., 1990, 37 pp. 69–77.

[50] Mac Lane, S. *Homology*, volume 114 of *Die Grundlehren der Math. Wissenschaften*. Springer–Verlag Berlin Heidelberg New York, 1975.

[51] Mal'cev, A. I. *On a class of homogeneous spaces*. Translations A.M.S., 1951, 39, pp. 1–33.

[52] Milnor, J. *On fundamental groups of complete affinely flat manifolds.* Adv. Math., 1977, 25 pp. 178–187.

[53] Moser, L. *Elementary surgery along torus knots and solvable fundamental groups of closed 3–manifolds.* Thesis, University of Wisconsin, 1970.

[54] Nisse, M. *Structure affine des infranilvariétés et infrasolvariétés.* C. R. Acad. Sci. Paris Série I, 1990, 310 pp. 667–670.

[55] Orlik, P. *Seifert Manifolds.* Lect. Notes in Math. 291. Springer–Verlag, 1972.

[56] Passman, D. S. *The Algebraic Structure of Group Rings.* Pure and Applied Math. John Wiley & Sons, Inc. New York, 1977.

[57] Ruh, E. A. *Almost flat manifolds.* J. Differential Geometry, 1982, 17, pp. 1–14.

[58] Scheuneman, J. *Affine structures on three-step nilpotent Lie algebras.* Proc. A.M.S., 1974, 46 (3), pp. 451–454.

[59] Segal, D. *Polycyclic Groups.* Cambridge University Press, 1983.

[60] Wolf, J. A. *Spaces of constant curvature.* Publish or Perish, Inc. Berkeley, 1977.

[61] Wolfram, S. *Mathematica.* Wolfram Research, Inc., 1993.

[62] Zassenhaus, H. *Über einen Algorithmus zur Bestimmung der Raumgruppen.* Commentarii Mathematici Helvetici, 1948, 21, pp. 117–141.

[63] Zieschang, H. *Finite Groups of Mapping Classes of Surfaces.* Lect. Notes in Math. 875. Springer–Verlag, 1981.

Index

Vol. 1594: M. Green, J. Murre, C. Voisin, Algebraic Cycles and Hodge Theory. Torino, 1993. Editors: A. Albano, F. Bardelli. VII, 275 pages. 1994.

Vol. 1595: R.D.M. Accola, Topics in the Theory of Riemann Surfaces. IX, 105 pages. 1994.

Vol. 1596: L. Heindorf, L. B. Shapiro, Nearly Projective Boolean Algebras. X, 202 pages. 1994.

Vol. 1597: B. Herzog, Kodaira-Spencer Maps in Local Algebra. XVII, 176 pages. 1994.

Vol. 1598: J. Berndt, F. Tricerri, L. Vanhecke, Generalized Heisenberg Groups and Damek-Ricci Harmonic Spaces. VIII, 125 pages. 1995.

Vol. 1599: K. Johannson, Topology and Combinatorics of 3-Manifolds. XVIII, 446 pages. 1995.

Vol. 1600: W. Narkiewicz, Polynomial Mappings. VII, 130 pages. 1995.

Vol. 1601: A. Pott, Finite Geometry and Character Theory. VII, 181 pages. 1995.

Vol. 1602: J. Winkelmann, The Classification of Three-dimensional Homogeneous Complex Manifolds. XI, 230 pages. 1995.

Vol. 1603: V. Ene, Real Functions – Current Topics. XIII, 310 pages. 1995.

Vol. 1604: A. Huber, Mixed Motives and their Realization in Derived Categories. XV, 207 pages. 1995.

Vol. 1605: L. B. Wahlbin, Superconvergence in Galerkin Finite Element Methods. XI, 166 pages. 1995.

Vol. 1606: P.-D. Liu, M. Qian, Smooth Ergodic Theory of Random Dynamical Systems. XI, 221 pages. 1995.

Vol. 1607: G. Schwarz, Hodge Decomposition – A Method for Solving Boundary Value Problems. VII, 155 pages. 1995.

Vol. 1608: P. Biane, R. Durrett, Lectures on Probability Theory. VII, 210 pages. 1995.

Vol. 1609: L. Arnold, C. Jones, K. Mischaikow, G. Raugel, Dynamical Systems. Montecatini Terme, 1994. Editor: R. Johnson. VIII, 329 pages. 1995.

Vol. 1610: A. S. Üstünel, An Introduction to Analysis on Wiener Space. X, 95 pages. 1995.

Vol. 1611: N. Knarr, Translation Planes. VI, 112 pages. 1995.

Vol. 1612: W. Kühnel, Tight Polyhedral Submanifolds and Tight Triangulations. VII, 122 pages. 1995.

Vol. 1613: J. Azéma, M. Emery, P. A. Meyer. M. Yor (Eds.), Séminaire de Probabilités XXIX. VI, 326 pages. 1995.

Vol. 1614: A. Koshelev, Regularity Problem for Quasilinear Elliptic and Parabolic Systems. XXI, 255 pages. 1995.

Vol. 1615: D. B. Massey, Lê Cycles and Hypersurface Singularities. XI, 131 pages. 1995.

Vol. 1616: I. Moerdijk, Classifying Spaces and Classifying Topoi. VII, 94 pages. 1995.

Vol. 1617: V. Yurinsky, Sums and Gaussian Vectors. XI, 305 pages. 1995.

Vol. 1618: G. Pisier, Similarity Problems and Completely Bounded Maps. VII, 156 pages. 1996.

Vol. 1619: E. Landvogt, A Compactification of the Bruhat-Tits Building. VII, 152 pages. 1996.

Vol. 1620: R. Donagi, B. Dubrovin, E. Frenkel, E. Previato, Integrable Systems and Quantum Groups. Montecatini Terme, 1993. Editors:M. Francaviglia, S. Greco. VIII, 488 pages. 1996.

Vol. 1621: H. Bass, M. V. Otero-Espinar, D. N. Rockmore, C. P. L. Tresser, Cyclic Renormalization and Auto-morphism Groups of Rooted Trees. XXI, 136 pages. 1996.

Vol. 1622: E. D. Farjoun, Cellular Spaces, Null Spaces and Homotopy Localization. XIV, 199 pages. 1996.

Vol. 1623: H.P. Yap, Total Colourings of Graphs. VIII, 131 pages. 1996.

Vol. 1624: V. Brînzănescu, Holomorphic Vector Bundles over Compact Complex Surfaces. X, 170 pages. 1996.

Vol.1625: S. Lang, Topics in Cohomology of Groups. VII, 226 pages. 1996.

Vol. 1626: J. Azéma, M. Emery, M. Yor (Eds.), Séminaire de Probabilités XXX. VIII, 382 pages. 1996.

Vol. 1627: C. Graham, Th. G. Kurtz, S. Méléard, Ph. E. Protter, M. Pulvirenti, D. Talay, Probabilistic Models for Nonlinear Partial Differential Equations. Montecatini Terme. 1995. Editors: D. Talay, L. Tubaro. X, 301 pages. 1996.

Vol. 1628: P.-H. Zieschang, An Algebraic Approach to Association Schemes. XII, 189 pages. 1996.

Vol. 1629: J. D. Moore, Lectures on Seiberg-Witten Invariants. VII, 105 pages. 1996.

Vol. 1630: D. Neuenschwander, Probabilities on the Heisenberg Group: Limit Theorems and Brownian Motion. VIII, 139 pages. 1996.

Vol. 1631: K. Nishioka, Mahler Functions and Transcendence.VIII, 185 pages.1996.

Vol. 1632: A. Kushkuley, Z. Balanov, Geometric Methods in Degree Theory for Equivariant Maps. VII, 136 pages. 1996.

Vol.1633: H. Aikawa. M. Essén, Potential Theory – Selected Topics. IX, 200 pages.1996.

Vol. 1634: J. Xu, Flat Covers of Modules. IX, 161 pages. 1996.

Vol. 1635: E. Hebey, Sobolev Spaces on Riemannian Manifolds. X, 116 pages. 1996.

Vol. 1636: M. A. Marshall, Spaces of Orderings and Abstract Real Spectra. VI, 190 pages. 1996.

Vol. 1637: B. Hunt, The Geometry of some special Arithmetic Quotients. XIII, 332 pages. 1996.

Vol. 1638: P. Vanhaecke, Integrable Systems in the realm of Algebraic Geometry. VIII, 218 pages. 1996.

Vol. 1639: K. Dekimpe, Almost-Bieberbach Groups: Affine and Polynomial Structures. X, 259 pages. 1996.

Printing: Weihert-Druck GmbH, Darmstadt
Binding: Theo Gansert Buchbinderei GmbH, Weinheim